Cambridge astrophysics series

X-ray emission from clusters of galaxies

FOR JANE

X-ray emission from clusters of galaxies

CRAIG L. SARAZIN

Department of Astronomy, University of Virginia

CAMBRIDGE UNIVERSITY PRESS

Cambridge

New York New Rochelle

Melbourne Sydney

CAMBRIDGE UNIVERSITY PRESS
Cambridge, New York, Melbourne, Madrid, Cape Town, Singapore, São Paulo, Delhi

Cambridge University Press
The Edinburgh Building, Cambridge CB2 8RU, UK

Published in the United States of America by Cambridge University Press, New York

www.cambridge.org
Information on this title: www.cambridge.org/9780521113137

First published 1988
This digitally printed version 2009

A catalogue record for this publication is available from the British Library

Library of Congress Cataloguing in Publication data

Sarazin, Craig L.

X-ray emission from clusters of galaxies.

(Cambridge astrophysics series)
Bibliography
1. Galaxies – Clusters. 2. X-ray astronomy.
I. Title. II. Series.
QB858.7.S27 1988 523.1′12 86-23322

ISBN 978-0-521-32957-6 hardback
ISBN 978-0-521-11313-7 paperback

CONTENTS

PREFACE

As this book was being completed, observational X-ray astronomy was in a relatively quiet period between the demise of the *Einstein* X-ray observatory, and the launch of the next generation X-ray observatory, AXAF. This seemed like a good time to summarize what we have learned about the X-ray emission of clusters of galaxies. This book is mainly devoted to a review of the observational properties of X-ray clusters of galaxies and of the theoretical understanding we currently have of the physical state, dynamics, and origin of the hot intracluster gas. The book also contains less complete reviews of the optical and radio properties of clusters of galaxies, and of their dynamics.

Much of the material in this book first appeared in a review article in *Reviews of Modern Physics* (Sarazin, 1986a). I should like to thank Ed Salpeter for suggesting this review, and for many helpful suggestions.

Andy Fabian, Richard Mushotzky, Paul Nulsen, Simon White and an anonymous referee very kindly provided detailed comments on an early draft of the manuscript and caught many significant errors. I really would like to thank them for all their help. I am also indebted to John Bahcall, James Binney, Pat Henry, Hernan Quintana, Yoel Rephaeli, Herb Rood, Hy Spinrad, Mel Ulmer, and Ray White for helpful comments, suggestions, and unpublished results. I should particularly like to thank John Bahcall for his encouragement. Figures for the original review paper and for this book were very kindly provided by Neta Bahcall, Jack Burns, Claude Canizares, Christine Jones, Andy Fabian, Bill Forman, Paul Gorenstein, Mark Henriksen, Roger Lynds, George Miley, Richard Mushotzky, Chris O'Dea, Herb Rood, Steve Strom, and Simon White. I should like particularly to thank Christine Jones and Bill Forman for producing special figures from the *Einstein* X-ray data.

Much of the original review article was written while I was a visitor at the Institute for Advanced Study in Princeton, N.J., and I would like to thank the

Institute and particularly John Bahcall for their hospitality. Karen Jobes at the Institute for Advanced Study helped considerably with the word processing. Part of the review was written during visits to the Aspen Center for Physics and the National Radio Astronomy Observatory, and I would also like to thank them.

The manuscript for the book was completed while I was a Visiting Fellow at the Joint Institute for Laboratory Astrophysics of the University of Colorado and the National Bureau of Standards. I should like to thank the JILA Fellows for their hospitality. During this sabbatical leave I was also supported in part by a Sesquicentennial Fellowship from the University of Virginia.

This research was supported in part by NSF Grant AST 81-20260 and NASA Astrophysics Theory Center Grant NAGW-764.

This work would have been impossible without the support and patient understanding of my wife Jane.

1

INTRODUCTION

Regular clusters of galaxies are the largest organized structures in the universe. They typically contain hundreds of galaxies, spread over a region whose size is roughly 10^{25} cm. Their total masses exceed 10^{48} gm. They were first studied in detail by Wolf (1906), although the tendency for galaxies to cluster on the sky had been noted long before this. A great advance in the systematic study of the properties of clusters occurred when Abell compiled an extensive, statistically complete catalog of rich clusters of galaxies (Abell, 1958). For the last quarter century, this catalog has been the most important resource in the study of galaxy clusters. Optical photographs of several of the best studied clusters of galaxies are shown in Figure 1.[1] The Virgo cluster (Figure 1a) is the nearest rich cluster to our own galaxy; the Coma cluster (Figure 1b) is the nearest very regular cluster.

In 1966, X-ray emission was detected from the region around the galaxy M87 in the center of the Virgo cluster (Byram *et al.*, 1966; Bradt *et al.*, 1967; Figure 1a). In fact M87 was the first object outside of our galaxy to be identified as a source of astronomical X-ray emission. Five years later, X-ray sources were also detected in the directions of the Coma (Figure 1b) and Perseus (Figure 1c) clusters (Fritz *et al.*, 1971; Gursky *et al.*, 1971a, b; Meekins *et al.*, 1971). Since these are three of the nearest rich clusters, it was suggested that clusters of galaxies might generally be X-ray sources (Cavaliere *et al.*, 1971). The launch of the *Uhuru* X-ray astronomy satellite permitted a survey of the entire sky for X-ray emission (Giacconi *et al.*, 1972) and established that this was indeed the case. These early *Uhuru* observations

[1] Unless otherwise indicated, all the figures in this book showing optical, X-ray, or radio brightness on the sky have north at the top and east at the left. When coordinates are given, the east–west coordinate is right ascension (in hours, minutes, and seconds) and the north–south coordinate is declination (in degrees, minutes, and seconds).

Fig. 1. Optical photographs of clusters of galaxies. (a) An optical photograph of the Virgo cluster of galaxies, an irregular cluster that is the nearest cluster to our galaxy. The galaxy M87, on which the X-ray emission is centered, is marked, as are the two bright galaxies M84 and M86. Photograph from the Palomar Observatory Sky Survey (Minkowski and Abell, 1963). (b) The Coma cluster of galaxies (Abell 1656), showing the two dominant D galaxies. Coma is one of the nearest rich, regular clusters. Photograph copyright 1973, AURA, Inc., the

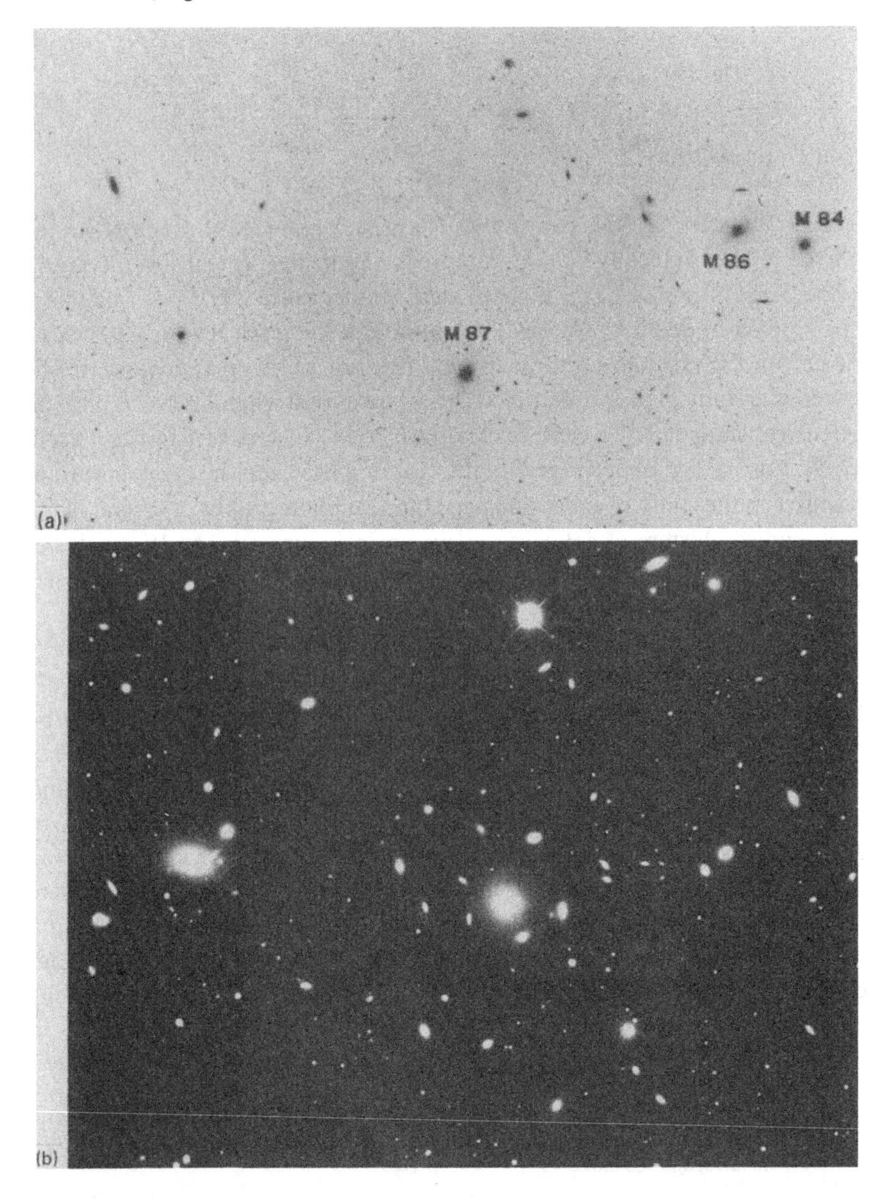

indicated that many clusters were bright X-ray sources, with luminosities typically in the range of 10^{43-45} erg/s. The X-ray sources associated with clusters were found to be spatially extended; their sizes were comparable to the size of the galaxy distribution in the clusters (Kellogg *et al.*, 1972; Forman *et al.*, 1972). Unlike other bright X-ray sources but consistent with their spatial extents, cluster X-ray sources did not vary temporally in their brightness (Elvis, 1976). Although several emission mechanisms were proposed, the X-ray spectra of clusters were most consistent with thermal bremsstrahlung from hot gas.

This interpretation requires that the space between galaxies in clusters be filled with very hot ($\approx 10^8$ K), low density ($\approx 10^{-3}$ atoms/cm^3) gas. Remarkably, the total mass in this intracluster medium is comparable to the total mass in all the stars in all the galaxies in the cluster. As to the origin of this gas, it was widely assumed that it had simply fallen into the clusters from

Fig. 1 *continued*
National Optical Astronomy Observatories, Kitt Peak. (c) The Perseus clusters of galaxies (Abell 476), showing the line of bright galaxies. NGC1275 is the brightest galaxy, on the east (left) end of the chain. NGC1265 is a head-tail radio galaxy. Photograph from Strom and Strom (1978c). (d) The irregular cluster Abell 1367. Photograph from Strom and Strom (1978c).

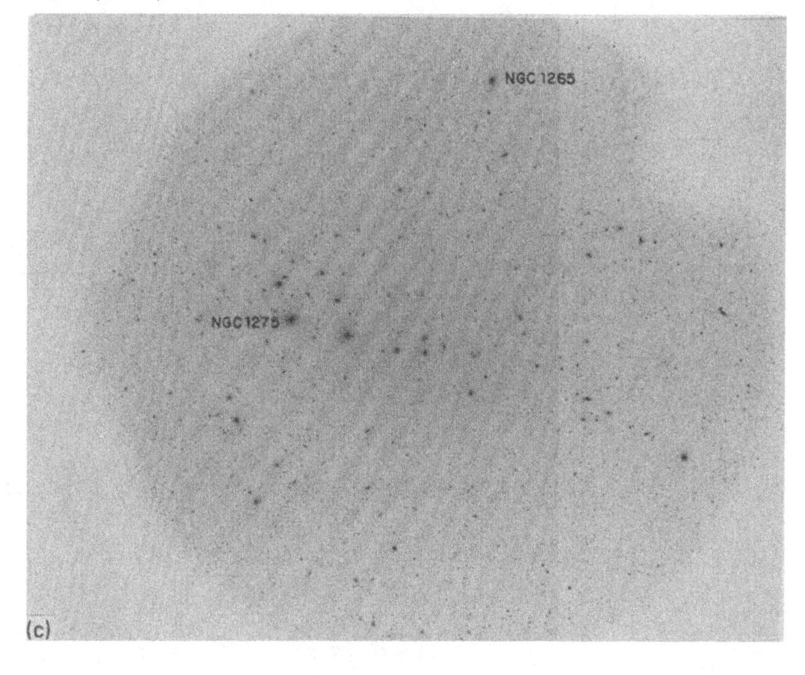

the great volumes of space between them, where it had been stored since the formation of the universe (Gunn and Gott, 1972).

In 1976, X-ray line emission from iron was detected from the Perseus cluster of galaxies (Mitchell *et al.*, 1976), and shortly thereafter from Coma and Virgo as well (Serlemitsos *et al.*, 1977). The emission mechanism for this line is thermal, and its detection confirmed the thermal interpretation of cluster X-ray sources. However, the only known sources for significant quantities of iron or any other heavy element in astronomy are nuclear reactions in stars, and no significant population of stars has been observed which does not reside in galaxies. Since the abundance of iron in the intracluster gas was observed to be similar to its abundance in stars, a substantial portion of this gas must have been ejected from stars in galaxies in the cluster (Bahcall and Sarazin, 1977). This is despite the fact that the total mass of intracluster gas is on the same order as the total mass of stars presently observed in the clusters. Obviously, these X-ray observations suggest that galaxies in clusters have had more interesting histories than might otherwise have been assumed.

Fig. 1 *continued*

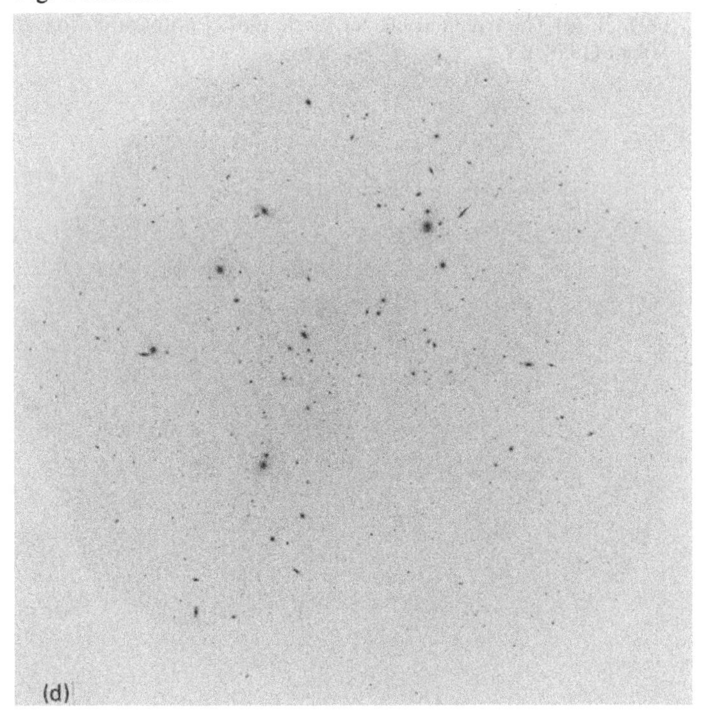

In this book, the X-ray observations of clusters of galaxies and the theories for the intracluster gas will be reviewed. Because clusters are still largely defined by their optical properties, I shall first review the optical observations of clusters (Chapter 2). I shall particularly emphasize information on their dynamical state, and the possibility that the galaxy population has been affected by the intracluster gas. Radio observations of clusters also provide information on the intracluster gas, which is summarized in Chapter 3. For example, certain distortions seen in radio sources in clusters are most naturally explained as arising from interactions with this gas. Moreover, extensive searches have been made for 'shadows' in the cosmic microwave radiation due to electron scattering by intracluster gas. Then the X-ray observations will be reviewed (Chapter 4), including the recent results of X-ray imaging and spectroscopy from the *Einstein* X-ray satellite. In Chapter 5 theories for the X-ray emission mechanism, the physical state, the distribution, the origin, and the history of the intracluster medium will be reviewed. Finally, I shall comment briefly on the prospects for further observations of X-ray clusters, particularly with the AXAF satellite, in Chapter 6.

Review articles on clusters of galaxies emphasizing their optical properties include Abell (1965, 1975), van den Bergh (1977b), Bahcall (1977a), Rood (1981), White (1982), and particularly Dressler (1984). Superclusters of galaxies are reviewed by Oort (1983). Some recent reviews which include the X-ray properties of clusters are Gursky and Schwartz (1977), Binney (1980), Cavaliere (1980), Cowie (1981), Canizares (1981), and Holt and McCray (1982). Fabian *et al.* (1984b) give an excellent review of cooling flows in X-ray clusters (see Sec. 5.7). Up-to-date reviews of the spectroscopic properties of X-ray clusters are given by Mushotzky (1984, 1985). Forman and Jones (1982) give a comprehensive review of the X-ray images of clusters. This book is based in large part on my review paper on X-ray clusters (Sarazin, 1986a).

2

OPTICAL OBSERVATIONS

2.1 Catalogs

The two most extensive and often cited catalogs of rich clusters of galaxies are those of Abell (1958) and Zwicky and his collaborators (Zwicky *et al.*, 1961–1968). As is conventional, in this book Abell clusters will be denoted by giving A and then the number of the cluster in Abell's list. Both of these catalogs were constructed by identifying clusters as enhancements in the surface number density of galaxies on the National Geographic Society – Palomar Observatory Sky Survey (Minkowski and Abell, 1963) and thus are confined to northern areas of sky (declination greater than $-20°$ for Abell and $-3°$ for Zwicky). Abell surveyed only clusters away from the plane of our galaxy.

As clustering exists on a very wide range of angular and intensity scales (Peebles, 1974), it is not possible to give a unique and unambiguous definition of a 'rich cluster'. Thus the membership of a catalog of clusters is determined by the criteria used to define a rich cluster. These criteria must specify the required surface number density enhancement for inclusion and the linear or angular scale of the enhancement. The scale is necessary in order to exclude small groupings of galaxies; for example, a close pair of galaxies can represent a very large enhancement above the background galaxy density on a small angular scale. Alternatively, specifying the surface density and scale is equivalent to specifying the number of galaxies (the 'richness' of the cluster) and the scale size. Because the number of galaxies observed increases as their brightness diminishes (Section 2.4), one must also specify the range of magnitudes of the galaxies included in determining the cluster richness. Finally, because galaxies grow fainter with increasing distance, the catalog can only be statistically complete out to a limiting distance or redshift, and only clusters within this distance range should be included in a statistically complete sample.

Abell's criteria were basically (1) that the cluster contain at least 50 galaxies in the magnitude range m_3 to $m_3 + 2$, where m_3 is the magnitude of the third brightest galaxy; (2) that these galaxies be contained within a circle of radius $R_A = 1.7/z$ minutes of arc or $3h_{50}^{-1}$ megaparsecs,[2] where z is the estimated redshift of the cluster (Section 2.2); (3) that the estimated cluster redshift be in the range $0.02 \leqslant z \leqslant 0.20$. R_A is called the Abell radius of the cluster. The Abell catalog contains 2712 clusters, of which 1682 satisfy all these criteria. The other 1030 were discovered during the search and were included to provide a more extensive finding list for clusters. The Abell catalog gives estimates of the cluster center positions (see also Sastry and Rood, 1971), distance, and richness of the clusters, as well as the magnitude of the tenth brightest galaxy m_{10}.

For the Zwicky catalog, the criteria were as follows: (1) the boundary (scale size) of the cluster was determined by the contour (or isopleth) where the galaxy surface density fell to twice the local background density; (2) this isopleth had to contain at least 50 galaxies in the magnitude range m_1 to $m_1 + 3$, where m_1 is the magnitude of the first-brightest galaxy. No distance limits were specified, although in practice, very nearby clusters such as Virgo (Figure 1a), were not included because they extended over several Sky Survey plates. Obviously, the Zwicky catalog criteria are much less strict than Abell's, and the Zwicky catalog thus contains many more clusters that are less rich. For each cluster, the Zwicky catalog gives a classification (Section 2.5), and estimates of the coordinates of the center, the diameter, the richness, and the redshift. Finding charts showing the cluster isopleths and positions of brighter galaxies and stars are also presented.

A number of smaller catalogs have been compiled, consisting of clusters in the southern sky or clusters at higher redshifts ($z \gtrsim 0.2$). Early southern cluster catalogs or lists include those of de Vaucouleurs (1956), Klemola (1969), Snow (1970), Sersic (1974), and Rose (1976). Until recently, the search for southern clusters was severely handicapped by the lack of deep survey plates. The first deep optical survey of the south, the European Southern Observatory Quick Blue Survey (ESO-B), was completed in 1978 (West, 1974; Holmberg *et al.*, 1974). A catalog of southern clusters from the first portion of this survey was prepared by Duus and Newell (1977) and a list of southern clusters near X-ray sources was given by Lugger (1978). More recently, portions of the ESO/SRC-J survey of very deep blue plates have been used to compile southern cluster catalogs by Braid and MacGillivray (1978) and White and Quintana (1985). Before his untimely death, Abell was preparing a southern

[2] For convenience, in this book the Hubble constant H_o will be parameterized as $H_o = 50h_{50}$ km/sec/Mpc.

continuation of the Abell catalog in collaboration with Corwin. The red portion of the ESO/SRC survey is currently being done, and these plates are being used to detect high redshift clusters (West and Frandsen, 1981).

The discovery of higher redshift clusters ($z \gtrsim 0.2$) is of great importance to cosmological studies; lists of such clusters include Humason and Sandage (1957), Gunn and Oke (1975), Sandage, Kristian, and Westphal (1976), Spinrad *et al.* (1985), and Vidal (1980). In addition, the lists of southern clusters from the ESO/SRC surveys discussed in the last paragraph contain many high redshift clusters.

2.2 Redshifts

The clusters in the Abell (1958) catalog were assigned to distance groups, based on the redshift estimated from the magnitude of the tenth brightest galaxy in the cluster. Leir and van den Bergh (1977) have given improved estimates of redshifts for 1889 rich Abell clusters, using the magnitudes of the first and tenth brightest galaxies, and an estimate of the cluster radius. Their distance scale is calibrated using measured redshifts for 101 clusters. Photometric distance estimators derived from a larger sample of redshifts have been given by Sarazin *et al.* (1982). Similarly, clusters in the Zwicky catalog were placed in distance groups, based on the magnitudes and sizes of the brightest cluster galaxies.

Spectroscopic redshifts are now available for about 500 Abell clusters. Sarazin (1986a, Table I) gives an extensive list of the redshifts for Abell clusters, taken primarily from Sarazin *et al.* (1982), Noonan (1981), Struble and Rood (1982, 1985), and Hoessel, Gunn, and Thuan (1980).

Of course, many redshifts are known for non-Abell clusters as well. The compilation of Noonan (1981) includes many such redshifts.

2.3 Richness – the number of galaxies in a cluster

The richness of a cluster is a measure of the number of galaxies associated with that cluster. Because of the presence of background galaxies, it is not possible to state with absolute confidence that any given galaxy belongs to a given cluster. Thus one cannot give an exact tally of the number of galaxies in a cluster. Richness is a statistical measure of the population of a cluster, based on some operational definition of cluster membership. The richness will be more useful if it is defined in such a way as to be reasonably independent of the distance to and morphology of a cluster.

Zwicky *et al.* (1961–1968) define the richness of their clusters as the total number of galaxies visible on the red Sky Survey plates within the cluster isopleth (Section 2.1); the number of background galaxies expected is subtracted from the richness. These richnesses are clearly very dependent on a cluster redshift (Abell, 1962; Scott, 1962). First, a wider magnitude range is counted for nearby clusters, because the magnitude of the first brightest galaxy is further from the plate limit. Second, a larger area of the cluster is counted for nearby clusters, because the point at which the surface density is twice that of the background will be farther out in the cluster.

Abell (1958) has divided his clusters into richness groups using criteria that are nearly independent of distance (see Section 2.1); that is, the magnitude range and area considered do not vary with redshift. Just (1959) has found a slight richness–distance correlation in Abell's catalog; however, the effect is small and is probably explained by a slight incompleteness (10%) of the catalog for distant clusters (Paal, 1964). When more accurate determinations have been made, it has been found that the Abell richness generally correlated well with the number of galaxies, but that the Abell richness may be significantly in error in some individual cases (Mottmann and Abell, 1977; Dressler, 1980a). Thus the Abell richnesses are very useful for statistical studies, but must be used with caution in studies of individual clusters. Note also that it is generally preferable to use the actual Abell counts rather than just the richness group (Abell, 1982).

2.4 Luminosity function of galaxies

The luminosity function of galaxies in a cluster gives the number distribution of the luminosities of the galaxies. The integrated luminosity function $N(L)$ is the number of galaxies with luminosities greater than L, while the differential luminosity function $n(L)dL$ is the number of galaxies with luminosities in the range L to $L+dL$. Obviously, $n(L) \equiv -dN(L)/dL$. Luminosity functions are often defined in terms of galaxy magnitudes $m \propto -2.5\, log_{10}(L)$; $N(\leqslant m)$ is the number of galaxies in a cluster brighter than magnitude m. Observational studies of the luminosity functions of clusters include Zwicky (1957), Kiang (1961), van den Bergh (1961a), Abell (1962, 1975, 1977), Rood (1969), Gudehus (1973), Bautz and Abell (1973), Austin and Peach (1974b), Oemler (1974), Krupp (1974), Austin *et al.* (1975), Godwin and Peach (1977), Mottmann and Abell (1977), Dressler (1978b), Bucknell *et al.* (1979), Carter and Godwin (1979), Thompson and Gregory (1980), and Kraan-Korteweg (1981). Figure 2 gives the observed integrated luminosity function for a composite of 13 rich clusters as derived by Schechter (1976) from Oemler's (1974) data.

Three types of functions are commonly used for fitting the luminosity function. Zwicky (1957) proposed the form

$$N(\leqslant m) = K(10^{0.2(m-m_1)} - 1), \tag{2.1}$$

where K is a constant and m_1 is the magnitude of the first brightest galaxy. In general, the Zwicky function fits the faint end of the luminosity function adequately, but does not fall off rapidly enough for brighter galaxies (see Figure 2). Clearly, equation (2.1) implies that $N(L) = K[(L_1/L)^{1/2} - 1]$, where L_1 is the luminosity of the first brightest galaxy, and K is the expected number of galaxies in the range $1/4L_1 \leqslant L \leqslant L_1$.

Abell (1975) has suggested that the luminosity function $N(L)$ be fit by two intersecting power laws, $N(L) = N^*(L/L^*)^{-\alpha}$, where $\alpha \approx 5/8$ for $L < L^*$, and $\alpha \approx 15/8$ for $L > L^*$. L^* is the luminosity at which the two power laws cross, and N^* is the expected number of galaxies with $L \geqslant L^*$. Of course, this form is intended only as a simple and practical fit to the data; the real luminosity function certainly has a continuous derivative $n(L)$, unlike Abell's function. The magnitude luminosity function corresponding to Abell's form is often written as

$$\log_{10} N(\leqslant m) = \begin{Bmatrix} K_1 + s_1 m & m \leqslant m^* \\ K_2 + s_2 m & m > m^* \end{Bmatrix} \tag{2.2}$$

Fig. 2. The luminosity function of galaxies in clusters. $N(\leqslant M_V)$ is the number of galaxies brighter than absolute magnitude M_V. The circles are the composite observed luminosity function derived by Schechter (1976). The solid circles exclude cD galaxies, while the open circles show the changes when they are included. The solid, dashed, and dash-dotted curves are the fitting functions of Schechter, Abell, and Zwicky, as discussed in the text.

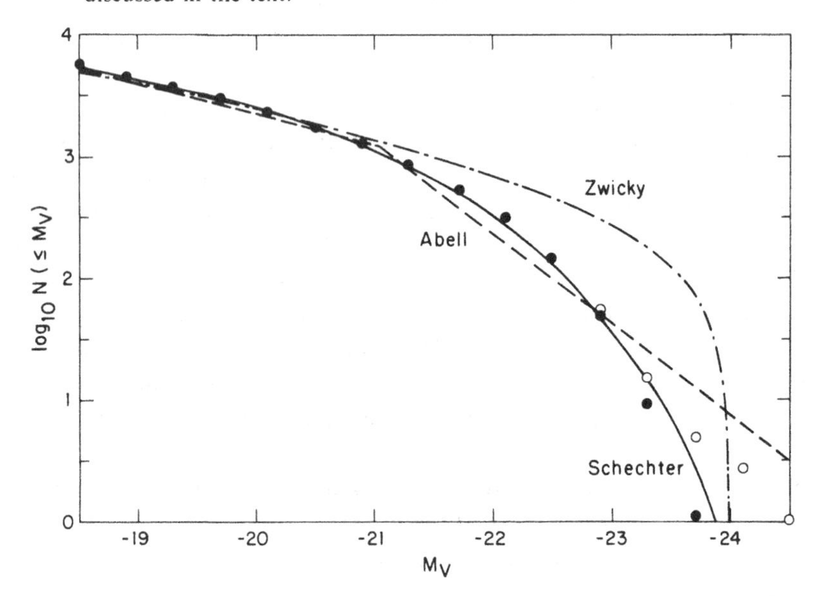

where K_1 and K_2 are constants. The slopes are approximately $s_1 \approx 0.75$ and $s_2 \approx 0.25$, and the power laws cross at $m = m^*$, so that $K_1 + s_1 m^* = K_2 + s_2 m^*$. As shown in Figure 2, the Abell form fits both the bright and faint ends of the luminosity function adequately.

Schechter (1976) proposed an analytic approximation for the differential luminosity function

$$n(L)dL = N^*(L/L^*)^{-\alpha} \exp(-L/L^*) d(L/L^*) \tag{2.3}$$

where L^* is a characteristic luminosity, $N^*\Gamma(1-\alpha, 1)$ is the number of galaxies with $L > L^*$, and $\Gamma(a, x)$ is the incomplete gamma function; Schechter derives a value for the faint end slope of $\alpha = 5/4$. The integrated luminosity function corresponding to equation (2.3) is $N(L) = N^*\Gamma(1-\alpha, L/L^*)$.

The advantages of the Schechter function are that it is analytic and continuous, unlike Abell's form, and that it is a real statistical distribution function, unlike Zwicky's form, which requires that the magnitude of the first brightest galaxy (which ought to be a random variable) be specified. An expression similar to the Schechter form was predicted by a simple analytical model for galaxy formation (Press and Schechter, 1974). The Schechter function is steeper than Abell's function at the bright end because it contains an exponential. The Schechter differential luminosity function decreases monotonically with luminosity, while the Abell function has a local maximum at $L = L^*$ (Abell, 1975); there is some weak evidence for such a peak, especially near the centers of rich clusters (Rood and Abell, 1973).

The Schechter function fits the observed distribution in many clusters reasonably well from the faint to the bright end (Figure 2), as long as the very brightest galaxies, the cD galaxies, are excluded (Schechter, 1976). These can have luminosities as large as $10L^*$, and thus are extremely improbable if equation (2.3) holds exactly. However, cD galaxies have a number of distinctive morphological characteristics which suggest that they were formed by special processes which occur primarily in the centers of rich clusters (see Section 2.10.1). In any case, they apparently can be excluded by morphological (as opposed to luminosity) criteria, and equation (2.3) can be taken to be the non-cD luminosity function.

The parameter N^* in the Abell or Schechter functions is a useful measure of the richness of the cluster. If these luminosity functions are adequate approximations, then fitting the luminosity function to determine N^* ought to give a more accurate measure of richness than counting galaxies in magnitude ranges. Note that while the total number of galaxies predicted by Abell's, Zwicky's, or Schechter's functions diverges at the faint end, the total luminosity is finite; for example, the total cluster optical luminosity is $L_{opt} =$

$N^*\Gamma(2-\alpha)L^*$ for the Schechter function, and N^* thus measures the total cluster luminosity.

Because of the break in the luminosity function near L^*, this parameter represents a characteristic luminosity of cluster galaxies. Moreover, L^* is nearly the same for many clusters; it corresponds to an absolute visual magnitude $M_V^* \approx -21.2 + 5\log_{10} h_{50}$ for Abell's form (Austin *et al.*, 1975), and $M_V^* \approx -21.9 + 5\log_{10} h_{50}$ for Schechter's form (Schechter, 1976). By comparing these values with the apparent magnitude m^* derived from the observed luminosity function of a cluster, one can derive a distance estimate for the cluster (Bautz and Abell, 1973; Schechter and Press, 1976). Unfortunately, this requires that magnitudes be available for a large number of galaxies.

The magnitude of the brightest galaxy in a cluster has often been used as a distance indicator in cosmological studies (see, for example, Hoessel, Gunn, and Thuan, 1980). It is obviously easier to determine observationally than the luminosity function. Moreover, the luminosities of the brightest cluster galaxies show a very small variation from cluster to cluster (Sandage, 1976) and depend only weakly on richness. There has been a controversy concerning whether the luminosities of these brightest galaxies are determined statistically by the cluster luminosity function, or whether the brightest galaxies are produced by some special physical processes operating in clusters (see Dressler, 1984, for a review). Under the statistical hypothesis, all of the galaxy luminosities are random variables drawn from the luminosity function (Geller and Peebles, 1976). Then, the observed number of galaxies in any luminosity range will have a Poisson distribution about the expectation given by the luminosity function, and the luminosity L_1 of the brightest galaxy will thus be distributed as $\exp[-N(L_1)]n(L_1)dL_1$. As $N(L_1)$ is proportional to richness, the statistical model predicts that L_1 increases with cluster richness, which is not really observed (Sandage, 1976). However, Schechter and Peebles (1976) have argued that the near constancy of L_1 results from a selection effect (that is, the observed sample is biased), and that the statistical hypothesis may still be valid.

Alternatively, the brightest cluster galaxies may be affected by special physical processes, such as the tidal interactions or mergers of galaxies (Peach, 1969; Ostriker and Tremaine, 1975; Hausman and Ostriker, 1978; Richstone, 1975). Evidence that this may indeed be the case is given in Section 2.10.1.

While the luminosity functions of many clusters are reasonably well represented by the Schechter or Abell form with a universal value of M^*, significant departures exist in a number of clusters (Oemler, 1974; Mottmann and Abell, 1977; Dressler, 1978b). These departures include variations in the value of M^*, variations in the slope of the faint end of the luminosity function (α in Schechter's form, equation (2.3)), and variations in the steepness of the bright end of the luminosity function (Dressler, 1978b). These variations are, in many cases, correlated with cluster morphology (Section 2.5). The variations in M^* and α probably reflect variations in the conditions in the cluster at its formation, while the variations in the bright end slope may result from evolutionary changes, such as the tidal interaction or merging of massive galaxies (Richstone, 1975; Hausman and Ostriker, 1978). In particular, the clusters with the steepest luminosity functions at the bright end often contain cD galaxies (Dressler, 1978b); this may indicate that the brighter galaxies were either eliminated by mergers to form the cD or diminished in brightness through tidal stripping (Section 2.10.1).

Turner and Gott (1976a) have shown that the luminosity function of galaxies in small groups is well represented by equation (2.3). In fact, Bahcall (1979a) has suggested that the luminosity function of all galaxian systems – from single galaxies (in or out of clusters) to the groups and clusters themselves – can be fit in a single function similar to the Schechter form (equation 2.3).

2.5 Morphological classification of clusters

A number of different cluster properties have been used to construct morphological classification systems for clusters. Somewhat surprisingly, these different systems are highly correlated, and it appears that clusters can be represented very crudely as a one-dimensional sequence, running from *regular* to *irregular* clusters (Abell, 1965, 1975). There is considerable evidence that the regular clusters are dynamically more evolved and relaxed than the irregular clusters. The various morphological classification schemes are described below, and the way in which they fit into the one-dimensional sequence is summarized in Table 1, which is adapted from Abell (1975) and Bahcall (1977a).

Zwicky *et al.* (1961–1968) classified clusters as *compact*, *medium compact*, or *open*. A *compact* cluster has a single pronounced concentration of galaxies, with more than ten galaxies appearing in contact as seen on the plate. A *medium compact* cluster has either a single concentration with ten galaxies separated by roughly their own diameters, or several concentrations. An *open* cluster lacks any pronounced concentration of galaxies.

Bautz and Morgan (1970) give a classification system based on the degree to which the cluster is dominated by its brightest galaxies. Bautz–Morgan *Type I* clusters are dominated by a single, central cD galaxy; cD galaxies have the most luminous and extensive optical emission found in galaxies (see Section

2.10.1). In *Type II* clusters, the brightest galaxies are intermediate between cD and normal giant ellipticals, while in *Type III*, there are no dominating cluster galaxies. *Type I–II* and *Type II–III* are intermediates. Leir and van den Bergh (1977) have classified 1889 rich Abell clusters on the Bautz–Morgan system, and some of the newer southern catalogs (e.g., White and Quintana, 1985) give Bautz–Morgan types for their clusters.

The original Rood–Sastry (1971) classification system is based on the nature and distribution of the ten brightest cluster galaxies. Basically, the six Rood–Sastry (RS) classes are defined as follows:

cD: the cluster is dominated by a central cD galaxy (example: A2199).
B: binary – the cluster is dominated by a pair of luminous galaxies (example: A1656 (Coma)).
L: line – at least three of the brightest galaxies appear to be in a straight line (example: A426 (Perseus)).
C: core – four or more of the ten brightest galaxies form a cluster core, with comparable galaxy separations (example: A2065 (Corona Borealis)).
F: flat – the brightest galaxies form a flattened distribution on the sky (example: A2151 (Hercules)).
I: irregular – the distribution of brightest galaxies is irregular, with no obvious center or core (example: A400).

Rood and Sastry (1971) give classifications for low redshift Abell clusters on this system. They show that these classifications form a bifurcated sequence,

Table 1. *Properties of morphological classes of clusters*

Property	Regular	Intermediate	Irregular
Zwicky Type	Compact	Medium–Compact	Open
Bautz–Morgan Type	I,I-II,II	II,II-III	II-III,III
Rood–Sastry Type	cD,B,L,C	L,C,F	F,I
Galactic Content	Elliptical-rich	Spiral-poor	Spiral-rich
E:S0:Sp	3:4:2	1:4:2	1:2:3
Morgan Type	ii	i-ii	i
Oemler Type	cD,Spiral-poor	Spiral-poor	Spiral-rich
Symmetry	Spherical	Intermediate	Irregular
Central Concentration	High	Moderate	Low
Subclustering	Absent	Moderate	Significant
Richness	Rich	Rich–Moderate	Rich–Poor
	$n^* \approx 10^2$	$n^* \gtrsim 10^1$	$n^* \gtrsim 10^0$

which can be represented by a 'tuning-fork' diagram (Figure 3a). This sequence is correlated with the sequence of *regular* to *irregular* clusters in the sense that clusters on the left of the diagram (*cD* and *B*) are regular and those to the right (*F* and *I*) are irregular. Rich clusters are more or less equally distributed among the three arms of the diagram.

Recently, Struble and Rood (1982, 1985) have proposed a revised version of the RS classification system. The definitions have been revised slightly, and a number of subclasses of the main RS classes have been proposed. More significantly, Struble and Rood have rearranged the tuning fork diagram into a 'split linear' diagram (Figure 3b), based on systematic trends in the galaxy distribution and content of clusters. This new scheme was devised in part from

Fig. 3. (a) The Rood–Sastry (1971) cluster classification scheme. (b) The revised Rood–Sastry classes from Struble and Rood (1982).

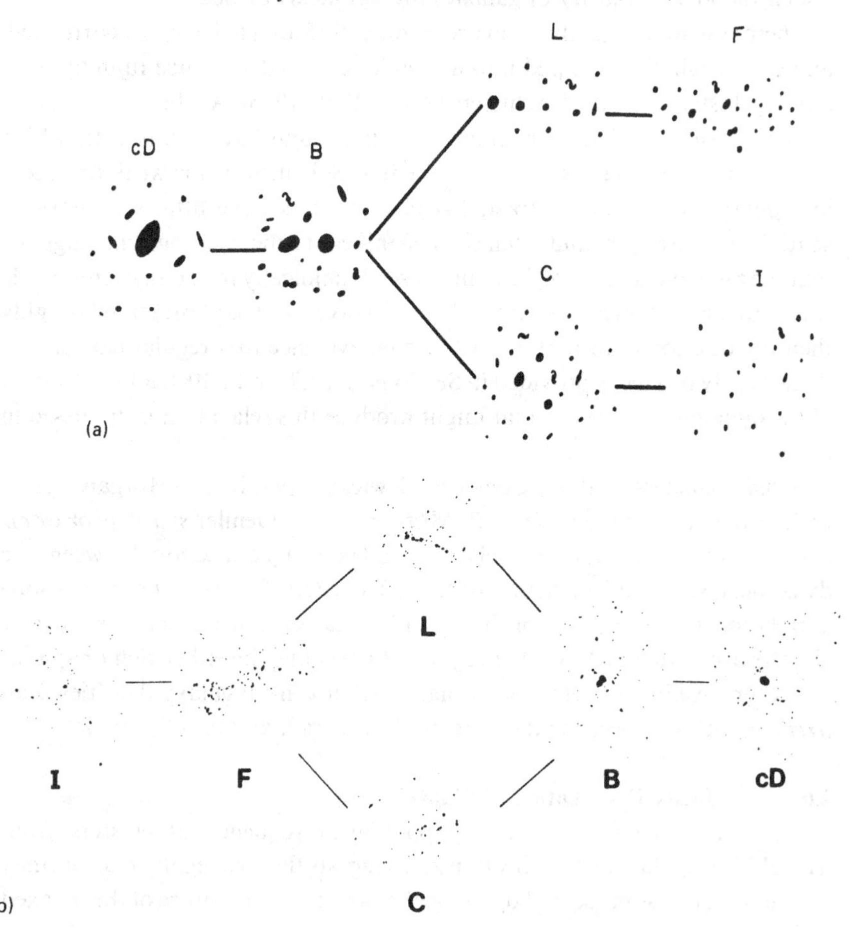

a comparison to numerical N-body simulations of the collapse of clusters (White, 1976c; Carnevali *et al.*, 1981; Farouki *et al.*, 1983; also see Figure 5). Struble and Rood propose that this sequence represents an evolutionary sequence of clusters from irregular *I* to *cD* clusters.

Morgan (1961) and Oemler (1974) have constructed classification systems based on the galactic content of clusters (that is, the fraction of cluster galaxies which are spirals (Sps), disk galaxies without spiral structure (S0s), or ellipticals (Es)). Morgan (1961) classified clusters as *type i* if they contained large numbers of spirals and as *type ii* if they contained few spirals. Oemler (1974) has refined this system, defining three classes of clusters: *spiral-rich* clusters, in which spirals (Sps) are the most common galaxies; *spiral-poor* clusters, in which spirals are less common and S0s are the most common galaxies; and *cD* clusters, which are dominated by a central cD galaxy and in which the great majority of galaxies are ellipticals or S0s.

These systems of classification are empirically found to be highly correlated, and can roughly be mapped into a one-dimensional sequence running from *regular* clusters to *irregular* clusters (Abell, 1965, 1975). As shown in Table 1, *regular* clusters are highly symmetric in shape, and have a core with a high concentration of galaxies toward the center. Subclustering is weak or absent in *regular* clusters. In contrast, *irregular* clusters have little symmetry or central concentration, and often show significant subclustering. This suggests that the *regular* clusters are, in some sense, dynamically relaxed systems, while the *irregular* clusters are dynamically less-evolved and have preserved roughly their distribution of formation. Additional evidence that regular clusters are dynamically relaxed is provided in Sections 2.6, 2.7, and 2.10.1, and the nature of the dynamical processes that might produce this relaxation is discussed in Section 2.9.

Regular clusters tend to be *compact* (Zwicky type), Bautz–Morgan *Type I* to *II*, Rood–Sastry types *cD* or *B*, Morgan *type ii*, Oemler *spiral–poor* or *cD* clusters. These last four correlations indicate a connection between the dynamical state and galactic content of clusters. There is no one-to-one correlation between the morphology of a cluster and its richness; *regular* clusters are always rich, while *irregular* clusters may be either rich or sparse. However, *regular* clusters tend to have higher central galaxy densities than *irregular* clusters, because they are at least as rich and more compact.

2.6 Velocity Distribution of Galaxies

The existence of the morphological sequence of clusters from irregular to regular clusters (Section 2.5) suggests that the regular clusters may have undergone some sort of dynamical relaxation. The nature of this relaxed distribution is examined in the next two sections through the distribution of cluster galaxy velocities and positions.

The redshifts in Section 2.2 are determined from the mean radial velocity of galaxies in a cluster; in fact, the radial velocities of individual galaxies are distributed around this mean. It has been conventional to characterize this distribution by the dispersion σ_r of radial velocities about the mean

$$\sigma_r = \langle (v_r - \langle v_r \rangle)^2 \rangle^{1/2} \tag{2.4}$$

where v_r is the radial velocity, which is the component of the galaxy velocity along the line-of-sight. The dispersion completely characterizes the radial velocity distribution function if it is Gaussian:

$$p(v_r)dv_r = \frac{1}{\sigma_r \sqrt{2\pi}} \exp(-(v_r - \langle v_r \rangle)^2 / 2\sigma_r^2) dv_r \tag{2.5}$$

Here, $p(v_r)dv_r$ is the probability that an individual cluster galaxy has a radial velocity in the range v_r to $v_r + dv_r$. While the Gaussian distribution has usually been adopted simply for convenience, statistical tests reveal that it is a consistent fit to the observed total distribution function in many clusters, at least if high velocities ($|v_r - \langle v_r \rangle| > 3\sigma_r$) are excluded (Yahil and Vidal, 1977). However, the velocity dispersion in a given cluster generally decreases with distance from the cluster center; in Coma and Perseus the decline is about a factor of two from the center to the outer edge ((Rood *et al.*, 1972; Kent and Gunn, 1982; Kent and Sargent, 1983). Moreover, the velocity dispersion can differ in different clumps of an irregular cluster showing subclustering (Geller and Beers, 1982; Bothun *et al.*, 1983).

A Gaussian distribution for a single component of the velocity obtains for a system of non-identical particles in thermodynamic equilibrium, in which case we identify the velocity dispersion with $\sigma_r \equiv (kT/m)^{1/2}$, where T is the galaxy 'temperature' and m the galaxy mass. While the Gaussian velocity distribution found in clusters suggests that they are at least partially relaxed systems, they are not fully relaxed to thermodynamic equilibrium. In thermodynamic equilibrium, all components of the cluster would have equal temperatures; what is observed is that it is the velocity dispersion (not temperature) which is nearly independent of galaxy mass and position (Rood *et al.*, 1972; Kent and Gunn, 1982; Kent and Sargent, 1983).

Some variation of the velocity dispersion with galaxy mass or cluster position is observed (Rood *et al.*, 1972; Kent and Gunn, 1982; Kent and Sargent, 1983). In the Coma and Perseus clusters, the most luminous galaxies have a somewhat smaller velocity dispersion than the less luminous galaxies, and the velocity dispersion decreases with increasing projected distance from the cluster center. In fact, the latter effect must necessarily occur if clusters are finite, bound systems. Then, the velocities of bound galaxies at any point in the cluster cannot exceed the escape velocity at that position. As the escape velocity decreases with increasing distance from the cluster center, the velocity dispersion must also decrease with projected distance from the cluster center.

The observed mean galaxy velocity $\langle v_r \rangle$ will depend on the projected position in the cluster if the cluster is rotating. The projected shapes of many clusters are substantially flattened. Of course, this is true of the L (line) and F (flat) clusters (Section 2.5); however, many regular (cD and B) clusters are also significantly flattened (Abell, 1965; Dressler, 1981). If these clusters are actually oblate in shape due to significant rotational support, the variation in $\langle v_r \rangle$ across the cluster would be expected to be comparable to σ_r. In fact, such large velocity gradients are not observed (Rood *et al.*, 1972; Gregory and Tifft 1976; Schipper and King, 1978; Dressler, 1981); apparently the flattening of clusters is not due to rotation.

Useful compilations of velocity dispersions for clusters have been given by Hintzen and Scott (1979), Danese *et al.* (1980), and particularly Noonan (1981).

2.7 Spatial Distribution of Galaxies

The most regular clusters show a smooth galaxy distribution with a concentrated core (Figure 4; Table 1). In general, models to describe the galaxy distribution in these clusters will possess at least five parameters, which can be taken to be the position of the cluster center on the sky, the central projected density of galaxies per unit area of the sky σ_o, and two distance scales r_c and R_h. The core radius r_c is a measure of the size of the central core, and is usually defined so that the projected galaxy density at a distance r_c from the cluster center is one half of the central density σ_o. The halo radius R_h measures the maximum radial extent of the cluster. Of course, the observed value of the central density σ_o must depend on the range of galaxy magnitudes observed, and the values of other parameters may also depend on galaxy magnitude. If the cluster is elongated, at least two more parameters are necessary; these can be taken to be the orientation of the semimajor axis of the cluster on the sky, and the ratio of semimajor to semiminor axes. However, spherically symmetric galaxy distributions will be discussed first, and r_c and R_h will be assumed to be independent of galaxy mass or luminosity. Then, one can write

$$\begin{aligned} n(r) &= n_o \mathrm{f}(r/r_c, r/R_h) \\ \sigma(b) &= \sigma_o \mathrm{F}(b/r_c, b/R_h), \end{aligned} \tag{2.6}$$

where $n(r)$ is the spatial volume density of galaxies at a distance r from the cluster center, n_o is the central ($r = 0$) density, $\sigma(b)$ is the projected surface density at a projected radius b, and f and F are two dimensionless functions. Obviously,

$$\sigma(b) = 2 \int_b^{R_H} \frac{n(r) r dr}{(r^2 - b^2)^{1/2}}. \tag{2.7}$$

A number of models have been proposed to fit the distribution of galaxies. Among the simplest are the isothermal models, which assume a Gaussian radial velocity distribution for galaxies (equation 2.5). If one further assumes

Fig. 4. The dots give the projected galaxy number density observed in 12 regular clusters, from Bahcall (1975). The observed number densities are normalized to the central surface number density σ_o and given as a function of the projected radius b divided by the core radius r_c. These parameters were determined by fitting equation (2.11) (the solid curve) to the observed distributions.

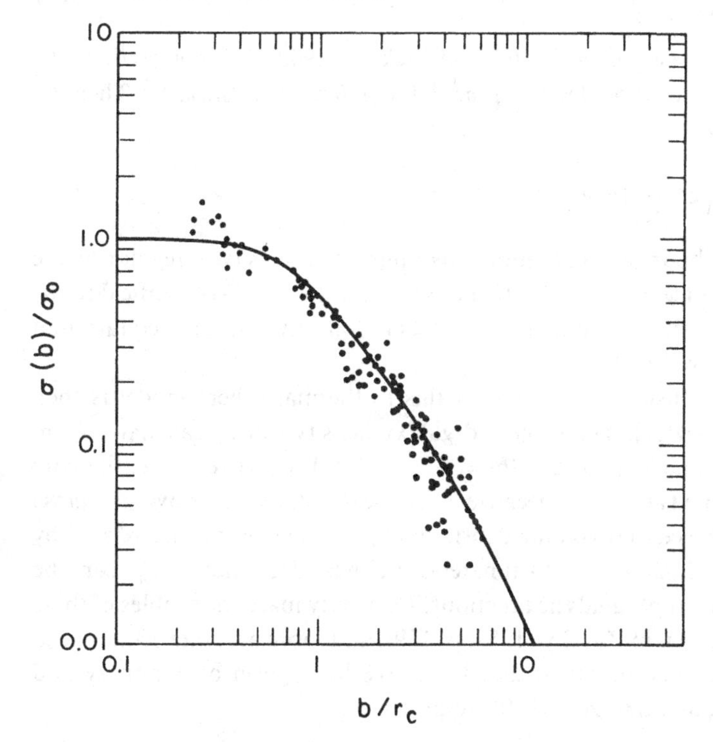

that the velocity distribution is isotropic and independent of position, that the galaxy distribution is stationary, and that galaxy positions are uncorrelated, then one can write the galaxy phase space density $f(\mathbf{r}, \mathbf{v})$ as

$$f(\mathbf{r}, \mathbf{v})d^3rd^3v = n(r)(2\pi\sigma_r^2)^{-3/2} \exp\left[-\frac{1}{2}\left(\frac{v}{\sigma_r}\right)^2\right]d^3vd^3r. \quad (2.8)$$

Now, the time-scale for two-body gravitational interactions in a cluster is much longer than the time for a galaxy to cross the cluster (see Section 2.9). Thus the galaxies can be considered to be a collisionless gas, and the phase space density f is conserved along particle trajectories (Liouville's theorem); f is then a function of only the integrals of the motion ('Jeans' theorem'). If the velocity distribution is isotropic, f does not depend on the orbital angular momentum of the galaxies, and can only depend on the energy per unit mass $\varepsilon = \frac{1}{2}v^2 + \phi(r)$, where $\phi(r)$ is the gravitational potential of the cluster. If we measure $\phi(r)$ relative to the cluster center, the galaxy spatial distribution is thus $n(r) = n_o \exp[-\phi(r)/\sigma_r^2]$. If the mass in the cluster is distributed in the same way as the galaxies, then Poisson's equation for the gravitational potential becomes

$$\frac{1}{r^2}\frac{d}{dr}\left(r^2\frac{d\phi}{dr}\right) = 4\pi G n_o m \exp(-\phi/\sigma_r^2), \quad (2.9)$$

where m is the mass per galaxy (Chandrasekhar, 1942). It is conventional to make the change of variables $\psi \equiv \phi/\sigma_r^2$, $\xi \equiv r/\beta$, $\beta \equiv \sigma_r/(4\pi G n_0 m)^{1/2}$. Then the equation for ψ is

$$\frac{1}{\xi^2}\frac{d}{d\xi}\left(\xi^2\frac{d\psi}{d\xi}\right) = e^{-\psi}, \quad (2.10)$$

subject to the boundary conditions (assuming the density is regular at the cluster center) of $\psi = 0$, $d\psi/d\xi = 0$ at $\xi = 0$. Equation (2.10) is identical to the equation for an isothermal gas sphere in hydrostatic equilibrium (Chandrasekhar, 1939).

The galaxy density distribution in this isothermal sphere model is then $n(\xi) = n_o \exp[-\psi(\xi)]$. The projected galaxy density can be calculated from equation (2.7) and written as $\sigma(b) = \sigma_o F_{isot}(b/\beta)$. For convenience, the core radius r_c is defined as $r_c \equiv 3\beta$, because $F_{isot}(3) \approx 0.502$, which is obviously close to one-half. The central volume density and projected density are related by $\sigma_o \approx 6.06 n_0 \beta = 2.02 n_o r_c$. Unfortunately, neither $\psi(\xi)$ nor F_{isot} can be represented by simple analytic functions. Relatively inaccurate tables of these functions are given in Zwicky (1957, p. 139), and more accurate economized analytic approximations to ψ and F_{isot} have been given by Flannery and Krook (1978), and Sarazin (1980), respectively.

At large radii $\xi \gg 1$, $n(\xi) \approx 2n_o/\xi^2$, and the total number of galaxies and total mass diverge in proportion to r. Thus the isothermal sphere cannot accurately represent the outer regions of a finite cluster. A number of methods to truncate the isothermal distribution have been used. If one is primarily concerned with representing the galaxy distribution near the core, and if one is not concerned with determining dynamically consistent velocity and spatial distributions, one can simply truncate the isothermal distribution at some radius. Zwicky (1957, p. 140) and Bahcall (1973a) have truncated the isothermal distribution with a uniform surface density cutoff C:

$$\sigma(b) = \frac{\sigma_o[\mathrm{F}_{isot}(b/\beta) - C/6.06]}{[1 - C/6.06]} \tag{2.11}$$

The cluster galaxy density then falls to zero at a radius R_h given by $\mathrm{F}_{isot}(R_h, \beta) = C/6.06$. Bahcall (1973a) also defines a modified central surface density parameter $\alpha \equiv \sigma_o/(6.06 - C)$; then, for small $C \ll 2\pi$, the central volume density is just $n_o \approx \sigma_o[(C/2\pi)^2]/(6.06\beta) \approx \alpha/\beta$. The solid curve for the surface density in Figure 4 is given by equation (2.11).

King (1966) has developed self-consistent truncated density distributions for clusters. The phase-space density he assumes is

$$f(\mathbf{r}, \mathbf{v})d^3r d^3v \propto \exp\left[\frac{\phi(0) - \phi(r)}{\sigma_{r\infty}^2}\right] \times \left\{\exp\left(-\frac{v^2}{2\sigma_{r\infty}^2}\right) - \exp\left(\frac{\phi(r)}{\sigma_{r\infty}^2}\right)\right\} d^3r d^3v \tag{2.12}$$

where $\sigma_{r\infty}$ is the radial velocity dispersion in an untruncated cluster. The velocity distribution is thus truncated at the escape velocity $v_e : f(\mathbf{r}, |\mathbf{v}| \geq v_e) = 0$ where $v_e^2 = -2\phi(r)$, and the potential goes to zero at infinity $\phi(\infty) = 0$. King showed that equation (2.12) gave an approximate solution to the Fokker–Planck equation for a finite cluster subject to two-body gravitational encounters. As shown in Section 2.9, galaxy clusters are nearly collisionless; however, it is possible that equation (2.12) is a reasonable approximation for the truncated phase-space density. The phase-space density f in equation (2.12) is a function of only the energy per unit mass ε and the parameter $\sigma_{r\infty}^2$, and thus satisfies Jeans' theorem. King integrates equation (2.12) over all velocities to give the density $n(r)$ as a function of the potential $\phi(r)$, and then solves Poisson's equation to give a self-consistent potential. The density $n(r)$ in these models falls continuously to zero at a finite radius R_h. The models can be scaled in distance and central density as with the unbounded isothermal models described earlier. The only characteristic parameter is $\sigma_r(0)/\sigma_{r\infty}$ or equivalently R_H/r_c (where r_c is again the core radius).

King prefers to use the potential difference between cluster center and edge, $W_o \equiv [\phi(R_h) - \phi(0)]/\sigma_{r\infty}^2$. These models predict that the velocity dispersion declines in the outer portions of the cluster, as is observed in Coma (Rood *et al.*, 1972).

Unfortunately, none of these bounded or unbounded isothermal models can be represented exactly in terms of simple analytic functions. However, King (1962) showed that the following analytic functions were a reasonable approximation to the inner portions of an isothermal function:

$$\begin{aligned} n(r) &= n_o[1 + (r/r_c)^2]^{-3/2}, \\ \sigma(b) &= \sigma_o[1 + (b/r_c)^2]^{-1}, \end{aligned} \tag{2.13}$$

where $\sigma_o = 2n_o r_c$. At large radii $r \gg r_c$, $n(r) \approx n_o(r_c/r)^3$ in the analytic King model; thus the cluster mass and galaxy number diverge as $\ln(r/r_c)$. Although this is a slower divergence than the unbounded isothermal model, this analytic King model also must be truncated at some finite radius R_h.

Another analytic model is that of de Vaucouleurs (1948a), which was proposed to fit the distribution of surface brightness in elliptical galaxies. However, this distribution also fits many regular clusters (de Vaucouleurs, 1948b). The projected density is

$$\sigma(b) = \sigma_o \exp[-7.67(b/r_e)^{1/4}] \tag{2.14}$$

where r_e is an effective radius such that one half of the galaxies lie at projected radii $b \leqslant r_e$. Accurate tables of the three-dimensional density and potential for this model have been given by Young (1976). The de Vaucouleurs form has several advantages over the isothermal function. It has only one distance scale, the effective radius r_e. It also converges to a finite total number of galaxies and cluster mass without a cutoff radius. Numerical simulations of the collapse of clusters seem to lead to distributions similar to this form (see Section 2.9.2). Unfortunately, the de Vaucouleurs form has not been widely used to fit galaxy distributions in clusters, and there have been few attempts to determine objectively whether it or the isothermal models give better fits to the actual distributions.

One major difference between the isothermal functions and the de Vaucouleurs law is that the latter has a density cusp at the cluster center; in fact, as can be seen from Figure 4, many clusters show these cusps, which must be removed in order to fit isothermal sphere models to the galaxy distributions. Beers and Tonry (1986) show that the galaxy distribution in clusters is very sensitive to the position chosen for the cluster center, and that many clusters have central number density spikes if the cluster center is assumed to correspond to the position of a cD galaxy (Section 2.10.1) or the maximum of the X-ray surface brightness (Section 4.4.1). They find that the surface density near this cusp varies roughly as $\sigma(b) \propto b^{-1}$, which is consistent with a singular isothermal sphere (one with $r_c \to 0$), or with an anisotropic galaxy velocity distribution, with an excess of radial orbits. The presence of these cusps is important to understanding the occurrence of multiple nuclei and companions about cD galaxies (Section 2.10.1).

Another useful fitting form is the Hubble (1930) profile, which is sometimes used to fit the light distribution in elliptical galaxies. It is

$$\sigma(b) = \sigma_o \left(1 + \frac{b}{r_c}\right)^{-2}, \tag{2.15}$$

which has the same asymptotic distribution as equation (2.13).

These models (equations 2.11–2.15) have been used to fit the projected distribution of galaxies. In most cases, the galaxy distributions have been fit to the truncated isothermal model (equation (2.11)). Figure 4 shows the surface number density distributions in 15 regular clusters, from Bahcall (1975), along with the fitting function (equation (2.11)). A compilation of the values of the core radii which have been derived for clusters is given in Table III of Sarazin (1986a), which includes values from Abell (1977), Austin and Peach (1974a), Bahcall (1973a, 1974a, 1975), Bahcall and Sargent (1977), Birkenshaw (1979), Bruzual and Spinrad (1978a, b), des Forets *et al.* (1984), Dressler (1978c), Havlen and Quintana (1978), Johnston *et al.* (1981), Koo (1981), Materne *et al.* (1982), Quintana (1979), Sarazin (1980), Sarazin and Quintana (1987), and Zwicky (1957).

Bahcall (1975) has suggested that the core radii of regular clusters are all very similar, with an average value

$$r_c = (0.25 \pm 0.004) h_{50}^{-1} \text{ Mpc}. \tag{2.16}$$

Sarazin and Quintana (1987) find that this may be true for the most compact, regular clusters. However, they also find that the core radius and galaxy distribution depend on the morphology of the cluster (Section 2.5).

The statistical uncertainty in the determination of the core radius or central density of a cluster tends to be rather large ($\gtrsim 30\%$), because even in a rich cluster only a small fraction of the galaxies are within the core. However, the errors in σ_o and r_c are highly anticorrelated, so that the product $(\sigma_o r_c)$ is relatively well-determined. The reason for this is that the number of galaxies within a projected radius b such that $r_c < b \ll R_h$ is roughly $N(b) \approx 2.17(\sigma_o r_c)b$ for an isothermal model. Thus the uncertainty in the product $(\sigma_o r_c)$ tends to be determined by Poisson statistics on the total number of cluster galaxies observed, and not just by the smaller number in the core (Sarazin and Quintana, 1987). Bahcall (1977b, 1981) has defined a related quantity $\bar{N}_o$ as the number of 'bright galaxies' with projected positions within 0.5 Mpc of the cluster center. Here, bright galaxies are those no more than two magnitudes fainter than the third brightest galaxy ($m \leqslant m_3 + 2$). Of course, the magnitude of the third brightest galaxy m_3 itself depends on richness; $\bar{N}_o$ is the number corrected for richness assuming a universal luminosity function (Section 2.4). From the discussion above, it is clear that $\bar{N}_o \propto (\sigma_o r_c)$, if σ_o is taken at a standard luminosity level (for example, L^* (Section 2.4)), and if $r_c \lesssim 0.5$ Mpc $\ll R_h$.

Because they are better determined statistically than the core radius r_c or center surface density σ_o, $(\sigma_o r_c)$ and $\bar{N}_o$ are often more useful as richness parameters when searching for correlations of integral properties of clusters in the optical, radio, and X-ray region. However, when comparing detailed spatial distributions the core radius is needed.

For example, from the arguments given above $\sigma_o r_c \propto \bar{N} \propto n_o r_c^2$, if σ_o and n_o are taken at a standard luminosity level. From equation (2.9), $r_c^2 = 9\sigma_r^2/4\pi G n_o m$, where m is the average galaxy mass $m \equiv \rho_o/n_o$, and ρ_o is the central mass density. As n_o is defined at a fixed luminosity level, this gives $(\sigma_o r_c) \propto \bar{N}_o \propto (M/L)^{-1}\sigma_r^2$, where (M/L) is the mass-to-light ratio of the cluster (Section 2.8). Bahcall (1981) finds the empirical correlation $\bar{N}_o \approx 21(\sigma_r/10^3\ \mathrm{km/s})^{2.2}$, which suggests that cluster mass-to-light ratios decrease with σ_r. This relationship may be useful for providing quick estimates of the velocity dispersions of clusters.

Several other size scales can be determined for clusters. The halo radius R_h gives the outermost limit of the cluster. Unfortunately, this is very poorly determined, because it depends critically on the assumed background. Moreover, clusters often have very extended haloes or are embedded in extended regions of enhanced density (superclusters). For Coma, the main isothermal distribution of galaxies extends to roughly $4h_{50}^{-1}$ Mpc; there is then a low-density halo extending to $\approx 10h_{50}^{-1}$ Mpc, which blends into the Coma supercluster which extends to a radius of about $35h_{50}^{-1}$ Mpc (Rood *et al.*, 1972; Rood, 1975; Chincarini and Rood, 1976; Abell, 1977; Gregory and Thompson, 1978; Shectman, 1982). However, studies of the galaxy covariance function (Peebles, 1974) suggest that there are no preferred scales for galaxy clustering and that the outer regions of clusters and superclusters represent a continuous distribution of clustering.

Other size scales for clusters have been measured that are intermediate between the core and halo size; they include the harmonic mean galaxy separation (Hickson, 1977), which is related to the gravitational radius R_G of a cluster (Section 2.8), the de Vaucouleurs effective radius r_e defined by equation (2.14), the mean projected distance from the cluster center (Noonan, 1974; Capelato *et al.*, 1980), and the Leir and van den Bergh radius (1977).

While the distribution functions for galaxies discussed above are spherically symmetric, most clusters appear to be at least slightly elongated, and some are highly elongated (Sastry, 1968; Rood and Sastry, 1972; Rood *et al.*, 1972; Bahcall, 1974a; MacGillivray *et al.*, 1976; Thompson and Gregory, 1978; Carter and Metcalfe, 1980; Dressler, 1981; Binggeli, 1982). Carter and Metcalfe (1980) and Binggeli (1982) give ellipticities and position angles for samples of Abell clusters. Their results suggest that clusters have average intrinsic ellipticities of ≈ 0.5–0.7; thus clusters are actually much more elongated on average than elliptical galaxies.

Carter and Metcalfe (1980) and Binggeli (1982) find that the position angles for the long axes of clusters are significantly aligned with the axis of the first-brightest cluster galaxy. In Sections 2.9.3. and 2.10.1, it is shown that such alignments may result if the first-brightest galaxies are produced by the merger of smaller galaxies through dynamical friction. They might also be produced during the collapse of the cluster.

Thompson (1976) has suggested that the axes of many of the elliptical galaxies in clusters may be aligned with the cluster axis; Adams *et al.* (1980) find a similar effect in two linear (L; see Section 2.5) clusters. Helou and Salpeter (1982) and Salpeter and Dickey (1985) do not find such alignments for the axes of spiral galaxies in the Virgo or Hercules clusters.

Binggeli (1982) finds that the long axes of Abell clusters tend to point at one another, even when the clusters are separated by as much as $\approx 30 h_{50}^{-1}$ Mpc. The alignments of nearby clusters were found to show evidence of a correlation even up to distances a factor of three larger.

In the previous discussion, it has been assumed that the galaxy density decreases monotonically with distance from the cluster center. However, Oemler (1974) found a plateau or local minimum in the projected distribution of galaxies in many clusters at a radius of about $0.4 R_G$ (where R_G is the gravitational radius). These features would imply the existence of significant oscillations in the three-dimensional galaxy density, although the process of deprojecting to counts is rather unstable (Press, 1976). Although these features are not statistically very significant in any one case, they do appear in a large number of clusters (Omer *et al.*, 1965; Bahcall, 1971; Austin and Peach, 1974a).

2.8 Cluster Masses – The Missing Mass Problem

The masses of clusters of galaxies can be determined if it is assumed that they are bound, self-gravitating systems. Cluster masses were first derived by Zwicky (1933) and Smith (1936). They found that the masses greatly exceed those which would be expected by summing the masses of all the cluster galaxies. Reviews of cluster mass determinations include Faber and Gallagher (1979) and Rood (1981).

If clusters were not bound systems, they would disperse rather quickly (in a crossing time $t_{cr} \approx 10^9$ years). Because many clusters appear regular and relaxed, and because their galactic content is quite different from that of the field, it is very unlikely that the regular compact clusters are flying apart. Assuming they are held together by gravity, one limit on the mass of clusters comes from the binding condition,

$$E = T + W < 0. \tag{2.17}$$

Here E is the total energy, T the kinetic energy, and W the gravitational potential energy:

$$T = \frac{1}{2} \sum_i m_i v_i^2,$$

$$W = -\frac{1}{2} \sum_{i \neq j} \frac{G m_i m_j}{r_{ij}}, \tag{2.18}$$

where the sums are over all galaxies, m_i and v_i are the galaxy mass and velocity, and r_{ij} is the separation of galaxies i and j.

A stronger mass limit results if we assume that the cluster distribution is stationary. The equations of motion of galaxies can be integrated to give

$$\frac{1}{2} \frac{d^2 I}{dt^2} = 2T + W, \tag{2.19}$$

where $I = \sum m_i r_i^2$ is the moment of inertia measured from the center of mass. For a stationary configuration, the left-hand-side of equation (2.19) is zero, and

$$W = -2T, \qquad E = -T. \tag{2.20}$$

This is the virial theorem. The total cluster mass is $M_{tot} = \sum m_i$; we can define a mass weighted velocity dispersion $\langle v^2 \rangle \equiv \sum m_i v_i^2 / M_{tot}$, and gravitational radius

$$R_G \equiv 2 M_{tot}^2 \left(\sum_{i \neq j} \frac{m_i m_j}{r_{ij}} \right)^{-1}. \tag{2.21}$$

Then, the virial theorem gives

$$M_{tot} = \frac{R_G \langle v^2 \rangle}{G}. \tag{2.22}$$

Now, $\langle v^2 \rangle$ and R_G can be evaluated from the radial velocity distribution and the projected spatial distribution of a fair sample of galaxies if one assumes that the positions of galaxies and the orientation of their velocity vectors are uncorrelated. Then, $\langle v^2 \rangle = 3\sigma_r^2$, where σ_r is the (mass-weighted) radial velocity dispersion, and $R_G = (\pi/2)b_G$ (Limber and Mathews, 1960), where

$$b_G = 2M_{tot}^2 \left(\sum \frac{m_i m_j}{b_{ij}} \right)^{-1} \tag{2.23}$$

and b_{ij} is the projected separation of galaxies i and j. Alternatively, R_G can be calculated from a fit to the galaxy distribution (as in Section 2.7) or from strip counts of galaxies (Schwarzschild, 1954). Combining these we have

$$M_{tot} = \frac{3R_G\sigma_r^2}{G} = 7.0 \times 10^{14} M_\odot \left(\frac{\sigma_r}{1000\ \text{km/s}} \right)^2 \left(\frac{R_G}{\text{Mpc}} \right). \tag{2.24}$$

Since $\sigma_r \approx 10^3$ km/s and $R_G \approx 1$ Mpc for an average rich cluster, one finds $M_{tot} \approx 10^{15} M$ typically.

The virial theorem requires that one sum over all the particles in a cluster; it can only be directly applied to galaxies alone if they contain all of the mass in the cluster. One can also derive the mass of a cluster by treating the galaxies as test particles in dynamical equilibrium in the gravitational potential of the cluster. Such calculations generally give results for the total mass which are consistent with the virial theorem determinations; usually, one must assume that the galaxies and the total mass have the same spatial distribution. The most accurate of these determinations involve comparing numerical or analytical models for the galaxy distribution in a cluster with the observed spatial and velocity distributions (White, 1976c; Kent and Gunn, 1982; Kent and Sargent, 1983). The best data exist for the Coma and Perseus clusters; the masses for both these clusters are about $3 \times 10^{15} M_\odot$ within a projected radius of 3° from the cluster center (Kent and Gunn, 1982; Kent and Sargent, 1983).

These analyses give surprisingly large masses for clusters, particularly when compared to the total luminosity of a cluster $L_{tot} \approx 10^{13}\ L_\odot$ The conventional method of quantifying this comparison is to calculate the mass-to-light ratio of a system in solar units $(M_{tot}/L_{tot})/(M_\odot/L_\odot)$. Obviously, any system composed solely of stars like our Sun would have a mass-to-light ratio of unity. Mass-to-light ratios have been derived for a number of clusters using the virial theorem and the measured magnitudes of the galaxies at visual (V) or blue (B) magnitudes. Typically (see Faber and Gallagher, 1979, and Rood, 1981, and references in these reviews) one finds

$$\begin{aligned} (M/L_V)_{tot} &\approx 250 h_{50} M_\odot / L_\odot, \\ (M/L_B)_{tot} &\approx 325 h_{50} M_\odot / L_\odot. \end{aligned} \tag{2.25}$$

The mass-to-light ratios found for the luminous portions of galaxies range from $(M/L_B)_{tot} \approx (1 - 12) h_{50} M_\odot / L_\odot$, with the large values corresponding to the E and S0 galaxies which predominate in compact regular clusters. *Thus only about* 10% *of the mass in clusters can be accounted for by material within the luminous portions of galaxies*. This is the so-called 'missing mass' problem, although this is something of a mixnomer. It is not the mass which is missing; what is missing is any other evidence for this invisible matter. The application of the virial theorem to clusters and the discovery that most of their mass was nonluminous was first made by Zwicky (1933) and Smith (1936).

How secure are these mass determinations for clusters? As long as a fair sample of galaxies are observed, the virial theorem masses are insensitive to velocity anisotropies, although this can be a problem if only galaxies near the cluster center are used. The dynamical model estimates for the masses are also not affected much by velocity anisotropies (Kent and Gunn, 1982). The virial theorem applies even if most galaxies are bound in binaries or subclusters, as long as the velocities and positions are measured for a fair sample of galaxies and R_G is computed directly from the galaxy positions. However, galaxies are not generally bound in binaries in regular clusters (Abell, Neyman, and Scott, 1964), and in any case typical binary or group velocities are too small to contribute significantly to σ_r (however, see Noonan, 1975a). If the clusters are bound but not relaxed, equation (2.17) would give a mass lower by only a factor of two than equation (2.20). The velocity dispersion and R_G of a cluster might be contaminated by field galaxies. This background contamination problem can be reduced by determining the mass-to-light ratio for the core of a regular cluster, using the core parameters of an isothermal model fit (Section 2.7). Again, one typically finds $(M/L_V)_{core} \approx 300 h_{50} M_\odot / L_\odot$ (Rood *et al.*, 1972; Bahcall, 1977a; Dressler, 1978b).

In principle, the missing mass could be associated with a large number of low luminosity, large M/L galaxies (see Noonan, 1975b; Section 2.4). Such galaxies could affect the mass-to-light ratio if they made a significant contribution to the total light of a cluster which was not included in the usual determinations of the total luminosity because the galaxies were too faint to be detected individually. However, these galaxies would then produce a significant amount of diffuse light between galaxies in a cluster. Observations of the diffuse light in the Coma cluster show that its luminosity is at most comparable to that in bright galaxies (de Vaucouleurs and de Vaucouleurs, 1970; Melnick *et al.*, 1977; Thuan and Kormendy, 1977). Thus the required mass-to-light ratios cannot be greatly affected by low mass galaxies. Abell (1977) included a large number of low mass galaxies in his determination of the mass-to-light ratio in Coma, and found a value about a factor of two smaller than that in equation (2.25).

There is now considerable direct and indirect evidence for invisible matter associated directly with individual galaxies as well as clusters (see Faber and

Gallagher, 1979, and Rood, 1981, for extensive reviews). It has been suggested that the invisible matter forms an extended halo around non-cluster galaxies, extending significantly further than the luminous material and having at least several times more mass (Einasto *et al.*, 1974; Ostriker *et al.*, 1974).

The nature of this missing mass component in clusters and galaxies remains one of the most important unsolved problems in astrophysics. Careful searches have been made for diffuse radiation in clusters, which might be produced if the missing mass were gaseous or stellar. These have included searches for radio free–free emission (Davidsen and Welch, 1974), optical and UV line emission (Crane and Tyson, 1975; Bohlin *et al.*, 1973), optical continuum emission (de Vaucouleurs and de Vaucouleurs, 1970; Melnick *et al.*, 1977; Thuan and Kormendy, 1977), and 21 cm radio emission from neutral hydrogen (Shostak *et al.*, 1980). The observations of X-ray emission from clusters, which are the primary subject of this book, also give limits on the mass of hot gas (Section 5.4.1). These studies and other similar searches have shown that the missing mass is not diffuse gas at any temperature, not dust grains of the type found in the interstellar medium, and not luminous stars.

In general, most suggestions as to the identity of the missing mass fall into three categories:

(1) substellar mass condensations, such as stars with mass $<0.1 M_{\odot}$ (Ostriker, Peebles, and Yahil, 1974), planetary size bodies, or comets (Tinsley and Cameron, 1974);
(2) invisible remnants of massive stars (black holes, neutron stars, or cool white dwarfs);
(3) stable, weakly interacting elementary particles, such as massive neutrinos, magnetic monopoles, axions, photinos, etc. (Blumenthal *et al.*, 1984).

Where is this missing mass located? A number of arguments suggest that the total mass distribution is similar to the galaxy distribution (Rood *et al.*, 1972). The fact that the galaxy distributions in compact, regular clusters are reasonably represented by isothermal spheres and that core and total mass-to-light ratios are equal within the errors supports this view. On the other hand, Smith *et al.* (1979b) and Smith (1980) have argued that the galaxy and mass distributions can be very dissimilar if the galaxies collapsed after the cluster potential was established. However, in this case violent relaxation (Section 2.9.2) is not effective, clusters could not have compact cores, and the mass-to-light ratio in the inner regions would not be the same as that for the entire cluster.

Is the missing mass bound to individual galaxies in clusters? Probably not. The rate of two-body relaxation (dynamical friction) of galaxies in clusters is proportional to galaxy masses (Section 2.9.1). If the missing mass were distributed among galaxies in proportion to their luminosity, very significant mass segregation would be expected (Rood, 1965; White, 1977b), which is not seen (Section 2.7). Moreover, if the missing mass were bound to galaxies in haloes having velocity dispersions similar to those in their visible portions, then the haloes would have to extend roughly 0.5 Mpc from the galaxy center (White, 1985). This is much larger than the typical separations of galaxies in the cores of clusters. Thus the missing mass probably forms a continuum, occupying the entire volume of the cluster, but with an overall density distribution fairly similar to that of the galaxies.

Although the missing mass is probably not currently bound to individual galaxies, it may initially have formed massive haloes around individual galaxies, which were stripped by tidal interactions during and after the formation of the cluster (Richstone, 1976, and Section 2.10.1). As mentioned above, there is considerable evidence that more isolated galaxies are immersed in extensive massive haloes (see reviews by Faber and Gallagher, 1979, and Rood, 1981). However, the time scales for tidal interactions and dynamical friction are similar; it is not clear, therefore, that tidal interactions will strip the massive haloes before galaxies merge. The problem of the competition between tidal forces and dynamical friction is the subject of a number of recent investigations (Richstone and Malumuth, 1983; Malumuth and Richstone, 1984; Merritt, 1983, 1984a, b, 1985; Miller, 1983), which unfortunately reach contradictory conclusions (Section 2.10.1). This problem is particularly serious for the massive binary galaxies which dominate B clusters (Section 2.5).

Alternatively, the missing mass in clusters may never have been associated with galaxies. For example, the missing mass may be more extensive than the luminous material; the luminous matter may only be the 'tip of the iceberg' of the real mass distribution. Rood *et al.* (1970) found that M/L increased with the scale size of the system (see also Rood, 1981), which supports this hypothesis; however, Turner and Sargent (1974) have argued that this was an artifact of calculating M from equation (2.22).

In many ways, a much better way to determine the masses of clusters of galaxies and the distribution of the missing mass is to apply the hydrostatic equation to the distribution of hot intracluster gas. This method, which is discussed in Section 5.5.5, avoids any assumption about the shapes of the orbits of galaxies, since the velocities of the particles in the gas are isotropic.

2.9 Dynamics of Galaxies in Clusters

In Sections 2.5, 2.6, and 2.7, evidence was presented indicating that the velocity and spatial distribution of galaxies in regular, compact clusters are in a relaxed, quasistationary state. In this section, the nature of the relaxation processes of galaxies in clusters will be briefly discussed.

If clusters result from the gravitational growth of initially small perturbations, we expect the initial state of material in the protocluster to be irregular. The relaxation of the cluster thus involves the spatial motion of the galaxies, and a lower limit to the relaxation time is the crossing time

$$t_{cr}(r) \equiv \frac{r}{v_r} \approx 10^9 \text{ yr}\left(\frac{r}{\text{Mpc}}\right)\left(\frac{\sigma_r}{10^3 \text{ km/s}}\right)^{-1}, \tag{2.26}$$

where the radial velocity dispersion σ_r has been substituted for the radial velocity v_r. An upper limit on the age of the cluster is the Hubble time (age of the universe)

$$t_H \approx (1.3 - 2.0) h_{50}^{-1} \times 10^{10} \text{ yr}, \tag{2.27}$$

where the numerical coefficient depends on the cosmological model. Thus the outer parts of a cluster or surrounding supercluster for which $r \gtrsim 10$ Mpc cannot possibly have relaxed and are expected to be irregular, as is observed (Sections 2.7 and 2.11.2).

2.9.1 *Two-body Relaxation*

The phase-space distribution of galaxies in the central parts of spherical regular clusters can be represented as $f(\mathbf{r}, \mathbf{v}) \propto \exp(-\varepsilon/\sigma_r^2)$, where $\varepsilon \equiv \frac{1}{2}v^2 + \phi(r)$ is the energy per unit mass in the cluster (Section 2.7). This is very similar to a Maxwell–Boltzmann distribution, except that it is the energy per unit mass (velocity dispersion) which is determined, and not the energy (temperature). We first consider the possibility that clusters have relaxed thermodynamically.

Thermodynamic equilibrium could result from elastic collisions between galaxies, and would be expected if the time scale for energy transfer in such collisions were shorter than the age of the cluster or the time-scale for the loss of kinetic energy through dissipative processes. The most important elastic two-body collisions in a cluster are gravitational. The resulting relaxation times have been calculated analytically by Chandrasekhar (1942) and Spitzer and Harm (1958), and numerically through N-body simulations, including a realistic luminosity function for galaxies by White (1976c, 1977b) and Farouki and Salpeter (1982). A useful characteristic time-scale (0.94 times the Spitzer 'reference time') is

$$t_{rel} = \frac{3\sigma_r^3}{4\sqrt{\pi}G^2 m m_f n_f \ln \Lambda} \tag{2.28}$$

for the relaxation of a galaxy of mass m in a background field of galaxies of mass m_f and number density n_f. Λ is the ratio of maximum to minimum impact parameters of collisions which contribute to the relaxation; the maximum impact parameter is on the order of half the gravitation radius of the cluster R_G (equation (2.21)), while the minimum impact parameter is roughly the larger of the galaxy radius r_g or the 'turning radius' $Gm_f/3\sigma_r^2$ (White, 1976b). Thus

$$\Lambda \approx \min\left[\left(\frac{3R_G\sigma_r^2}{2Gm_f}\right), \left(\frac{R_G}{2r_g}\right)\right]. \tag{2.29}$$

Usually, the second value applies. Thus, for $r_g \approx 20$ kpc, $R_G \approx 1$ Mpc, $\ln \Lambda \approx 3$.

In order to give a lower limit to the relaxation time, we assume for the moment that all the mass in a cluster (including the missing mass (Section 2.8)) is bound to individual galaxies; we later show this is unlikely to be the case. Then, we define an average density $\langle \rho \rangle \equiv 3M_{tot}/(4\pi R_G^3)$, where M_{tot} is the total mass of the cluster. If we assume a Schechter luminosity function (equation (2.3)), and a fixed galaxy mass-to-light ratio, we find

$$t_{rel}(m, r) \gtrsim 0.24 t_{cr}(R_G) N^* \left[\left(\frac{m}{m^*}\right)\left(\frac{\rho(r)}{\langle \rho \rangle}\right) \ln \Lambda\right]^{-1}, \tag{2.30}$$

where m^* is the characteristic galaxy mass (corresponding to L^*), N^* is the characteristic galaxy number (richness), t_{cr} is the cluster crossing time (equation (2.26)), and $\rho(r)$ is the total cluster density at r. Equation (2.30) is a lower limit because it assumes all the cluster mass is bound to individual galaxies. Typically, $R_G \approx 1$ Mpc, $\sigma_r \approx 1000$ km/s, $N^* \approx 1000$, and $\ln \Lambda \approx 3$, giving $t_{cr}(R_G) \approx 10^9$ yr and $t_{rel}(m^*, R_G) \gtrsim 3 \times 10^{11}$ yr. This is much longer than a Hubble time (equation (2.27)); it is therefore unlikely that the apparently relaxed state of regular clusters results from two-body collisions. However, two-body relaxation processes can affect the more massive galaxies ($m \gg m^*$) near the cluster center ($\rho(r) \gg \langle \rho \rangle$), as is discussed below.

2.9.2 *Violent Relaxation*

The fact that clusters exhibit a nearly constant velocity dispersion (rather than kinetic temperature) suggests that the relaxation is produced by collective gravitational effects. Lynden-Bell (1967) showed that collective relaxation effects can result in a very rapid quasirelaxation ('violent relaxation'). The existence of these effects involves a somewhat subtle point; collective relaxation develops through collisionless interactions, and thus the detailed ('fine-grain') phase-space distribution of galaxies is conserved. However, if the relaxation is sufficiently violent (for example, the energy of particles is changed by a significant fraction), then initially adjoining units of phase-space will be widely separated in the final state ('phase mixing'). Thus, if one averages the fine-grain phase-space density over any observable volume to give a 'coarse-grain' distribution, this coarse-grain distribution can be an equilibrium distribution and independent of the details of the initial state.

If clusters were formed by the growth and collapse of initial perturbations, then during the collapse the gravitational potential ϕ fluctuated violently. This would cause a change in the energy per unit mass $\varepsilon \equiv \frac{1}{2}v^2 + \phi(r)$ of

$$\frac{D\varepsilon}{Dt} \approx \frac{\partial \phi}{\partial t}. \tag{2.31}$$

For example, if the cluster collapsed from a stationary state ($v = 0$) to a virialized final state (Section 2.8), then $\Delta\varepsilon \approx \Delta\phi \approx \varepsilon$. The time scale for collapse t_{coll} is roughly a dynamical time scale or crossing time

$$t_{coll} \approx (G\langle\rho\rangle)^{-1/2} \approx (R_G^3/GM_{tot})^{1/2} \approx t_{cr}(R_G). \tag{2.32}$$

As the energy of a galaxy changes by $\approx 100\%$ during a collapse time, violent relaxation can be completed during the collapse of the cluster, after which time the potential is constant, and the galaxy distribution is in stationary virial equilibrium.

Since the equation of motion of a particle in the cluster's mean gravitational field is independent of mass (equation (2.31)), the equilibrium is independent of mass. Lynden-Bell (1967) derived an equilibrium state by assuming that the system relaxed to the most probable coarse-grained phase-space state subject to conservation of energy, particle number, and fine-grain phase density. This equilibrium state was found to be a Fermi–Dirac distribution that reduces to the Maxwell–Boltzmann distribution for the appropriate number densities in clusters (see also Shu, 1978),

$$f(\mathbf{r}, \mathbf{v}) \propto \exp(-\varepsilon/\sigma_r^2). \tag{2.33}$$

This phase-space density produces a Gaussian velocity distribution and isothermal spatial distribution of galaxies, which are roughly consistent with the observed distributions in the inner parts of clusters (Sections 2.6 and 2.7). The distribution is also independent of mass, roughly as observed. If equation (2.33) held for all radii, the cluster mass would be infinite (Section 2.7). However, galaxies at large radii or with large energies never reach equilibrium because their crossing times (equation (2.26)) are longer than the Hubble or collapse times (equations (2.27) and (2.32)). Because the relaxation occurs only while the cluster is collapsing, it is not clear that this equilibrium state can ever be achieved in any real collapsing system.

The state of equilibrium of a collisionless, gravitationally collapsing system can be studied directly through numerical N-body experiments. These experiments do show that isolated collapsing or merging systems relax rapidly (in a few crossing times), and they agree roughly on the nature of the equilibrium state. In general, they *do not* find that the systems become isothermal spheres; instead they find spatial distributions that are reasonably represented by the de Vaucouleurs (1948a, b; equation (2.14)) or Hubble (1930; equation (2.15)) forms (White, 1979; Villumsen, 1982; van Albada, 1982). These distributions are more centrally condensed and fall off more rapidly at large radii than the isothermal sphere. However, it is possible that the subsequent collapse of surrounding material can increase the density at large distances; Gunn (1977) has shown that for some initial conditions this process can lead to nearly isothermal mass distributions.

The idea that the distribution of galaxies in clusters is determined by violent relaxation during the formation of the cluster provides a simple explanation for the one-dimensional morphological sequence of clusters running from irregular to regular (Table 1; Gunn and Gott, 1972; Jones *et al.*, 1979; Dressler, 1984). The regular clusters are those old enough to have collapsed and relaxed, while the irregular ones have not. As the collapse time is $t_{coll} \approx (R_h^3/GM_{tot})^{1/2} \approx (G\rho_i)^{-1/2}$, where ρ_i is the initial density, higher density protoclusters will collapse more rapidly. Since the age of clusters is limited by the Hubble time t_H, *regular clusters will be produced by higher density protoclusters, and irregular clusters by lower density protoclusters.* Thus we expect regular clusters to have higher densities than irregular clusters, as is observed. Moreover, violent relaxation and phase mixing will eliminate subclustering and produce a centrally condensed, symmetric distribution, as observed in regular clusters.

White (1976c) has followed the collapse of a model cluster numerically, with an N-body code; the results are shown in Figure 5. He finds that the cluster first forms irregular subcondensations around massive galaxies (like an I cluster). These continually merge until the galaxy distribution is elongated and has two large clumps (like an F cluster), which merge to form a smooth, regular cluster with a prominent core (a C cluster) (see also Henry *et al.*, 1981; Forman and Jones, 1982). Recently, Cavaliere *et al.* (1983) have produced many similar models for cluster collapse.

The other aspects of the morphological sequence have to do with galactic content – the fractions of spiral, S0, and elliptical galaxies, or domination by supergiant (B, D, and cD) galaxies. In Section 2.10 evidence is presented which suggests that the galactic content of clusters is also determined by the density of the cluster, although the mechanism is still controversial.

2.9.3 *Ellipsoidal Clusters*

The distribution in equation (2.33) is isotropic in velocity space and spherically symmetric in real space. However, many of the most regular clusters have observed galaxy distributions that are not symmetric on the sky (Section 2.7). In principle, this asymmetry could be due to rotation; however, the velocity fields in clusters show no evidence for dynamically significant rotation (Section 2.6).

Numerical N-body simulations of the formation of clusters show that if the

Fig. 5. The projected galaxy distributions in White's (1976c) N-body calculations of the evolution of a collapsing cluster. Each symbol represents a galaxy. In each figure, the bar represents a fixed length scale, one-half of the gravitational radius R_G (equation 2.21). The times t are given in units of the initial collapse time of the cluster. (a) The initial configuration ($t=0$). (b) An irregular distribution with subclustering at $t=0.19$. (c) A bimodal distribution in the cluster at $t=0.97$. (d) The final relaxed configuration at $t=2.66$.

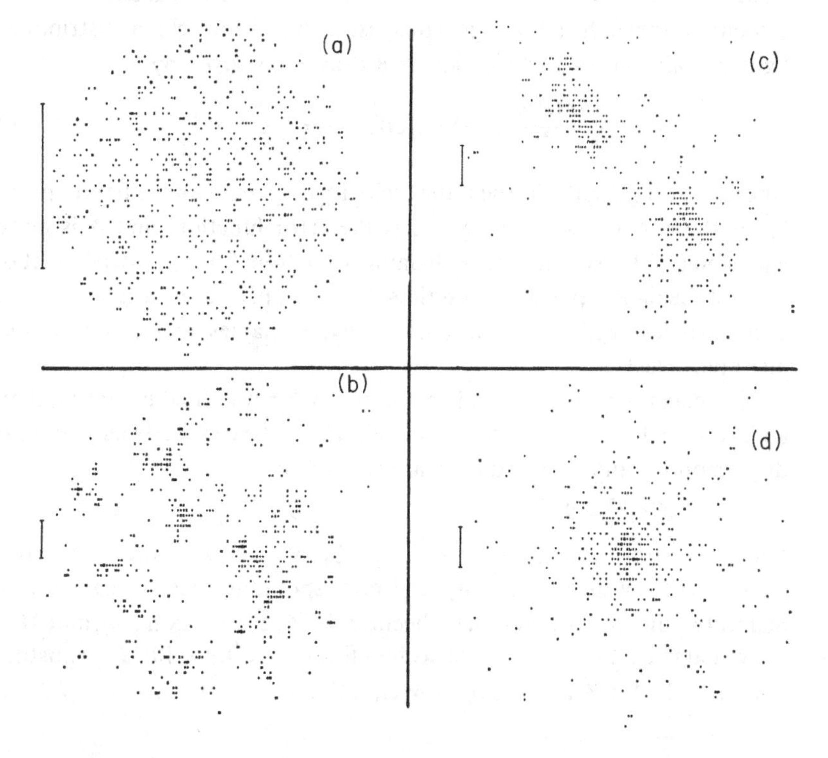

initial distribution of galaxies is aspherical, the final distribution after violent relaxation will be aspherical (Figure 5d; Aarseth and Binney, 1978). The most general final configurations have compact cores and are regular, but the surfaces of constant density are basically triaxial ellipsoids rather than spheres. The triaxial shape is maintained by an ellipsoidal Gaussian velocity distribution $p(v) \propto \exp[-(2\Sigma_{ij})^{-1} v_i v_j]$, where Σ is the velocity dispersion tensor. The principal axes of Σ and the spatial figure of the galaxy are parallel and nearly proportional to one another (Binney, 1977), as expected from the tensor virial theorem (Chandrasekhar, 1968). Thus the velocity dispersion is largest parallel to the longest diameter of the cluster.

2.9.4 *Dynamical Friction*

Once the collapse of a cluster is completed, violent relaxation is ineffective, and further relaxation occurs through two-body interactions. As shown above, these effects are probably not important for a typical galaxy at an average position in the cluster, but they can be significant for a more massive galaxy ($m \gg m^*$) in the cluster core ($\rho \gg \langle \rho \rangle$; see equation (2.30)). Chandrasekhar (1942) showed that a massive object of mass m moving at velocity $\mathbf{v}$ through a homogeneous, isotropic, Maxwellian distribution of lighter, collisionless particles suffers a drag force given by

$$\frac{d\mathbf{v}}{dt} = -\mathbf{v}\left\{\frac{4\pi G^2 m}{v^3} \ln(\Lambda)\rho_{tot}[\mathrm{erf}(x) - x \times \mathrm{erf}'(x)]\right\} \tag{2.34}$$

where $x \equiv v/(\sqrt{2}\sigma_r)$, σ_r is the radial velocity dispersion of the lighter particles, ρ_{tot} is their total mass density, erf is the error function, and Λ is given by equation (2.29). Note that this 'dynamical friction' force is independent of the mass of the lighter particles, and thus depends only on the total density ρ_{tot} of such particles (regardless of whether they are galaxies or collisionless missing mass particles).

The relaxation time in the cluster core can be evaluated by noting that, for an isothermal cluster distribution (Section 2.7), the central density ρ_o, velocity dispersion σ_r, and core radius r_c are related by

$$4\pi G \rho_o r_c^2 = 9\sigma_r^2. \tag{2.35}$$

Let $m = (m/m^*)m^* = (m/m^*)(M/L_V)_{gal} L_V^*$, where L_V^* and m^* are the characteristic visual luminosity and corresponding mass of galaxies in the Schechter luminosity function (Schechter, 1976, and Section 2.4), and M/L_V is the visual mass-to-light ratio. Schechter finds $L_V^* \approx 4.9 \times 10^{10} L_\odot$ Substituting this value, $\ln \Lambda \approx 3$, and the ρ_o from equation (2.35) into equation (2.28) gives for the dynamical friction time-scale

$$t_{rel}(m, r=0) \approx 6 \times 10^9\,\mathrm{yr}\left(\frac{\sigma_r}{1000\,\mathrm{km/s}}\right)$$
$$\times\left(\frac{r_c}{0.25\,\mathrm{Mpc}}\right)^2\left(\frac{m}{m^*}\right)^{-1}\left[\frac{(M/L_V)_{gal}}{10M_\odot/L_\odot}\right]^{-1}. \qquad (2.36)$$

Dynamical friction slows down the more massive galaxies near the center of a spherical cluster; they spiral in toward the cluster center (Lecar, 1975). The kinetic energy removed from the massive galaxies is transferred to the lighter particles (galaxies or missing mass components), which then expand. Thus dynamical friction produces mass segregation in a cluster; more massive galaxies are found preferentially at smaller radii. Moreover, at a fixed radius the velocity dispersion of the more massive galaxies will be lower, as they have been slowed down.

If the cluster is aspherical because of an anisotropic velocity dispersion, the dynamical friction force will not be parallel to the velocity (Binney, 1977). Because the force depends inversely on velocity, it is strongest along the shortest axis of the cluster. Thus dynamical friction increases the anisotropy of the most massive galaxies in a cluster. This can explain the formation of L clusters such as Perseus, in which the brightest galaxies form a narrow chain along the long axis of the cluster (Binney, 1977).

The massive galaxies that spiral into the cluster center will eventually merge to form a single supergiant galaxy if they are not tidally disrupted first. In Section 2.10.1 evidence is given which suggests that this is the mechanism by which cD galaxies form; however, there are also strong arguments that galaxies lose mass due to tidal effects so rapidly that dynamical friction is not important (Merritt, 1983, 1984a, 1985). The merger of the most massive and therefore brightest galaxies in the cluster core might cause the luminosity function to fall off more rapidly at high luminosities, as may have been observed in some cD clusters (Dressler, 1978b), and might produce a deficiency of brighter galaxies (other than the cD) in the cluster core (White, 1976a).

The fact that the dynamical friction time scale depends inversely on galaxy mass provides a method to measure the mass bound to individual galaxies in a cluster, as opposed to the total virial mass of the cluster. If the mass-to-light ratio derived for the cluster as a whole (typically $(M/L_V) \approx 250 M_\odot/L_\odot$; equation (2.25)) is substituted in equation (2.36), $t_{rel} \approx 2 \times 10^8\,\mathrm{yr}(m/m^*)^{-1}$, which is much smaller than the probable age of a cluster of about a Hubble time $t_H \approx 10^{10}$ yr (for typical parameters for a compact, regular cluster). Thus, if the missing mass in clusters were bound to individual galaxies, the massive galaxies $m \gtrsim m^*$ would be highly relaxed (at least in the cluster core). All of these galaxies would have segregated into a very small core and possibly merged to form a single galaxy. Observations of the degree of mass segregation, luminosity function, and luminosity of the brightest galaxies in well studied clusters are all inconsistent with this much relaxation. From the observations one can limit the mass-to-light ratios of massive galaxies in cluster cores to $(M/L_V) \lesssim 30 M_\odot/L_\odot$. This indicates that the missing mass cannot be bound to individual, massive galaxies, but rather must form a continuum. This important result was first given by Rood (1965), and has been verified by numerical integration of equation (2.34) and N-body models by White (1976a, 1977b) and Merritt (1983).

2.10 Galactic Content of Clusters

In the discussion of the morphological classification of clusters (Section 2.5), it was noted that the regular, compact clusters have a galactic content that differs markedly from that of the field. First, these clusters are often dominated by a single very luminous (cD) galaxy, or by a pair of very bright galaxies. Second, elliptical and S0 galaxies predominate over spiral galaxies in regular, compact clusters, where the opposite is true in the field. In this section possible origins for these differences are described and are related to the overall picture of cluster evolution given in the preceding section.

2.10.1 cD Galaxies

cD galaxies were defined by Mathews, Morgan, and Schmidt (1964) as galaxies with a nucleus of a very luminous elliptical galaxy embedded in an extended amorphous halo of low surface brightness. They are usually found at the center of regular, compact clusters of galaxies (Morgan and Lesh, 1965; Bautz and Morgan, 1970), and about 20% of all rich clusters contain cD galaxies. However, some galaxies that appear to be cDs have been found in poor clusters and groups (Morgan, Kayser, and White, 1975; Albert, White, and Morgan, 1977).

cD galaxies are extremely luminous; if one excludes nuclear sources (Seyfert galaxies, N galaxies, and quasars), they are, as a class, the most luminous galaxies known. Sandage (1976) and Hoessel (1980) find average absolute magnitudes $\langle M_V \rangle_{cD} \approx -23.7 + 5 \log h_{50}$ and $-22.7 + 5 \log h_{50}$ for apertures of $43/h_{50}$ kpc and $19/h_{50}$ kpc, respectively. Since the galaxies extend well beyond this, total luminosities are at least a magnitude brighter. Moreover, the magnitudes of cDs show a rather small dispersion (≈ 0.3 mag) and are only weakly dependent on cluster richness (Sandage, 1976). cD galaxies are usually considerably brighter (often by a full magnitude) than other galaxies in the same cluster.

The question naturally arises as to whether cD galaxy luminosities simply represent the high end of the normal galaxy luminosity function, or whether cDs are 'special'. The distribution of brightest galaxy luminosities L_1 in a cluster would be $p(L_1)dL_1 = \exp(-N(L_1))dN(L_1)$ if the brightest galaxies were drawn at random from the integrated luminosity function $N(L)$ (Section 2.4). This distribution would produce a large dispersion in L_1 and a strong dependence on cluster richness (neither of which are observed), unless the luminosity function were much steeper than is observed at the bright end (Sandage, 1976). However, cD luminosities themselves are too high to be drawn from the galaxy luminosity function (Schechter, 1976), unless it were less steep than is observed. While selection effects may account for the small dispersion of brightest elliptical and S0 galaxies in groups and clusters (Schechter and Peebles, 1976), it does not appear that this can explain the high luminosities of cD galaxies. A number of statistical tests (Sandage, 1976; Tremaine and Richstone, 1977; Dressler, 1978a) indicate that the magnitudes of cD galaxies cannot be drawn statistically from a general galaxy luminosity function.

cD galaxies are more extended than the other giant elliptical galaxies in two ways (see Figure 6a and 6b where the surface brightness of the cD in A1413 is compared to the giant elliptical NGC1278 in the Perseus cluster). First, the core regions of cDs are apparently larger. Hoessel (1980) finds that cD galaxies have core radii that average $\langle r_c(gal)\rangle \approx 4/h_{50}$ kpc, while typical giant elliptical galaxies have $\langle r_c(gal)\rangle \approx 0.4/h_{50}$ kpc (here, the surface brightness of the galaxy at projected radius b is assumed to vary as $\{1 + [b/r_c(gal)]^2\}^{-1}$). Unfortunately, these core radii represent fairly small angles and their determination may be ambiguous (Dressler, 1984). If one fits galaxy surface brightnesses to a de Vaucouleurs (1948a) form (equation (2.14)), the effective radii r_e of cD galaxies are roughly a factor of two larger than the effective radii of normal giant elliptical galaxies. Alternatively, one can define the integrated luminosity slope

$$\alpha \equiv \frac{d \ln L(b)}{d \ln b}, \tag{2.37}$$

where $L(b)$ is the luminosity observed within the projected radius b. Hoessel (1980) finds $\langle\alpha\rangle_{cD} \approx 0.59$ for $b = 19/h_{50}$ kpc, which is about twice the value found for typical giant ellipticals.

In cD galaxies in rich clusters, the giant elliptical-like core of the galaxy is embedded in a very extended low surface brightness halo (Oemler, 1973, 1976;

Carter, 1977; Dressler, 1979). In Figure 6 the surface brightness of the very extended cD in the cluster A1413 is compared to that of a typical giant elliptical galaxy NGC1278 in the Perseus cluster (Oemler, 1976). The solid line shows a de Vaucouleurs fit (equation (2.14)) to each galaxy. This profile fits the inner parts of either galaxy reasonably well, but the cD in A1413 has a halo of low surface brightness extending to beyond 1 Mpc from the galactic center. The surface brightness in cD haloes generally falls off as roughly the 3/2 power of projected distance from the galaxy center.

Masses of cD galaxies have been estimated by a measurement of the stellar velocity dispersion in the outer parts of the cD in A2029 (Dressler, 1979) and by measurements of the velocities of smaller companion galaxies which are assumed to be bound to the cD (Wolf and Bahcall, 1972; Jenner, 1974). Typically, one finds $M \approx 10^{13} M_{\odot}$, but it is difficult to separate the galaxy and cluster mass distributions in the outermost parts of the cD (Dressler, 1979). For the same central velocity dispersion, cD galaxies are about 60% brighter than other giant ellipticals (Malumuth and Kirshner, 1981).

cD galaxies are usually found very near the centers of compact clusters (Morgan and Lesh, 1965; Leir and van den Bergh, 1977; White, 1978a). They

Fig. 6. The surface photometry of cD galaxies. The surface brightness in magnitudes per square second of arc is plotted against the one-fourth power of the radius. The dots are the observed points, and the straight lines are de Vaucouleurs fits to the inner points. NGC1278 (Oemler, 1976) is a normal giant elliptical galaxy, showing no extended halo. The middle figure shows the cD in the cluster A1413 (Oemler, 1976), with a more extended central de Vaucouleurs profile and a very extended halo. The right panel is the cD in the poor cluster AWM4 (Thuan and Romanishin, 1981), which has an even more extensive de Vaucouleurs profile, but no apparent halo.

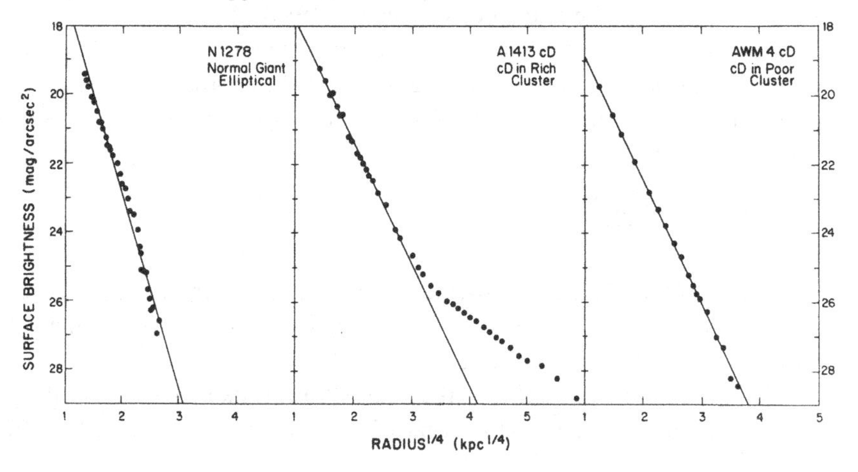

also have velocities very near the mean velocity of galaxies in the cluster (Quintana and Lawrie, 1982), and may in fact give a better estimate of the cluster mean velocity than the average value for a few bright galaxies. These results suggest that cDs are usually sitting at rest at the bottom of the cluster gravitational potential well.

cD galaxies often have double or multiple nuclei (Minkowski, 1961; Morgan and Lesh, 1965; Hoessel, 1980; Schneider and Gunn, 1982); that is, there are several peaks in the surface brightness within the central part of the cD. Many cDs appear to be in binary or multiple galaxy systems (Leir and van den Bergh, 1977; Rood and Leir, 1979; Struble and Rood, 1981); when two cD nuclei appear to be surrounded by a common halo, these are referred to as 'dumbbell' galaxies. One important problem with these multiple nuclei is whether they are physically associated with and bound to the cD, or whether they are chance projections. In many cases, the multiple nuclei have rather large velocities relative to the cD, and cannot be bound to it (Jenner, 1974; Tonry, 1984, 1985a, b; Hoessel *et al.*, 1985). These multiple nuclei must then be chance projections. Such projections would be unlikely if galaxies in clusters have isotropic velocity dispersions and constant surface number density cores. However, they are much more likely if galaxies have radial orbits, and the cluster density has a cusp near the center (Tonry, 1985a; Merritt, 1984b, 1985), as appears to be the case in many clusters (Beers and Tonry, 1986; Section 2.7).

The special structural and kinematic properties of cD galaxies suggest that they have been formed or modified by dynamical processes in clusters. Gallagher and Ostriker (1972) and Richstone (1975, 1976) have suggested that cDs consist of the debris from galaxy collisions. In a rich cluster, the outer envelopes of galaxies will be stripped by tidal effects during these collisions. The rate of mass loss due to tidal collisions in a cluster was derived by Richstone (1976); if his rate is integrated over a Schechter luminosity function (Section 2.4) and an isothermal galaxy density function with core radius r_c (Section 2.7), one finds

$$\left(\frac{dM}{dt}\right)_{tidal} \approx 3 \times 10^4 M_\odot \, \mathrm{yr}^{-1} \left(\frac{\sigma_r}{10^3 \, \mathrm{km/s}}\right)^4 \times \left(\frac{\sigma_*}{200 \, \mathrm{km/s}}\right)^{-1} \left(\frac{r_t}{r_c}\right), \tag{2.38}$$

where σ_r is the cluster line-of-sight velocity dispersion (Section 2.6), and r_t and σ_{gal} are the tidal (outermost) radius and line-of-sight stellar velocity dispersion of a typical galaxy ($L = L^*$; see Section 2.4). Equation (2.38) assumes that all the cluster mass is in galaxies; if this is not the case, then the rate of mass loss from galaxies is reduced by the square of the fraction of mass in galaxies.

Knobloch (1978a, b) and Da Costa and Knobloch (1979) have argued that Richstone's expression overestimates the rate of mass loss, because the tidal stripping of a galaxy halo is limited by the rate of diffusion of stars into the halo. However, they take as their basic model a galaxy in which the halo mass is very small; this does not agree with the determinations of the mass distributions in galaxies (see, for example, Faber and Gallagher, 1979). For realistic mass models, it appears that the claimed discrepancy is a factor of roughly (σ_*/σ_r).

If most of the mass in a cluster were initially in extended ($r_t \approx r_c \approx 300$ kpc) haloes of dark material bound to individual galaxies, equation (2.38) shows that these dark haloes could have been stripped by tidal interactions within the age of a cluster, as was suggested in Section 2.8.

The outer portions of the luminous material would also be stripped (Strom and Strom, 1978a, 1979); if we assume a tidal radius of $r_t \approx 0.1 r_c \approx 30$ kpc, we would expect $\gtrsim 10^{12} L_\odot$ of luminous material to be stripped from galaxies in a compact cluster. The stripped material would settle to the center of the cluster gravitational potential; because of the dependence of the collision rate on the square of galaxy density, the stripped material would probably be somewhat more centrally condensed than the galaxy distribution in the cluster. This tidal debris might thus be observed as the extended halo about a central cD galaxy. The stripped material would be centered on the cluster center, be at rest (on average) relative to the cluster center-of-mass, and have a mass and spatial distribution similar to that observed for cD haloes. This model for the formation of cD haloes (Gallagher and Ostriker, 1972; Richstone, 1975, 1976) predicts that they have high velocity dispersions, as has been observed for the cD in A2029 by Dressler (1979). This tidal debris model may explain the observed properties of cD haloes; however, it cannot naturally explain the giant elliptical galaxy nucleus of the cD.

Ostriker and Tremaine (1975), Gunn and Tinsley (1976), and White (1976a, 1977a) have suggested that cD galaxies are produced by the merger of massive galaxies within the core of a cluster. As discussed in Section 2.9, dynamical friction causes the orbits of massive cluster galaxies to decay. As such galaxies reach the cluster center, they merge to form a single supergiant galaxy, which swallows any galaxies that subsequently pass through the cluster center. The merger hypothesis (called 'galactic cannibalism' by Ostriker and Hausman, 1977) provides an attractive explanation for the formation of cD galaxies in a cluster.

First of all, since a cD galaxy would be produced by the merger of many of the more luminous galaxies in a cluster, the high luminosities of cDs (in excess of that expected from the galaxy luminosity function) is naturally explained. Once a massive galaxy reaches the cluster center, it will swallow other galaxies. Initially, its luminosity will increase at the rate (Ostriker and Tremaine, 1975)

$$\frac{d(L_{cD}/L^*)}{d(t/10^{10}\,\mathrm{yr})} \approx 13h_{50}^{-1}\left(\frac{r_t}{100\,\mathrm{kpc}}\right)^3\left(\frac{r_c}{250\,\mathrm{kpc}}\right)^{-4} \times\left(\frac{\sigma_r}{10^3\,\mathrm{km/s}}\right)\left(\frac{M/L_V}{30M_\odot/L_\odot}\right), \tag{2.39}$$

where r_t is the tidal radius of the central galaxy, r_c and σ_r are the core radius and velocity dispersion of the cluster, and M/L_V is the visual mass-to-light ratio of all galaxies in the cluster. (Very recently, Merritt (1985) has shown that equation (2.39) greatly overestimates the actual merger rate at early times in a cluster.) Once most of the massive galaxies within the core have been accreted, the luminosity grows as $t^{1/2}$ rather than as t (Ostriker and Tremaine, 1975):

$$\left(\frac{L_{cD}}{L^*}\right) \approx 8\left(\frac{\sigma_r}{10^3\,\mathrm{km/s}}\right)^{3/2}\left(\frac{M/L_V}{30M_\odot/L_\odot}\right)^{1/2}\left(\frac{t}{10^{10}\,\mathrm{yr}}\right)^{1/2}. \tag{2.40}$$

Thus luminosities of $L_{cD} \approx 10L^*$ can be produced during the lifetime of a compact cluster.

The merger product should be larger than the initial galaxy, because the kinetic energy of the merging galaxies 'heats' and inflates the final product (Ostriker and Hausman, 1977; Hausman and Ostriker, 1978). Hausman and Ostriker (1978) argue that the galaxy core radius $r_c(gal)$ will increase by roughly a factor of ten, and the structure parameter α by a factor of at least two, in rough agreement with the observations (Hoessel, 1980).

With the large increase in the luminosity of the first-brightest galaxy through mergers (equation (2.40)), it might seem surprising that these galaxies show a rather small dispersion in absolute magnitude (Sandage, 1976). The observed magnitudes are for fixed measuring apertures, which are generally much smaller than the halo size of the cD; depending on the amount of swelling the cD undergoes, the observed magnitude may either increase or decrease (Gunn and Tinsley, 1976). In the simulations of Hausman and Ostriker (1978), the apparent magnitude within an aperture of 16 kpc remains roughly constant once the massive galaxies in the core of the cluster have been swallowed. This may explain the small dispersion in observed cD absolute magnitudes.

The first-brightest galaxy grows in luminosity by swallowing other massive galaxies; this increases the contrast between the first-brightest and the other bright galaxies (Gunn and Tinsley, 1976; Hausman and Ostriker, 1978), as is observed (Sandage, 1976; Tremaine and Richstone, 1977; Dressler, 1978a; McGlynn and Ostriker, 1980).

Of course, the process of dynamical friction and merging leaves the merger product nearly at rest relative to the average cluster galaxy and nearly at the center of the cluster, as is observed for cDs.

If a cluster is elongated, then the merger product at its center will tend to be elongated in the same direction. In general, cD galaxies are more highly flattened than other ellipticals (Leir and van den Bergh, 1977; Dressler, 1978c), and the axes of the cDs align with those of their clusters, as predicted (Sastry, 1968; Rood and Sastry, 1972; Dressler, 1978c, 1981; Carter and Metcalfe, 1980; Binggeli, 1982).

If cD galaxies are produced by mergers of massive galaxies, one expects mergers to take place roughly every 10^9 years (Ostriker and Hausman, 1977). The actual merger will occur over roughly an orbital period in the cD galaxy $\approx 3 \times 10^8$ years (White, 1978b), and for this period the nucleus of the merging galaxy will be visible within the cD envelope. Thus one would expect about 1/4 of cD galaxies to have multiple nuclei, roughly as is observed (Hoessel, 1980). However, a major problem with this interpretation of multiple nuclei in cDs is that these nuclei, in many cases, have rather large velocities relative to the cD (Jenner, 1974; Tonry, 1984, 1985). This suggests that they are not bound to the cD, but might only be cluster galaxies seen in projection against the cD.

Rood and Leir (1979) find that about 1/4 of all cD galaxies are binaries (actually dumbbells); they argue that this fraction is much larger than is expected due to dynamical friction and merging. The reason is that in these systems the two galaxies in the binary differ by less than a magnitude in apparent luminosity. However, the primary component of the system would have swallowed roughly five other galaxies in the merger picture, and therefore should be much brighter than its companion. Since one magnitude corresponds to a factor of 2.5 in brightness, and magnitudes at a fixed aperture do not increase directly with total luminosity because of swelling, it is not clear how serious a discrepancy this is (see also Tremaine, 1981).

The dynamical friction and merger theory may explain most of the observed properties of cD galaxies except for their very extended haloes. The tidal debris theory explains the extended haloes, but not the properties of the inner elliptical galaxy. It seems natural to suggest that cD galaxies result from the action of both types of processes.

However, Merritt (1983, 1984a, b, 1985) has recently pointed out some very serious problems with this simple picture of cD formation. He argues that the tidal effect of the global cluster potential (as opposed to that due to individual galaxy interactions) dominates the evolution of cluster galaxies. Tidal stripping of galaxies would then lower their masses and prevent any mergers. If this argument is correct and if mergers formed cDs, then the mergers must have occurred before the cluster collapsed, perhaps in smaller subclusters or groups (Carnevali *et al.*, 1981; van den Bergh, 1983a; Dressler, 1984).

Theories of the origin of cD galaxies can be tested through observations of the apparent cD galaxies in 23 poor clusters which were discovered by Morgan, Kayser, and White (1975) (hereafter MKW) and Albert, White, and Morgan (1977) (hereafter AWM). These clusters range in Abell richness (Section 2.3) from < 10 to ≈ 50 galaxies; none are richer than Abell richness class zero (Bahcall, 1980). These clusters have velocity dispersions σ_r (Stauffer and Spinrad, 1978) and core radii r_c (Bahcall, 1980) which are probably about 1/2 of those of rich, compact clusters. Since the dynamical friction rate (when summed over all galaxies) varies as σ_r/r_c (equation (2.30)) and the merger rate varies as σ_r/r_c^4 (equation (2.39)). one expects mergers to form remnants in these poor clusters even larger than those in rich clusters. On the other hand, the tidal stripping rate (equation (2.38)) varies as σ_r^4/r_c, and is thus much lower in poor clusters. If mergers form the extended elliptical galaxy body of a cD and tidal debris makes up the very extended halo, the cDs in poor clusters should lack such haloes. Surface photometry of the cDs in poor clusters (Oemler, 1976; Stauffer and Spinrad, 1980; Thuan and Romanishin, 1981) indicates that this is indeed the case. In Figure 6c, the surface photometry of the cD in the poor cluster AWM4 as observed by Thuan and Romanishin (1981) is shown.

While the process of merging and tidal stripping, perhaps in preexisting subclusters and groups, seem to provide a possible explanation of many of the unusual properties of cD galaxies, the core of a rich compact cluster is a very active physical environment in which many other processes may be important. For example, recent X-ray and optical observations suggest that central, dominant galaxies are accreting vast quantities of gas, as much as $400 M_{\odot}$/yr in the case of NGC1275 in the Perseus cluster (Sections 4.3.3 and 5.7). The accreting galaxies include a number of cDs. There are several arguments which suggest that the accreted gas is being converted into low mass stars (Cowie and Binney, 1977; Fabian *et al.*, 1982a; Sarazin and O'Connell, 1983); if so, accretion can significantly increase the core luminosities of cDs (Fabian *et al.*, 1982a; Sarazin and O'Connell, 1983).

Finally, I would like to comment on a semantic issue that has arisen concerning cDs and other similar galaxies. It should be clear from the discussion above that cDs represent the extreme result of a number of dynamical processes which must be occurring continuously in clusters and groups. There are a number of related types of galaxies which share many but not all of the properties of cDs; they may be extended, and be the brightest galaxy in a cluster (or nearly so), and be located at the spatial and velocity center of the cluster. In Section 4.4.2 we shall see that such galaxies may affect the X-ray morphology of clusters, even if they do not satisfy the technical definition of a cD galaxy; examples are M87 in the Virgo cluster and NGC 1275 in the Perseus cluster. It has sometimes been argued that the definition of cD galaxies should be extended in some way to include all such objects. It seems to me that this is not a good idea; it is more useful to retain the degree of specificity in the original definition of cD galaxies, based on their optical properties as defined by Morgan and his collaborators (Mathews, Morgan, and Schmidt, 1963). I suggest that a new term, 'central dominant galaxy' (abbreviated cd?) be used for galaxies that are the brightest cluster member or within 0.5 magnitudes of the brightest, and that appear to be at rest at the cluster center.

2.10.2 Proportion of Spiral, S0, and Elliptical Galaxies

Elliptical and S0 galaxies are more common than spiral galaxies in the inner portions of regular compact clusters, while the opposite is true in irregular clusters and in the field (Table 1). Many explanations have been proposed for the origin of this systematic variation in galactic content. In general, these theories fall into two broad classes. In the first class, the proportion of galaxy types is set by the conditions when the galaxies form, and once the galaxies form they do not alter their morphology. Thus, in regions that are or will become regular clusters, more ellipticals are formed. In the second group of theories, galaxies may form with the same distribution of morphological types everywhere, but physical processes that depend on environment cause galaxies to alter their morphology. That is, in compact regular clusters, spirals become S0s or ellipticals, and S0s become ellipticals. Often an analogy is made between the generation of variations in the character of galaxy populations and human populations; in the first case, galaxy morphology is determined at conception (the 'heredity' hypothesis), while in the second case it is influenced primarily by the 'environment' in later life.

From the point of view of the environment theories, the primary difference between spiral galaxies and S0 or elliptical galaxies is that spirals contain much more gas. As a result, spirals have active star formation, and the gas allows shocks, which delineate the spiral structure. The basic idea behind most of the environment theories, first suggested by Spitzer and Baade (1951), is that spiral galaxies would become S0 galaxies if their gas were removed. A number of mechanisms have been proposed to remove the gas from spiral galaxies in clusters; these are reviewed in detail in Section 5.9.

Spitzer and Baade (1951) suggested that collisions between spiral galaxies in the cores of compact clusters remove the interstellar medium from the disks of these galaxies. Subsequent increases in the estimates of the distance scale to clusters have had the effect of seriously reducing the efficiency of this process (Section 5.9). It certainly could strip some of the spirals in a cluster. However, if the cluster has a significant amount of intracluster gas, there are several other processes which are more efficient.

Gunn and Gott (1972) suggested that S0 galaxies are formed when spiral galaxies lose their interstellar medium through ram pressure ablation on intracluster gas. This process has now been studied fairly extensively, and is reviewed in Section 5.9. In general, these studies indicate that a galaxy will be stripped almost completely during a single passage through the core of a cluster if the intracluster gas density exceeds roughly 3×10^{-4} atoms/cm^3.

Another mechanism for removing gas from galaxies, which can operate even when the galaxies are moving slowly through the intracluster gas, is evaporation (Cowie and Songaila, 1977; Sections 5.9 and 5.4.2). Heat is conducted into the cooler galactic gas from the hotter intracluster gas, and if the rate of heat conduction exceeds the cooling rate, the galactic gas will heat up and flow out of the galaxy (equation (5.118)). The evaporation rate can be significantly reduced if the conductivity saturates (Section 5.4.2), or because of the magnetic field (Section 5.4.3). If the conductivity is not suppressed by these effects, this mechanism can play an important role in stripping gas from galaxies.

The X-ray observations that are the primary subject of this book have shown that many clusters have intracluster gas densities high enough to make ram pressure stripping effective (Section 5.9; equation (5.115)). In addition, there is a strong inverse correlation between the X-ray luminosity and the fraction of spiral galaxies in a cluster (Bahcall, 1977c; Melnick and Sargent, 1977; Tytler and Vidal, 1979; equation (4.9)). The spiral fraction also decreases as the velocity dispersion of the cluster increases, as required for ram pressure ablation, and the spirals that are observed in X-ray clusters are on average at large projected distances from the cluster center (Gregory, 1975; Melnick and Sargent, 1977), where the intracluster gas density is presumably much lower than in the cluster core. Although not directly connected with stripping from spiral galaxies, the X-ray observations of M86 (an elliptical galaxy in the Virgo cluster) suggest that it is currently undergoing ram pressure ablation (Forman *et al.*, 1979; Fabian *et al.*, 1980).

Ram pressure ablation or evaporation removes the gas from cluster galaxies, and this prevents ongoing star formation. Thus one would expect that even when spiral galaxies do occur in a cluster, they will have less gas than field spirals and less active star formation. Studies of the 21 cm radio line of hydrogen from spiral galaxies in clusters indicate that they have considerably less gas than field spirals, and that the amount of gas present increases with distance from the cluster center (Section 3.7). Spirals in clusters also show weaker optical line emission than those in the field, which also suggests they have less gas (Gisler, 1978; but see Stauffer, 1983); the same is apparently true of cluster ellipticals and S0s (Davies and Lewis, 1973; Gisler, 1978). Moreover, many of the spirals in clusters have poorly defined spiral arms; they are classed as 'anemic' spirals by van den Bergh (1976). These anemic cluster spirals are intermediate in appearance and in color between field spirals and S0s, which suggests that they have less active star formation than the field spirals. Cluster spirals may also be smaller than field spirals (Peterson *et al.*, 1979). Finally, Gallagher (1978) and Kotanyi *et al.* (1983) present possible examples of spiral galaxies currently undergoing stripping.

Of course, the most direct test of the hypothesis that spiral galaxies evolve to become elliptical galaxies is the observation of spiral galaxies in high redshift clusters which may be undergoing this transformation. Unfortunately, the image sizes of galaxies in high redshift clusters (with ground based telescopes) are so small that the galaxies cannot be directly classified. However, Butcher and Oemler (1978a, 1984a, b) showed that a number of moderately high redshift ($z \approx 0.4$) clusters apparently contained a high proportion of blue galaxies. The blue galaxies lie at larger projected distances from the cluster center than the redder galaxies. When redshift effects were removed, these blue galaxies had colors indistinguishable from those of nearby spiral galaxies. No such population of blue galaxies occurs in nearby compact clusters (Butcher and Oemler, 1978b). These blue galaxies in high redshift clusters probably contain substantial quantities of gas and may be undergoing star-formation; they may indeed be spiral galaxies. If so, we would have fairly direct evidence that galactic populations evolve in rich clusters.

There are, however, a number of problems associated with this interpretation of the 'Butcher–Oemler effect'. First, Mathieu and Spinrad (1981) and van den Bergh (1983b) have argued, based on the photometry and positions of the blue galaxies in the Butcher–Oemler cluster about 3C295, that the cluster is actually relatively poor, and that the blue galaxies are primarily background and foreground field galaxies; they claim that the actual cluster members have a more normal color distribution. Redshifts and spectra now exist for a reasonably large sample of galaxies in the Butcher–Oemler clusters (Dressler and Gunn, 1982; Dressler, 1984; Dressler *et al.*, 1985); they show that many of the blue galaxies are cluster members, although many are foreground galaxies. The blue galaxies that are cluster members do have colors which might suggest that they are spirals. However, many of them have spectra and surface brightnesses that are very different from those of present day normal spiral galaxies; they appear to consist of Seyfert-like active galaxies (Henry *et al.*, 1983; Dressler *et al.*, 1985) and galaxies that have recently undergone or are undergoing very large bursts of star formation (Butcher and Oemler, 1984b). There is some evidence from deep optical images that the blue galaxies are indeed disk galaxies (Thompson, 1986). The present rather uncertain situation concerning the Butcher–Oemler effect has been reviewed recently by Dressler (1984). Second, the effect may not be universal; high redshift clusters are known which appear not to contain such a blue population (see, for example, Koo, 1981). Third, X-ray observations show that the Butcher–Oemler clusters contain significant quantities of intracluster gas (Section 4.8). The ram pressure ablation timescales are thus rather short ($\lesssim 10^9$ years). It seems unlikely that the clusters formed such a short time before they were observed; how then have the spirals survived? It is possible, given the fact that the blue galaxies lie in the outer parts of the clusters, that they have not passed through the cluster core yet. Another possibility, suggested by Gisler (1979), is that the rate of formation of interstellar gas in galaxies was higher at these redshifts. This could make it considerably more difficult to strip the galaxies through ram pressure ablation. (However, also see Livio *et al.* 1980). Finally, the redshifts of the clusters observed by Butcher and Oemler are not that large; why is it that similar clusters are not found at the present time?

Norman and Silk (1979) and Sarazin (1979) have suggested that nearly all the gas in a cluster was initially in the form of gaseous haloes around galaxies. These haloes would be stripped rather slowly by collisions until sufficient gas built up in the cluster core for ram pressure ablation to become effective. In this way, the stripping of spiral galaxies could be delayed for a significant fraction of a Hubble time, in agreement with the Butcher–Oemler effect. Larson *et al.* (1980) argued that the gaseous disks of galaxies are supplied by the continual infall of gas from the gaseous haloes proposed by Norman and Silk (1979) and Sarazin (1979). In this scenario, a galaxy will be transformed from a spiral into an S0 following the removal of its gaseous halo; star formation will then exhaust the interstellar medium in the disk in a few billion years. The Butcher–Oemler clusters might be in the midst of this transformation.

The stripping mechanisms described above (ram pressure ablation, evaporation, or the Spitzer–Baade effect) remove the gas from galaxies, but probably would not seriously affect the distribution of the stars because the mass fraction of gas in typical spirals is not terribly large (Farouki and Shapiro, 1980). Unfortunately, simply removing the gas from a spiral would leave behind a thin stellar disk; thus a stripped spiral galaxy might resemble an S0 galaxy, but never an elliptical galaxy. As the fraction of elliptical galaxies is higher in the centers of compact clusters than in the field (Table 1), the difference in galactic population between these two environments cannot result solely from the transformation of spirals into S0s. Moreover, S0s appear to have thicker disks than spirals (Burstein, 1979b). If galactic populations have evolved since formation, the stellar distribution in galaxies must have been modified during this evolution.

Of course, the process of gas removal could directly affect the stellar distribution if, during the removal of the gas, a significant portion were converted into stars which remained bound to the galaxy. This might produce the 'thick disk' component seen in S0s by Burstein (1979b).

Tidal gravitational effects during galaxy collisions can alter the stellar and mass distributions in galaxies. The possibility that massive haloes around galaxies might be removed by tidal collisions in clusters has already been discussed in Sections 2.8 and 2.10.1. Such tidal encounters might also puff up the disks of spiral galaxies and transform them into S0 or elliptical galaxies (Richstone, 1976). Unfortunately, detailed numerical simulations suggest that tidal interactions are not capable of transforming disk galaxies into ellipticals unless they do so before the cluster collapses (Da Costa and Knobloch, 1979; Farouki and Shapiro, 1981).

Another possibility is that elliptical galaxies are formed by the merger of spiral galaxies in clusters (Toomre and Toomre, 1972; White, 1979). There are a number of serious problems with this hypothesis, which have been summarized by Ostriker (1980) and Tremaine (1981). Most of these objections disappear if the mergers occur in subclusters before the formation of the cluster (see White, 1982).

Evidence given in support of the 'heredity hypothesis' (the theory that galaxy morphology is determined at the time of galaxy formation) is primarily in the form of evidence against the 'environment hypothesis'. Specifically, most of these arguments attack the extreme suggestion that the differences in galaxy morphology are determined solely by some environmental mechanism which removes the gaseous content of galaxies.

First, the total fraction of disk galaxies (Sp + S0)/(Sp + S0 + E) is not fixed, but varies from regular clusters to the field (Faber and Gallagher, 1976; Table 1). If the only change in galaxy morphology were the conversion of spirals into S0s, this ratio would be a constant. Second, some S0 and E galaxies are sometimes found in low density regions (the field), where ram pressure ablation and other environmental influences should be very weak (Sandage and Visvanathan, 1978; Dressler, 1980b), and these field S0s and Es have the same color distribution as the cluster S0s and Es. Third, Dressler (1980b) has found that distribution of galaxy morphologies correlates most strongly with the local galaxy density, and not as strongly with the global environment (which probably determines the density of intracluster gas). Fourth, the properties of S0 galaxies (colors, bulge-to-disk ratios, gaseous content, etc.) are generally intermediate between Sp and E galaxies (Sandage *et al.*, 1970; Faber and Gallagher, 1976). This would not necessarily be true if S0s were stripped Sps, but would be explained under the 'heredity hypothesis' if the nature of galaxy formation were characterized by a single parameter, which is often taken to be the density of the region in which galaxy formation occurs (see below). Fifth, S0 galaxies have larger ratios of bulge-to-disk luminosity than spirals (Burstein, 1979a), and also absolutely larger and more luminous bulges (Dressler, 1980b). This would not be expected if S0s were simply spirals with their gas removed. Finally, S0 galaxies appear to have a thick, boxy component to the disk, which is not present in spirals (Burstein, 1979b). Of course, one could argue that these thick disks actually arise during the process of stripping the gaseous disks from spirals; for example, the stripping process might induce star formation in the gas while it is being stripped, and produce a thick disk of stars supported by large velocity components perpendicular to the disk.

If galaxy morphology is determined by the conditions at the time of galaxy formation, what is the mechanism and to what conditions is it responsive? Most theories of galaxy formation assume that galaxies form by the collapse of initially gaseous matter (Eggen *et al.*, 1962; Gott, 1977). The stellar bulge components of galaxies all have distributions similar to the de Vaucouleurs profile (equation (2.14)), which suggests that they have relaxed violently (Section 2.9) during the collapse. As discussed in Section 2.9, this implies that the collapse takes place on nearly a free-fall time (equation (2.32)). If, prior to this collapse, most of the gas in the galaxy were converted to stars, these stars would act as a collisionless, dissipationless fluid, and only violent relaxation would occur. This would produce an ellipsoidal distribution of stars – that is, an elliptical galaxy. On the other hand, if star formation were ineffective, and most of the collapsing material remained gaseous, it would dissipate through radiation much of its energy, while maintaining its net angular momentum, and collapse to form a disk. With this hypothesis one can understand why the galaxies that contain significant quantities of gas are the disk-dominated spirals, and why the galaxies that lack gas are ellipticals and S0s.

Moreover, if galaxies initially have little gas compared to their stellar content, it is easier for them to remain free of gas. First, the stripping processes discussed above (ram pressure ablation, evaporation, etc.) are more effective if the density of interstellar gas is low. Second, even in the absence of such external mechanisms for removing gas, a galaxy with a high ratio of the density of stars to gas can clean itself of interstellar medium through the formation of a galactic wind (Mathews and Baker, 1971). The interstellar gas in a galaxy is heated by the stars, through supernovae, stellar winds, ionizing radiation, and the motion of mass-losing stars at high velocity through the ambient gas. If the gas density is sufficiently high compared to the stellar density, the gas will be able to cool efficiently and the energy input from stars will be radiated away. Conversely, if the gas density is low compared to the stellar density, the gas will heat up until thermal velocities in the gas exceed the escape velocity from the galaxy, and the gas leaves the galaxy in a transonic wind. Thus, if a galaxy starts with a small proportion of gas to stars, it can keep itself free of gas. Since the standard hypothesis is that E and S0 galaxies start with little gas, one can understand how they have managed to stay relatively gas free.

If it is the efficiency of star formation during the collapse of a protogalaxy which determines its morphology, what determines the efficiency of star formation? Why does it depend on the location of the protogalaxy? Two attempts to answer these questions have received particular attention. First, Sandage, Freeman, and Stokes (1970) argued that the efficiency of star formation in a protogalaxy is determined by the specific angular momentum content of the gas. If the angular momentum content of the gas is high, the collapse of protostars will be halted or delayed by centrifugal forces. Thus one would expect protogalaxies with a high angular momentum content to form spirals, and those with low angular momentum to form ellipticals. This hypothesis is in agreement with the observation that the specific angular momentum of disks significantly exceeds that of ellipsoidal components of galaxies.

Alternatively, Gott and Thuan (1976) have argued that the efficiency of star formation during the collapse of a galaxy is set by the density in the protogalaxy. Star formation requires that gases cool. Cooling processes generally involve two-body collisions. Therefore it is reasonable to assume that the cooling rate increases with density (see equation (5.23) for example).

In protogalaxies with a sufficiently high initial density, the gas will be largely converted to stars during the collapse. If the initial density is sufficiently low, star formation is not effective and the gas collapses to a disk.

In Section 2.9 evidence was presented which indicated that the sequence of cluster morphology, from the field to irregular clusters to regular clusters, was a dynamical sequence resulting from increasing initial density. Thus regular, compact clusters formed from regions of high density, and irregular clusters from lower density regions. If density is also the factor that determines galaxy morphology, the relationship of galaxy morphology and cluster morphology can be understood.

In summary, it remains controversial whether galaxy morphology is determined primarily by conditions at the time of galaxy formation, or whether galaxy morphology evolves in response to the environment after formation. It seems unlikely that *all* galaxies have identical forms at birth, and that *all* the variation in galaxy morphology is due to environment. It seems reasonable that environment has played *some* role in determining galaxy morphology; surely, somewhere at least one spiral galaxy has blundered into a core of a rich, compact cluster and been stripped. Thus I believe both mechanisms have probably significantly affected galaxy morphology. Note that two other possibilities further obscure the distinction between these two hypotheses. First, galaxies may have formed before clusters; then, the galaxies might have evolved in an environment different from that observed today (i.e., Roos and Norman, 1979). Second, the formation of disks in galaxies might be a slow and ongoing process (Larson *et al.*, 1980); then, there is no real distinction between the heredity and environment hypotheses, at least to the origin of disks.

Ultimately, the Hubble Space Telescope (Hall, 1982) will permit structural studies of galaxies in and out of clusters at large redshifts. These studies will show whether morphological evolution has occurred in galaxies, at least over the last half of the age of the universe.

2.11 Extensions of Clustering

Rich clusters of galaxies represent only a portion of a spectrum of clustering (Peebles, 1974), which ranges from individual galaxies and binary galaxies to enormous regions of enhanced density ('superclusters') or reduced density ('voids').

2.11.1 Poor Clusters

Lists of poor clusters and groups have been given by Sandage and Tammann (1975), de Vaucouleurs (1975), Turner and Gott (1976b), Hickson

(1982), Beers *et al.* (1982), and Huchra and Geller (1982). Of particular interest are the poor clusters containing possible cD galaxies which have been catalogued by Morgan, Kayser, and White (1975, MKW) and Albert, White, and Morgan (1977, AWM). In Section 2.10.1 observations of the cDs in these poor clusters were used to constrain models for the formation of cD galaxies. Bahcall (1980) has studied the optical properties (richness, galaxy distribution, and galactic content) and finds that they represent a smooth continuation to lower richness of the properties of the Abell clusters. Recent optical studies of these clusters include Beers *et al.* (1984) and Malumuth and Kriss (1986).

2.11.2 Superclusters and Voids

Clustering does not stop at the well-defined rich clusters, but extends to a much larger scale (Peebles, 1974). These superclusters appear in the distribution of galaxies, or in the distribution of clusters of galaxies. Recent reviews of superclustering include those of Rood (1981) and Oort (1983). Our own galaxy appears to lie within the Local Supercluster, a flattened system dominated by the Virgo cluster (de Vaucouleurs, 1953).

The recognition of distinct superclusters and of voids (nearly empty regions between superclusters) has largely become possible as larger samples of redshifts have become available for galaxies and clusters. With redshifts, one can study the three-dimensional distribution of galaxies, rather than just their angular distribution on the sky, and the confusing effects of projection can be reduced. Such studies have shown that Coma (A1656) and A1367 form part of a large ($\gtrsim$30 Mpc) Coma supercluster (Rood *et al.*, 1972; Chincarini and Rood, 1976; Tifft and Gregory, 1976; Gregory and Thompson, 1978), and that the Perseus cluster is embedded in the Perseus supercluster (Gregory *et al.*, 1981). Other possible superclusters identified as groupings of clusters have been catalogued by Abell (1961), Rood (1976), Murray *et al.* (1978), and Thuan (1980).

Similar large scale clustering is observed directly in the distribution of galaxies in redshift surveys in small regions of the sky by Kirshner, Oemler, and Schechter (1978) and by the more extensive survey of Davis *et al.* (1982). Many of the superclusters appear to be highly elongated.

Large voids have also been found to lie between the superclusters. Such voids appear in front of the Coma, Perseus, and Hercules superclusters (Rood, 1981). An extremely large void (100–200 Mpc in size) in the galaxy distribution in the direction of Bootes has been discovered by Kirshner *et al.* (1981). A similarly large void in the distribution of Abell clusters was discovered recently by Bahcall and Soneira (1982).

3

RADIO OBSERVATIONS

3.1 General Radio Properties

The association between radio sources and clusters of galaxies was first made by Mills (1960) and van den Bergh (1961b). Radio emission from clusters has been reviewed recently by Robertson (1983). A brief review of the radio properties of clusters relevant to their X-ray emission will be given in this section. First, the general radio luminosities and spectra will be summarized. Second, possible correlations between the radio and X-ray emissions of clusters will be discussed. Third, two classes of radio source morphology (head-tail radio sources and cluster halo sources) which are unique to the cluster environment will be described. Fourth, possible observations of the effect of the intracluster medium on radio emission due to the cosmic blackbody background or background source will be reviewed. Finally, the use of the 21 cm hyperfine line to detect neutral hydrogen gas in clusters will be briefly discussed.

Figure 7 shows a contour plot of the radio emission in the Perseus cluster, superimposed on an optical photograph. The very strong radio emission from the central galaxy NGC1275 (Section 4.5.2) and the two head-tail radio galaxies NGC1265 and IC310 are shown.

The radio emission from clusters of galaxies (as well as most other extragalactic objects) is synchrotron emission due to the interaction of a nonthermal population of relativistic electrons (with a power-law energy distribution) with a magnetic field (Robertson, 1983). Such nonthermal synchrotron emission generally has a spectrum in which the intensity I_ν (erg/cm^2 Hz s) is well represented as a power-law over a wide range of frequencies ν,

$$I_\nu \propto \nu^{-\alpha_r}, \tag{3.1}$$

where α_r is the radio spectral index. Typical extragalactic radio sources have

$\alpha_r \approx 0.8$. Most of the radio emission from clusters is due to discrete sources, which can be associated with individual galaxies within the cluster. The properties of the nonthermal radio emission from radio galaxies have been reviewed recently by Miley (1980).

Radio emission from Abell clusters at a frequency of 1400 MHz has been surveyed by Owen (1975), Jaffe and Perola (1975), and Owen *et al.* (1982). At lower frequencies there are a number of older surveys (e.g., Fomalont and Rogstad, 1966), as well as an extensive list based on the 4C Cambridge survey (Slingo, 1974a, b; Riley, 1975; McHardy, 1978b). These observations suggested that only sources detected in the inner portions (within 1/3 Abell radii) of the Abell clusters were likely to belong to the cluster, the rest being background objects. Observations at high frequencies have been made by Andernach *et al.* (1980, 1981), Haslam *et al.* (1978), and Waldthausen *et al.*

Fig. 7. A radio map of the Perseus cluster of galaxies from Gisler and Miley (1979). Contours of constant radio surface brightness at 610 MHz are shown superimposed on the optical image of the cluster. Note the very strong source associated with the galaxy NGC1275 (the highest contours associated with this source have been removed), and the two head-tail radio sources associated with NGC1265 and IC310. The rings are diffraction features due to NGC1275 and are not real.

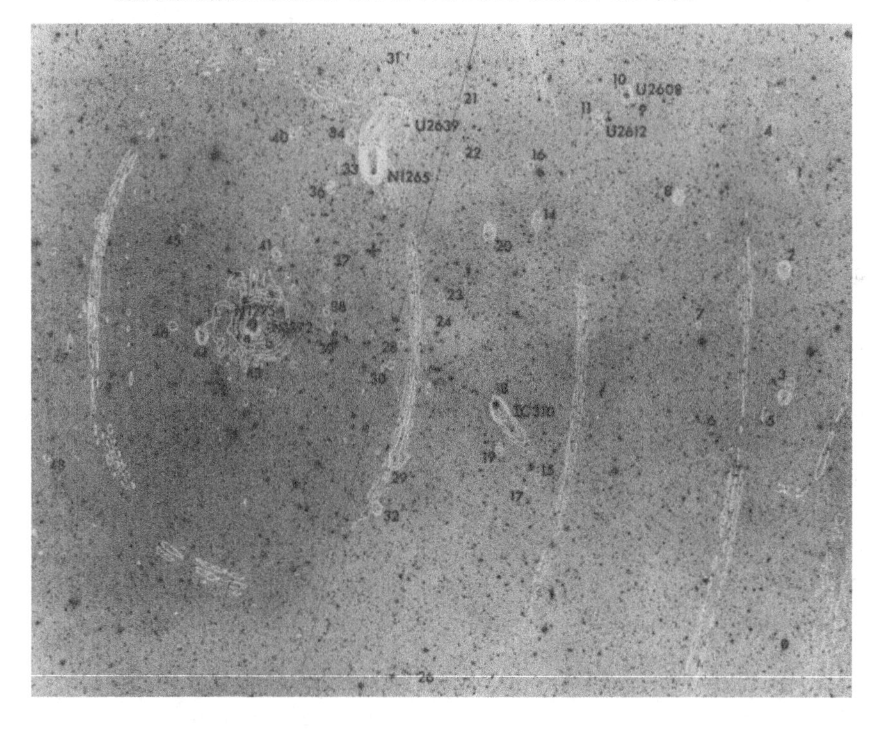

(1979). The more distant Abell clusters were searched for radio emission by Fanti *et al.* (1983), while the richest Abell clusters were observed by Birkinshaw (1978). Jaffe (1982) surveyed a sample of high redshift clusters, and found evidence of evolution in the radio luminosity of clusters for the range of redshifts $0.25 < z < 0.95$.

In discussing the radio luminosity functions of clusters, it is important to distinguish the luminosity function of galaxies in clusters from the luminosity function of the cluster as a whole. The radio emission from clusters is mainly due to sources associated with individual radio galaxies. About 20% of the nearby strong radio galaxies are located in rich clusters of galaxies (McHardy, 1979). This appears to be mainly due to the fact that strong radio emission is primarily associated with giant elliptical galaxies, which occur preferentially in clusters. A galaxy of a given morphology (elliptical, for example) and optical luminosity apparently has the same radio luminosity function whether inside or outside of a cluster (Jaffe and Perola, 1976; Auriemma *et al.*, 1977; Guindon, 1979; McHardy, 1979). The radio luminosity function of the whole cluster can be fit as the result of superposing the luminosity function of an average of ≈ 5 radio galaxies per cluster (Owen, 1975). The cluster radio luminosity function does not appear to depend strongly on richness for the Abell clusters (Riley, 1975; Owen, 1975; McHardy, 1979).

There does appear to be a correlation between cluster radio emission and cluster morphology. Owen found that the more evolved RS types (cD, B, C, and L; see Section 2.5) have stronger radio emission. McHardy (1979) found that more evolved BM types (I, I–II; see Section 2.5) tend to contain stronger radio galaxies, at least at low frequencies.

Powerful radio sources are found most often near the cluster center (McHardy, 1979). They are usually associated with optically dominant galaxies, which often have multiple nuclei and are often cD galaxies (Guthie, 1974; McHardy, 1974, 1979; Riley, 1975).

Cluster radio sources generally have steeper radio spectra (values of $\alpha_r \gtrsim 1$) than radio sources in the field (Costain *et al.*, 1972; Baldwin and Scott, 1973; Slingo, 1974a, b; McHardy, 1979). The steepness of the spectrum (the value of α_r) increases with cluster richness and decreases as the BM type increases (Roland *et al.*, 1976; McHardy, 1979). The steepest radio spectra ($\alpha_r \approx 1.3$) in clusters are associated with radio sources in optically dominant galaxies (Riley, 1975).

De Young (1972) claimed that cluster radio galaxies were generally smaller than those in the field; this claim was not supported by larger surveys (Hooley, 1974; McHardy, 1979). Owen and Rudnick (1976a) found that cluster radio sources generally had extended emission; unresolved sources seen towards clusters are usually background objects. Guindon (1979) found that, while the average size of double or triple radio sources in clusters was the same as those in the field, clusters did lack the very largest sources.

Differences in the morphology of cluster radio sources and field sources are discussed in Sections 3.3 and 3.4 below.

Recently, the radio emission properties of poor clusters have been studied. In general, the sources in poor clusters also have steep spectra and show some of the same morphological distortions found in rich cluster sources (Burns and Owen, 1977). As in rich clusters, the strongest radio emission is associated with optically dominant galaxies near the cluster center, and these sources have especially steep spectra (White and Burns, 1980; Burns *et al.*, 1981b; Hanisch and White, 1981). There is no apparent difference between the radio emission of poor and rich cD clusters beyond the direct effect of richness (Burns *et al.*, 1980).

3.2 Correlations Between X-ray and Radio Emission

Based on the small sample of X-ray clusters known at that time, Owen (1974) argued that there was a strong correlation between the radio and X-ray luminosity of clusters. Rowan-Robinson and Fabian (1975) did not find this correlation. Bahcall (1974b, 1977a) found that the X-ray emission of a cluster was increased if a strong radio source was located near the center of the cluster. McHardy (1978a) found that strong radio sources were more likely to be located in luminous X-ray clusters than in other clusters.

As discussed above, clusters generally contain steep-spectrum radio sources. It appears that radio sources in X-ray clusters have even steeper spectra, and that α_r correlates with X-ray luminosity (Erickson *et al.*, 1978; McHardy, 1978a; Cane *et al.*, 1981; Dagkesamansky *et al.*, 1982). This suggests that there might be a strong correlation between low frequency radio flux and X-ray luminosity (Erickson *et al.*, 1978; Cane *et al.*, 1981). Such a correlation is not found in larger samples of X-ray clusters (Mitchell *et al.*, 1979; Ulmer *et al.*, 1981). X-ray selected cluster samples from the *Einstein* observatory do not show a strong X-ray–radio correlation (Feigelson *et al.*, 1982; Johnson, 1981).

A possible correlation between X-ray emission and radio emission in poor clusters has also been found (Burns *et al.*, 1981c).

There appears to be a strong correlation between radio emission by central dominant galaxies in clusters and the presence of cooling flows, as evidenced by soft X-ray line emission or central spikes in the X-ray surface brightness (Burns *et al.*, 1981a; Valentijn and Bijleveld, 1983; Jones and Forman, 1984; Section 5.7.2).

To summarize, at present there may be a weak correlation between X-ray and radio luminosities, and there appears to be stronger correlation between the radio spectral index α_r and the X-ray luminosity L_x. Because of the many interrelationships between cluster properties, it is difficult to decide whether these possible correlations are primary or reflect other correlations (see Section 4.6). For example, the relationship between L_x and α_r may be a result of the fact that L_x correlates with cluster optical morphology (Section 4.6), as does α_r. Moreover, it is difficult to establish a clear causal basis for these correlations.

Costain *et al.* (1972) and Owen (1974) argued that the correlation between X-ray and radio emission implies that the same population of relativistic electrons produced both radio emission and X-ray emission. The radio emission is synchrotron emission; in this model the X-ray emission would be inverse Compton scattering of cosmic radiation photons by the same relativistic electrons. This 'inverse Compton' (IC) theory is described in more detail in Section 5.1.1. The IC model does require that cluster radio sources have steep spectra, as observed. However, the evidence against the IC model is now overwhelmingly strong (Sections 4.3 and 5.1). Thermal emission by diffuse gas provides the main X-ray emission from clusters.

Another direct connection between the radio emitting electrons and the X-ray emitting thermal gas would be established if the thermal gas were heated by the relativistic electrons and/or any associated 'cosmic ray' nuclei. Such heating would occur through Coulomb interactions (Lea and Holman, 1978) and might be enhanced by plasma interactions (Scott *et al.*, 1980; see Section 5.3.5. for more details). The heating is strongest for the lower energy electrons which produce very low frequency radio emission, and thus the heating requires that the radio sources in clusters have steep spectra, as observed. However, it is not clear that any ongoing heating of the thermal gas either is needed or does occur in the majority of X-ray clusters. First of all, there exist a reasonable number of strong X-ray clusters that do not have strong steep-spectrum radio sources. Second, the thermal energy per unit mass in the hot gas is roughly the same as the kinetic energy per unit mass in the galaxies. Thus the gas could have been heated initially by thermalizing its kinetic energy when it either was ejected from galaxies or fell into the cluster. In typical clusters, the cooling time in most of the gas is longer than the probable age of the cluster (the Hubble time), and no further heating of the gas would necessarily be required (Section 5.3).

A connection between the X-ray and radio luminosity of clusters might be produced if the radio emission were powered by accretion of gas which was initially part of the hot intracluster medium. In Sections 4.3 and 5.7 evidence is presented indicating that intracluster gas is cooling and being accreted by central dominant galaxies in many X-ray clusters. As mentioned previously, there is a correlation between radio emission by central dominant galaxies in clusters and the presence of these cooling flows (Burns *et al.*, 1981a; Valentijn and Bijleveld, 1983; Jones and Forman, 1984). The further accretion of this cooling gas onto a central massive object in the galaxy might produce the radio emission.

It is likely that the most important connection between the X-ray emitting gas in clusters and the relativistic electrons that produce the radio emission is dynamical. The hot gas provides pressure forces that can control the dynamics of the plasma of relativistic electrons. The current evidence suggests that radio emission from galaxies occurs when streams of blobs of relativistic nonthermal plasma are ejected from the nucleus of the galaxy (Miley, 1980). If the pressure and density of any surrounding medium is sufficiently large, the bulk motion and expansion of the radio emitting plasma will be retarded. This could explain the absence of very large radio galaxy sources associated with clusters. Moreover, the intracluster gas may confine the radio emitting plasma and prevent its adiabatic expansion. Expansion and synchrotron emission provide two competing energy loss mechanisms for the relativistic plasma. If expansion occurs, it weakens the radio emission but generally will not affect its spectrum. If the expansion is retarded by the pressure of a confining medium, synchrotron losses become important. These losses are most important for the highest energy electrons, and thus they cause the spectrum of the radio source to steepen. This mechanism probably provides the most plausible explanation of the steep spectrum of cluster radio sources.

Additional evidence for the dynamical effect of the hot intracluster gas on radio galaxies comes from distortions in the structure of the radio source produced if the radio galaxy is moving relative to the intracluster medium, as discussed below.

3.3 Head-Tail and Other Distorted Radio Structures

There are two general types of radio source structures that occur predominantly in clusters of galaxies. In this section, distortions in the structures of single radio galaxies are discussed. In Section 3.4 cluster sized radio halo sources are reviewed.

A large proportion of relatively isolated radio galaxies have a fairly simple and symmetrical radio structure; for a review of the properties of radio galaxies see Miley (1980). Many of these galaxies have a compact radio source associated with the nucleus of the galaxy. There is also extended radio emission, generally in the form of double radio 'lobes', which lie on either side of the galaxy. These lobes are often of comparable brightness and projected distance from the nucleus, and most importantly, lie on a line through the nucleus of the galaxy. Recently observations have detected 'jets' of radio emission originating at the nucleus and extending out to the radio lobes; in some cases only one jet on one side of the galaxy has been found. The conventional theoretical scenario for the origin and energetics of these radio galaxies contains the following elements (Miley, 1980). First, the ultimate energy source for the radio emission (and all other nonthermal galaxy emission) is thought to be a very compact object in the nucleus of the galaxy. Energy is carried from this nonthermal engine out to the radio lobes by twin 'beams' of plasma; this plasma probably contains a mixture of thermal gas and relativistic nonthermal particles, which cause the beams to be observable in some cases as radio jets. The beams may be more or less continuous, or may consist of blobs of plasma ('plasmoids'). The beams probably move outwards until they encounter a sufficient quantity of intergalactic (or intragalactic) gas, at which point their bulk kinetic energy of motion is converted into thermal energy and into the disordered relativistic motion of particles of nonthermal plasma. This nonthermal plasma produces the emission from the radio lobes.

Radio galaxies in clusters show more complex radio structures, which generally tend to lack the symmetrical, aligned double structure of standard radio galaxies. These range from double lobed radio sources in which the lobes are not aligned with the galaxy nucleus ('bent-doubles' or 'wide-angle-tails') to sources in which all the radio emission lies in a tail on one side of the galaxy, and the galaxy itself forms the head of the tail ('head-tail' or 'narrow-angle-tail' radio galaxies).

The first head-tail (HT) radio galaxies discovered were NGC1265 and IC310 in the Perseus cluster (Ryle and Windram, 1968; see Figure 7), followed by the discovery of head-tail radio galaxies in Coma (Willson, 1970) and the 3C129 cluster (MacDonald *et al.*, 1968). Figure 8 shows a radio map of NGC1265, which is the archetypical head-tail radio galaxy. Some lists of head-tail radio galaxies and other distorted radio sources are those of Rudnick and Owen (1976a, b) and Simon (1978, 1979), and Valentijn (1979a), and other observations of head-tail radio galaxies in rich clusters are given in Hill and Longair (1971), Vallee and Wilson (1976), Miley and Harris (1977), Gisler and Miley (1979), Burns and Ulmer (1980), Hintzen and Scott (1980), Bridle and Vallee (1981), Gavazzi *et al.* (1981), Vallee *et al.* (1981), and Dickey and Salpeter (1984).

NGC1265 and IC310, the first two head-tail radio galaxies discovered, are both in the Perseus cluster and have tails that lie on the line from the galaxy to the powerful radio galaxy NGC1275 at the cluster center (Figure 7). Ryle and

Fig. 8. A low resolution radio map at a frequency of 5 GHz of the head-tail radio source associated with the galaxy NGC1265 in the Perseus cluster, from Wellington *et al.* (1973). Contours of constant radio surface brightness are shown superimposed on the optical image of the galaxy.

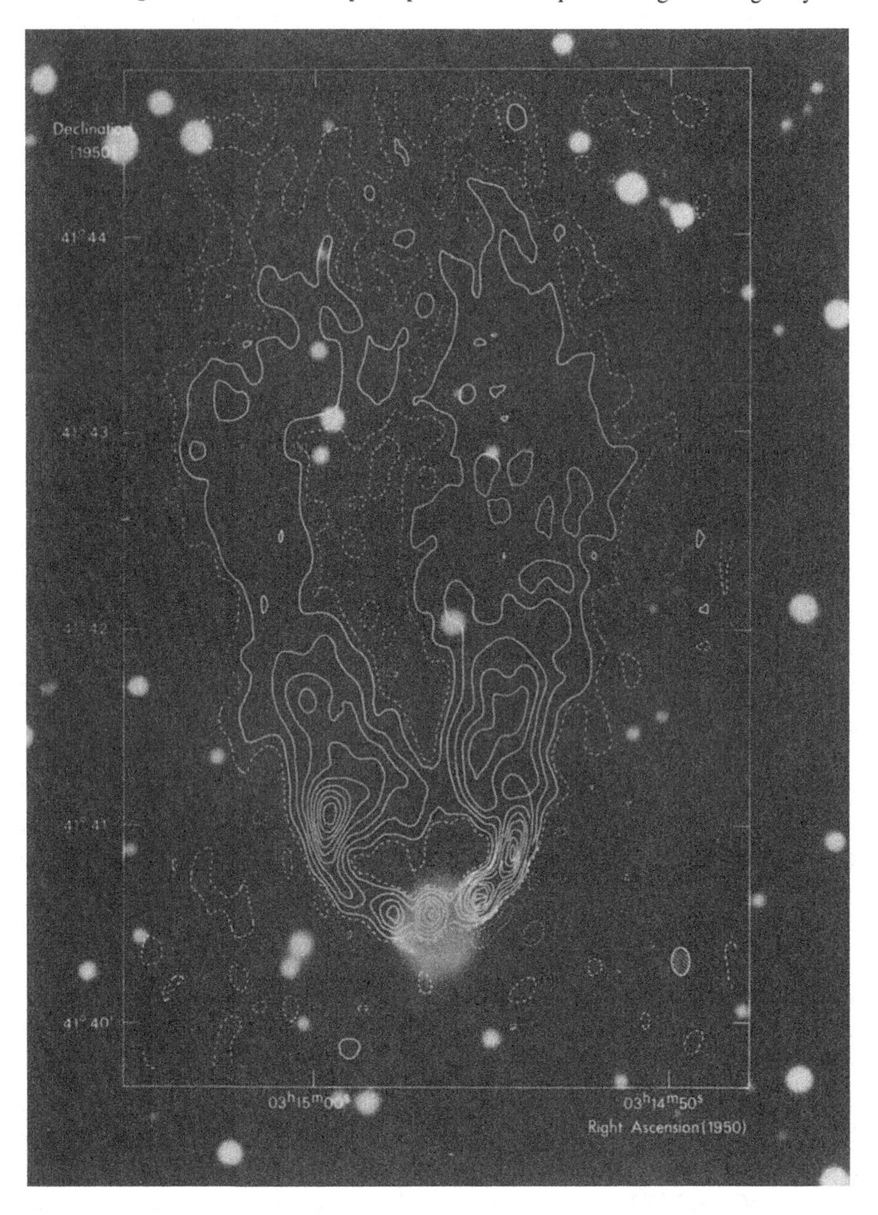

Windram (1968) suggested that the radio emission from these galaxies was activated by a wind of relativistic particles from NGC1275, which also determined the directions of the tails. However, subsequent head-tail radio galaxies have not been found to show any alignment with the direction to powerful radio galaxies or the cluster center. The accepted explanation of head-tail radio galaxies, originally due to Miley *et al.* (1972), is that they are conventional radio galaxies moving at a high velocity through a static intracluster gas. The radio emitting beams or plasmoids are decelerated by the ram pressure of the intracluster gas and form a wake behind the galaxy. The high velocity of the galaxy is a result of the gravitational potential of the cluster (that is, the velocity v is comparable to the cluster velocity dispersion). The ram pressure acting on the radio blobs or beams is then

$$P_r = \rho_g v^2, \tag{3.2}$$

where ρ_g is the intracluster gas density (Section 5.9).

Miley (1973) found that the spectral index α_r and the fractional polarization of the radio emission increased with distance away from the galaxy along the tail. Synchrotron emission energy losses will steepen a radio spectrum, so the spectral variations are consistent with injection of particles at the galaxy. The polarization indicates that the magnetic field is highly ordered and directed along the direction of the tail, probably by the sweeping of the radio emitting plasma behind the galaxy. There are some indications that head-tail radio galaxies are more rapidly moving than typical cluster galaxies (Guindon and Bridle, 1978), but the effect is not very large (Ulrich, 1978), and in any case, only one component of the velocity can be measured. Head-tail radio galaxies are never cD galaxies and are seldom among the most luminous galaxies in a cluster, either at radio or at optical wavelengths (Rudnick and Owen, 1976a, b; Simon, 1978; McHardy, 1979; Valentijn, 1979c). Since cD galaxies are nearly at rest in the cluster potential (Section 2.10.1), they would not be expected to form head-tail radio galaxies.

Of course, the intracluster gas that produces the head-tail radio galaxies also produces X-ray emission; in general, the densities of intracluster gas ρ_g derived from X-ray observations are consistent with those needed to give a ram pressure sufficient to produce the observed radio structures (see, for example, Simon, 1979).

The first detailed theoretical work on head-tail radio galaxies was done by Jaffe and Perola (1973). They suggested two models; in the first, blobs of plasma were ejected in opposite directions (taken to be perpendicular to the direction of motion of the galaxy). They gave several arguments against this model. First, the adiabatic expansion of the blobs would produce large losses in their energy; if these losses exceeded losses due to synchrotron emission, the spectrum would not vary as observed. Second, they argued that the magnetic field was too well ordered to be an initially disordered field. Thus they proposed a second model, in which the radio galaxy possessed an extensive magnetosphere, which was swept behind the galaxy. The magnetosphere provided an ordered magnetic field, and they argued that it could confine the adiabatic expansion of the radio emitting blobs. This second argument was shown to be incorrect by Cowie and McKee (1975) and Pacholczyk and Scott (1976). Cowie and McKee showed that the large adiabatic losses found by Jaffe and Perola were due in part to the assumption of high Mach number flow, which is not correct for the temperatures of the intracluster gas derived from X-ray observations (Section 4.3). As a result, subsequent theoretical work has largely been devoted to models for the interaction of free plasmoids (the first Jaffe–Perola model) or twin beams with intracluster gas (Cowie and McKee, 1975; Pacholczyk and Scott, 1976; Begelman *et al.*, 1979; Christiansen *et al.*, 1981).

One important problem with this model is that in many sources the age of the tail (projected length divided by the estimated galaxy velocity) is much

Fig. 9. A high resolution Very Large Array radio map at a frequency of 4.9 GHz of the head-tail radio galaxy NGC1265, from O'Dea and Owen, 1986. Note the twin radio jets leading from the nucleus of the galaxy out to the radio tails.

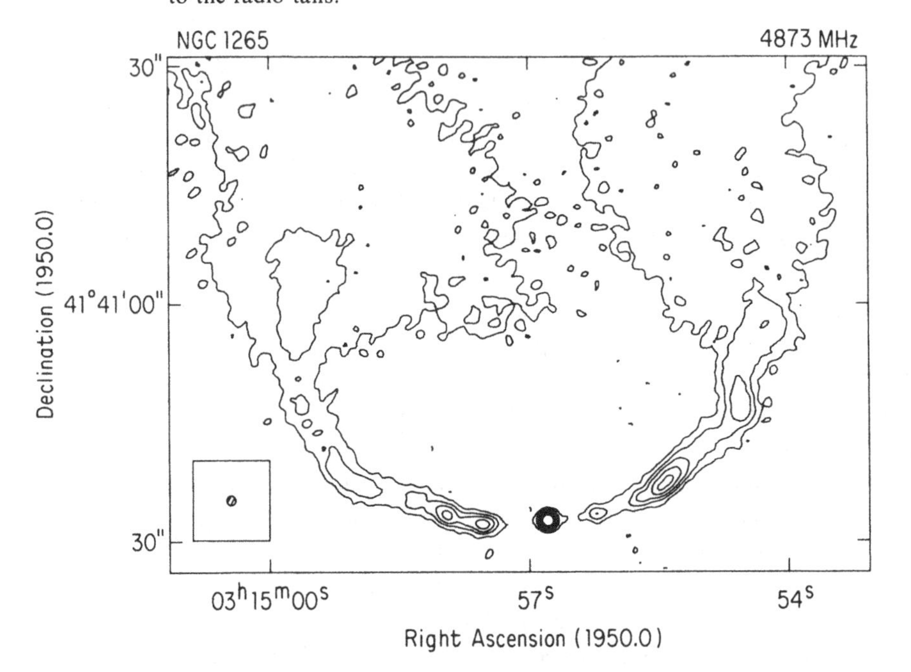

longer than the lifetimes of the emitting electrons against synchrotron losses (Wilson and Vallee, 1977). Pacholczyk and Scott (1976) and Christiansen *et al.* (1981) have argued that these particles are reaccelerated by turbulence in the tails.

Recent high resolution radio observations have detected well defined radio jets leading out from the nucleus of the head-tail radio galaxy to the start of the tails (Owen *et al.*, 1978, 1979; Burns and Owen, 1980). Figure 9 shows these jets in NGC1265. Begelman *et al.* (1979) presented a twin-jet theory for the formation of HTs. Because the observed jets are narrow and follow well defined curved paths, it is unlikely that they consist of independent plasmoids, because these plasmoids would most likely have different masses and surface areas and would be bent by different amounts by the ram pressure force. The transition from a collimated jet into the swept back tail is fairly abrupt. Jones and Owen (1979) argued that the jets in the inner parts of the galaxies are protected from the ram pressure of the intracluster gas. They suggested that the head-tail radio galaxies contain regions of gas that are bound to the galaxy and too dense to be stripped by ram pressure (Section 5.9). The jets propagate through and are confined by this gas until they reach the intracluster gas and are swept back. However, the jets are curved in this inner region (Figure 9), which is difficult to understand if they propagate through static intragalactic gas (although the curvature could be due to buoyancy force if the intragalactic gas is compressed by ram pressure from the intracluster gas). In any case, Begelman *et al.* (1979) are able to fit the observed shape of the jets assuming ram pressure from intracluster gas and no intragalactic gas.

In some cases the tails in head-tail radio galaxies are observed to be curved or bent. Jaffe and Perola (1973), Miley and Harris (1977), and Vallee *et al.* (1979, 1981) suggested that the curvature reflects nonlinear galaxy motion, due to the orbit of the head-tail radio galaxies in the cluster potential, or due to a binary orbit. Jaffe and Perola (1973) also suggested that there might be an intracluster gas wind with shear. Cowie and McKee (1975) suggested that the bending might be due to buoyancy forces; the bendings in a number of head-tail radio galaxies are consistent with this (Gisler and Miley, 1979).

Another common morphological type of cluster radio source is 'wide-angle-tails' or WATs. These resemble classical double radio sources in which the two radio lobes and/or jets are not oppositely aligned, but lie at an angle. Figure 10 shows the radio image of the WAT 1919 + 479, which is in a Zwicky cluster. Some lists of these sources include those of Owen and Rudnick (1976b), Simon (1978), Guindon and Bridle (1978), Valentijn and Perola (1978), Valentijn (1979b, c), and van Breugel (1980). WATs are generally associated with optically dominant cluster galaxies, often cD galaxies (Owen and Rudnick, 1976b; Simon, 1978; Valentijn, 1979c), and tend to be more luminous radio sources than head-tail radio galaxies. They tend to be slower moving galaxies than those containing head-tail radio galaxies (Guindon and Bridle, 1978). A particularly fascinating case is the radio source 3C75 in the cluster A400, which has two intertwined pairs of jets which merge into a WAT (Owen *et al.*, 1985).

A major theoretical question about WATs is whether they are the result of the same physical process (ram pressure by intracluster gas) as head-tail radio galaxies. Owen and Rudnick (1976b) suggested that ram pressure was the mechanism behind WATs; since the WATs occur in slow moving (cD) galaxies, the ram pressure is less, and the radio structure is bent at a smaller

Fig. 10. A Very Large Array radio image at a frequency of 1420 MHz of the wide-angle-tail (WAT) radio galaxy 1919 + 479 associated with a cD galaxy in a Zwicky cluster, from Burns *et al.* (1986).

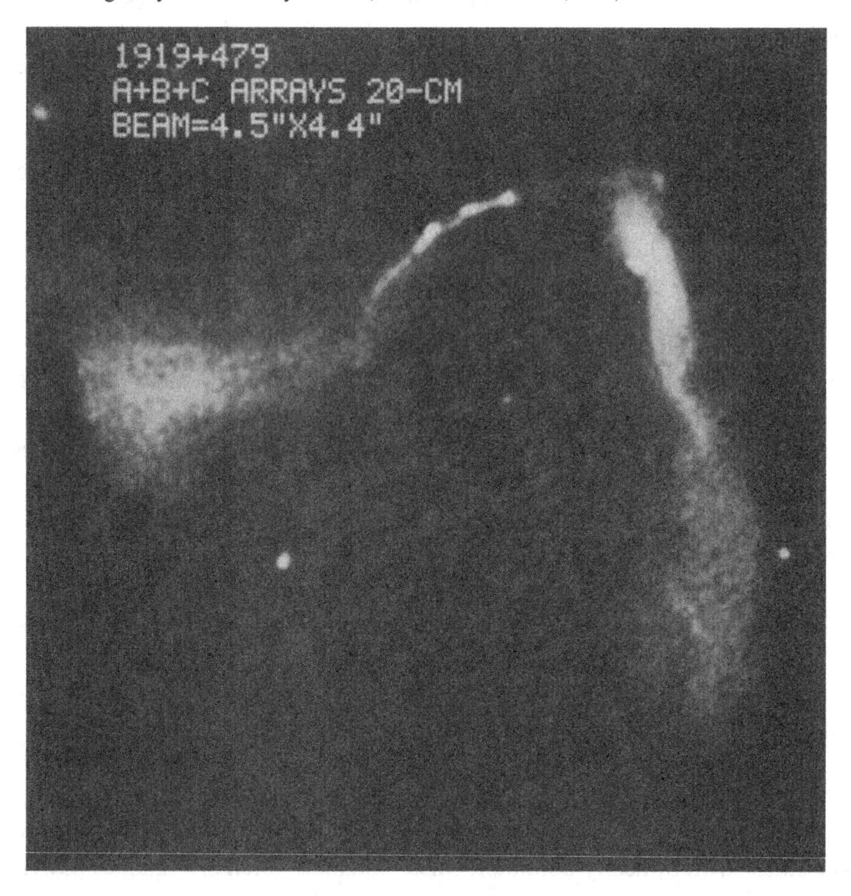

angle. Another possible reason that WATs might be less bent is that they are associated with stronger radio sources. The plasmoids or beams associated with these sources may have more momentum than those in weaker radio sources, and may be harder to deflect. Valentijn and Perola (1978) and Valentijn (1979b) suggested that WATs were similar to HTs, except that the ejection angle of the beams or plasmoids was not nearly perpendicular to the velocity of the galaxy. This would not, by itself, explain why WATs occur in galaxies that are more prominent optically and in radio emission. Valentijn (1979b) also invokes the smaller galaxy velocity and more rigid beams discussed above, and suggests that these brighter galaxies may have more intragalactic gas, which shields the beams from the intracluster gas.

Burns (1981) and Burns *et al.* (1982) suggest that the WATs are bent mainly by buoyancy forces and *not* ram pressure forces. Of course, if the galaxy associated with the WAT were a cD galaxy located at the center of the cluster potential and if the intracluster gas density were spherically symmetric about this center, no bending due to buoyancy could occur. Burns *et al.* (1982) argue that the cD associated with a WAT may not be exactly at the cluster center, and that accretion by the slowly moving cD may produce an aspherical density distribution in the intracluster gas.

Sparke (1983) argues that WATs and other distorted radio sources associated with cDs are indications that the clusters are in the process of collapsing. She notes that many of the WATs are associated with irregular clusters with clumpy X-ray emission and low X-ray temperatures (the irregular X-ray clusters of Table 3 below).

The association between these distorted radio morphologies and clusters of galaxies containing intracluster gas has been used to detect previously unobserved clusters and cluster X-ray sources. Burns and Owen (1979) found a number of distorted radio sources associated with poor Zwicky clusters, and suggested they might be X-ray sources. This conjecture was confirmed by Holman and McKee (1981). Fomalont and Bridle (1978) discovered a number of WATs in groups of galaxies.

Hintzen and Scott (1978) suggested that quasars with distorted radio structure were likely to be members of clusters of galaxies, and that radio observations of quasars could be used to detect high redshift clusters of galaxies and cluster X-ray sources. In general, optical studies have not found that quasars are associated with rich clusters, and any method that selected quasars in clusters or any type of high redshift clusters would be very useful. This method has in fact been used to detect clusters around several quasars (Hintzen *et al.*, 1981; Harris *et al.*, 1983a) and to provide a list of other candidates (Hintzen *et al.*, 1983).

3.4 Cluster Radio Haloes

The second type of radio morphology distinctly associated with clusters of galaxies is the cluster radio halo. They are diffuse, extended radio sources whose sizes are generally considerably larger than the cluster galaxy core radius, and smaller than the overall cluster size (for example, an Abell radius). The best studied example is Coma C, the halo source in the Coma cluster, which was first shown to be diffuse by Willson (1970), and which was studied further by Jaffe *et al.* (1976), Jaffe (1977), Valentijn (1978), Hanisch *et al.* (1979), Hanisch (1980), and Hanisch and Erickson (1980). Other clusters in which halo sources have probably been detected include A401 (Harris *et al.*, 1980; Roland *et al.*, 1981; but see Hanisch and Erickson, 1980), A1367 (Gavazzi, 1978; Hanisch, 1980; but this halo is considerably weaker and smaller than the others), A2255 (Harris *et al.*, 1980), A2256 (Bridle and Fomalont, 1976; Bridle *et al.*, 1979; but this is a very messy radio cluster), and A2319 (Grindlay *et al.*, 1977; Harris and Miley, 1978; Andernach *et al.*, 1980). While Ryle and Windram (1968) reported a rather large radio halo in the Perseus cluster, it does not appear to be a real source (Gisler and Miley, 1979; Jaffe and Rudnick, 1979; Birkinshaw, 1980; Hanisch and Erickson, 1980), although a smaller halo source has been reported recently (Noordam and de Bruyn, 1982).

Radio haloes do not appear to be very common, as a large number of surveys of clusters have failed to find them (Jaffe and Rudnick, 1979; Cane *et al.*, 1981; Andernach *et al.*, 1981; Hanisch, 1982a).

The cluster radio halos have very steep power-law radio spectra, with $\alpha_r \approx 1.2$. The power-law spectrum and some indications of polarization suggest that the emission process is synchrotron emission by nonthermal relativistic electrons, as in radio galaxies. The halo in Coma has a diameter (FWHM) of ≈ 1 Mpc/h_{50}, which is typical. In Coma, the spectrum of the halo is relatively uniform spatially (Jaffe, 1977). Although the sample of known radio haloes is small, they appear to be associated with clusters of intermediate optical morphology (BM II; RS B, C, L) (Hanisch, 1982b). These clusters are relaxed, but do not have a dominant cD galaxy (an exception may be A2319). The haloes are generally associated with clusters having a regular nXD X-ray morphology (Vestrand, 1982); Vestrand notes that many of these clusters have particularly luminous and extended X-ray emission and may have unusually high X-ray temperatures (see also Forman and Jones, 1982). In making these comparisons, the unusually weak and small halo associated with A1367 has not been included.

There is currently no consensus as to the origin of these haloes. Jaffe (1977) discussed observational and theoretical constraints on the origin of the nonthermal electrons producing the emission in the Coma cluster, and proposed that the electrons originate at strong radio sources in the cluster and diffuse out to form the halo. The observed spectral index α_r is about 0.5 larger than the spectral indices of strong cluster radio sources; such an increase occurs if there is a steady-state between the input of relativistic nonthermal electrons and synchrotron losses. Moreover, the number of electrons produced in strong radio sources is sufficient to explain the halo radio emission if the magnetic field in the cluster is $B_c \approx 1\mu$G, which is consistent with limits on the hard X-ray emission from clusters (Section 4.3.1). However, the halo radio emission is less strongly peaked at the cluster center than the distribution of galaxies, particularly of strong radio galaxies (Jaffe, 1977). Thus the nonthermal electrons must be transported out from the cluster core. In order that synchrotron losses should not affect the spectrum of the electrons and cause the halo radio spectrum to steepen dramatically with radius (which is not observed; Jaffe, 1977), the particles must be transported at a velocity which is $\gtrsim 2000$ km/s. Convective fluid motions of this order would be supersonic and would involve a very high rate of energy dissipation. Thus Jaffe argued that the relativistic electrons must diffuse out into the cluster.

As discussed extensively by Jaffe, a diffusion velocity $\gtrsim 2000$ km/s would greatly exceed the Alfvén velocity

$$v_A \equiv \left(\frac{B_c^2}{4\pi\rho_g}\right)^{1/2} \tag{3.3}$$

in the intracluster plasma. (Here, ρ_g is the density of intracluster gas.) For typical values of the gas density from X-ray observations and the required magnetic field discussed above, $v_A \lesssim 100$ km/s. Particles that diffuse through a plasma faster than the Alfvén velocity excite plasma waves, which rapidly slow down the diffusion of the particles, and thus Jaffe argued that the Alfvén velocity acts as an upper limit on the diffusion speed of the relativistic electrons in radio halos. This velocity is much too small to allow the particles to diffuse without losses. A possible solution to this Alfvén speed limit problem, suggested by Holman *et al.* (1979), is that the plasma waves generated by electrons diffusing at speeds greater than v_A may be damped by ions in the background thermal plasma. This would allow diffusion at speeds up to the speed of these background ions, essentially the sound speed in the intracluster gas.

Another solution to the Alfvén speed problem was suggested by Dennison (1980b). He noted that the flux of relativistic, nonthermal particles at the Earth (cosmic rays) is dominated by protons, and suggested that this might also be true in radio sources. The protons would diffuse away from cluster radio galaxies at the Alfvén speed, but suffer no significant synchrotron losses because of their rigidity. In the cluster they would collide with thermal protons and produce secondary electrons by a number of processes. These relativistic, nonthermal secondaries would then produce the observed radio haloes in this model.

Harris and Miley (1978) suggested that the radio haloes are remnants of previous head-tail radio galaxies, whose spectra have steepened due to synchrotron losses. One problem with this idea is that HTs are typically not very luminous, and the clusters with radio haloes therefore would be required to have had a large number of bright radio galaxies in the past. However, radio haloes are rare, so this may not be a serious objection.

Jaffe (1977) considered the possibility that the nonthermal electrons are accelerated to relativistic energies within the cluster by turbulence in the intracluster gas. Roland *et al.* (1981) suggested that the turbulence was generated by the wakes of galaxies moving through the intracluster medium. Based on the small available sample, they suggested that the luminosity of radio haloes increases with the cluster X-ray luminosity L_x (a measure of the amount of gas in the cluster) and the velocity dispersion of galaxies σ_r, with $L_{halo} \propto L_x \sigma_r^2$. There are several problems with this hypothesis; first, unless the acceleration of relativistic electrons is very efficient, the rate of dissipation of the turbulent energy is unacceptably large (Jaffe, 1977). Second, no galactic wake has been detected as a radio source. They ought to appear as tailed galaxies without heads (no radio source in the nucleus of the galaxy).

The cluster A401 has been observed to possess a radio halo. With A399, this cluster forms a possible merging double system (Ulmer and Cruddace, 1981; Section 4.4). Harris *et al.* (1980) suggested that radio haloes form during the coalescence of subclusters, possibly by the acceleration of relativistic particles in shocks which form in the intracluster gas. However, there are a reasonable number of other double clusters that do not show radio haloes.

Observations of radio haloes are important to the understanding of X-ray cluster emission because the nonthermal radio-emitting electrons and X-ray emitting thermal plasma coexist and may interact. Initially, it had been suggested that the X-ray emission might be inverse Compton emission from the nonthermal electrons (Section 5.1.1). However, the frequency of occurrence of X-ray emission and rarity of radio haloes is one of the many arguments against this theory. On the other hand, the nonthermal electrons may heat the thermal plasma and contribute to the X-ray emission indirectly (Lea and Holman, 1978; Rephaeli, 1979; Scott *et al.*, 1980; see also Sections 3.2 and 5.3.5). Vestrand (1982) has pointed out that radio halo clusters have extended X-ray emission and may have higher X-ray temperatures than nonhalo clusters; he attributes this difference to the heating of intracluster gas by nonthermal electrons.

3.5 Cosmic Microwave Diminution (Sunyaev–Zel'dovich Effect)

X-ray observations indicate that clusters of galaxies contain significant amounts of diffuse, hot gas. In this section and the next, the effect of this gas on sources of radio emission lying behind the cluster will be discussed. The free electrons in this intracluster plasma will have an optical depth for scattering low frequency photons given by

$$\tau_T = \int \sigma_T n_e dl, \tag{3.4}$$

where $\sigma_T = (8\pi/3)[e^2/m_e c^2)]^2$ is the Thompson electron scattering cross section, n_e is the electron density, and l is the path length along any line of sight through the gas. For a typical X-ray cluster, $n_e \approx 10^{-3}\,\mathrm{cm}^{-3}$ and $l \approx 1$ Mpc, and thus $\tau_T \approx 10^{-(2-3)}$. A fraction τ_T of the photons from any radio source behind a cluster will be scattered as the radiation passes through the cluster.

One 'source' of radio emission which lies 'behind' everything is the cosmic radiation which is a relic of the 'big bang' formation of the universe (Sunyaev and Zel'dovich, 1980a). This radiation has a spectrum that is nearly a blackbody, with a temperature of $T_r \approx 2.7$ K. Because this radiation is nearly isotropic, simply scattering the radiation would not have an observable effect. However, because the electrons in the intracluster gas are hotter than the cosmic radiation photons, they heat the cosmic radiation photons and change the spectrum of the cosmic radiation observed in the direction of a cluster of galaxies. This effect was first suggested by Zel'dovich and Sunyaev (1969) (Sunyaev and Zel'dovich, 1972). Reviews of the theory and current observational status of this Sunyaev–Zel'dovich effect have been given recently by Sunyaev and Zel'dovich (1980a, 1981).

During an average scattering, a photon with frequency ν has its frequency changed by an amount $\Delta\nu/\nu = 4kT_g/m_e c^2$, where T_g is the electron temperature of the intracluster gas. In calculating the effect this has on the radiation spectrum, it is conventional to measure the intensity in terms of a 'brightness temperature' T_r; this is defined as the temperature of a blackbody having the same intensity. Then, the change in the brightness temperature of the cosmic radiation due to passage through the intracluster gas is given by

$$\frac{\Delta T_r}{T_r} = \frac{\Delta I_\nu}{I_\nu} \frac{d \ln T_r}{d \ln I_\nu} = \tau_T \frac{kT_g}{m_e c^2} [x \coth(x/2) - 4], \tag{3.5}$$

where I_ν is the radiation intensity and $x \equiv h\nu/kT_r$. This expression is actually derived in the diffusion limit and is valid only for sufficiently small $x \lesssim 10$

(Sunyaev, 1981). The change in brightness temperature or intensity is negative for low frequencies $x < 3.83$ and positive for higher frequencies. This change occurs at a wavelength $\lambda_o = 0.14$ cm $(2.7\,\mathrm{K}/T_r)$. It is somewhat paradoxical that heating the background radiation lowers its brightness temperature at low frequencies. This is because Compton scattering conserves the number of photons, and shifting lower frequency photons to higher energies lowers the intensity at low frequencies.

Nearly all of the measurements that have been made of this effect have been at relatively low frequencies; taking the limit $x \to 0$ in equation (3.5) gives

$$\frac{\Delta T_r}{T_r} = -\int \frac{2kT_g}{m_e c^2}\, d\tau_T. \tag{3.6}$$

This reduction in the cosmic radiation in the direction of clusters of galaxies in the microwave region is often referred to as the 'microwave diminution'. Calculations of ΔT_r and its variation with projected position for a large set of models for the intracluster gas have been given by Sarazin and Bahcall (1977), and specific predictions of the size of the effect for the Coma, Perseus, and Virgo/M87 clusters are given in Bahcall and Sarazin (1977). Models for the Coma cluster have been given by Gould and Rephaeli (1978) (but note that their basic model for the intracluster gas is not physically consistent) and Stimpel and Binney (1979) (these models included the effect of cluster ellipticity).

Unfortunately, it has proved to be very difficult to make reliable measurements of this very small microwave diminution effect. Pariiskii (1973) claimed to have detected a microwave diminution from the Coma cluster. Gull and Northover (1976) claimed to have detected both Coma and A2218, and found very small diminutions towards four other clusters. The claimed detections of Coma were at a level much too high to be consistent with models for the X-ray emission from this well-studied cluster (Bahcall and Sarazin, 1977; Gould and Rephaeli, 1978), and subsequent observations have not confirmed the microwave diminution in Coma. Rudnick (1978) gave upper limits for five clusters, including Coma; his limits were consistent with models for the X-ray emission, but inconsistent with the previously claimed detections of Coma. Lake and Partridge (1977) claimed three detections of very rich clusters, but later withdrew the claim, saying that the measurements were undermined by systematic problems with the telescope. In a later survey of 16 clusters (Lake and Partridge, 1980), they detected only A576 at a level of $\Delta T_r = -1.3 \pm 0.3$ mK. (Note that $1\,\mathrm{mK} \equiv 10^{-3}$ K). Birkinshaw *et al.* (1978, 1981b) surveyed 10 clusters, detecting microwave diminutions in A576 ($\Delta T_r = -1.12 \pm 0.17$ mK), A2218 ($\Delta T_r = -1.05 \pm 0.21$ mK), and possibly A665 and

A2319. A2218 was *not* detected by Lake and Partridge (1980); in fact, their measurements have the opposite *sign*. Perrenod and Lada (1979) made measurement at higher frequencies ($\lambda = 9$ mm $\gg \lambda_o$) in order to reduce the effects of contamination by radio galaxies and beam smearing. They detected A2218 at the same level as Birkinshaw *et al.*, and also had a marginal detection of A665 at $\Delta T_r = -1.3 \pm 0.6$ mK (A665 is the richest cluster in the Abell catalog). Lasenby and Davies (1983) did not detect either A576 or A2218.

The apparent microwave diminutions from A576 and A2218 require very large masses of gas (comparable to the virial masses) at very high temperatures $T_g \gtrsim 3 \times 10^8$ K. X-ray observations of A576 are completely inconsistent with this much gas at these temperatures (Pravdo *et al.*, 1979; White and Silk, 1980), and thus the measured microwave reductions must be due to some other effect. While earlier low spatial resolution studies of A2218 suggested that it was too weak an X-ray source to produce the claimed microwave diminution (Ulmer *et al.*, 1981), a detailed high spatial resolution study of the X-ray emission from A2218 with the *Einstein* observatory (Boynton *et al.*, 1982) indicates that the required amounts of gas are present in this cluster at the required temperatures $T_g = 10$–30 keV.

From equation (3.6), the microwave diminution effect is independent of distance as long as the cluster can be resolved. In fact, Birkinshaw *et al.* (1981a) have measured a diminution of $\Delta T_r = -1.4 \pm 0.3$ mK from the distant ($z = 0.541$) cluster 0016 + 16. Optically, this is a rich cluster (Koo, 1981), although the field is somewhat confused by a foreground cluster at $z = 0.30$. There is a strong X-ray source towards 0016 + 16, which would imply a very high X-ray luminosity if it is associated with the more distant cluster (White *et al.*, 1981b). In some ways, microwave diminution observations of distant clusters are more straightforward than observations of nearby clusters, because the reference positions are further outside the cluster core.

Recently, a new set of observations of the microwave diminution were published by Birkinshaw *et al.* (1984) (see also Birkinshaw and Gull, 1984). These observations used the Owens Valley Radio Observatory and are apparently less subject to systematic effects than earlier observations. They confirm the detections of 0016 + 16 ($\Delta T_r = -1.40 \pm 0.17$ mK), A665 ($\Delta T_r = -0.69 + 0.10$ mK), and A2218 ($\Delta T_r = -0.70 \pm 0.10$ mK). Several of these detections have also been confirmed by Uson and Wilkinson (1985). Since there are now several confirming observations of the microwave diminution in these three clusters, it may be that the effect has finally been observed unambiguously. However, in view of the disagreements between different observers in the past, the withdrawal of previously claimed detections, and the inconsistency of some of the radio results with X-ray measurements of the amount of gas present, I do not feel completely confident that the current microwave diminution results are conclusive. It is clear that the major sources of errors in the measurements are not statistical but systematic. These include very low level systematic problems with the response of the radio telescopes used (Lake and Partridge, 1980).

One major source of problems is the possible presence of radio sources in the cluster. If these are concentrated at the cluster core, they will increase the radio brightness of the cluster and mask the microwave diminution. All of the observations are corrected for the presence of strong radio sources, and the observers generally avoid observing clusters, such as Perseus, which contain very strong radio sources. There is still the danger that a larger number of harder to detect, weaker radio sources will make a significant contribution to the cluster radio brightness. Birkinshaw (1978) surveyed six clusters, including A576 and A2218, for weak radio source emission, and concluded that it was unlikely to affect the microwave diminution measurements. Schallwich and Wielebinski (1978) detected a weak radio source in the direction of A2218, and corrected the microwave diminution measurement of Birkinshaw *et al.* (1978) for this cluster. Unfortunately, this correction would destroy the agreement of this measurement with the shorter wavelength measurement of Perrenod and Lada (1979), because the radio source and microwave diminution have different spectral variations. Tarter (1978) has suggested that if clusters contained a small amount of ionized gas at a cooler temperature than the X-ray emitting gas, the free–free radio emission from this gas could mask the microwave diminution. All of these radio source problems would generally mask the microwave diminution and might explain why some clusters that are predicted to have very strong diminutions, such as A2319, are in fact observed to have *positive* ΔT_r.

What about A576, in which a strong microwave diminution was initially observed, although very little gas is observed in X-rays? The microwave diminution measurements are generally relative measurements in which one compares the cosmic microwave brightness in the direction of the cluster core with the brightness at one or more positions away from the cluster core. A negative ΔT_r at the cluster core cannot be distinguished from a positive ΔT_r in these reference positions, which are generally not far outside the cluster core. Thus the observation of a negative ΔT_r in A576 may indicate that there is excess radio emission in the outer parts of the cluster. Cavallo and Mandolesi (1982) have suggested that this radio emission is produced by the stripping of gas from spiral galaxies in the outer parts of the cluster.

The microwave and X-ray observations of a cluster can be used to derive a distance to the cluster which is independent of the redshift (Cavaliere *et al.*,

1977, 1979; Gunn, 1978; Silk and White, 1978). From equation (3.6), ΔT_r depends on the electron density n_e, the gas temperature T_g, and the size of the emitting region. The X-ray flux f_x from the cluster depends on all of these, but also decreases with the inverse square of the distance to the cluster. Thus the distance can be determined by comparing the X-ray flux and the microwave diminution:

$$D_A \propto \left(\frac{\Delta T_r}{T_r}\right)^2 f_x^{-1}[T_g(0)]^{-3/2}\theta_c(1+z)^{-4}, \tag{3.7}$$

where D_A is the angular diameter distance (Weinberg, 1972), z is the redshift, θ_c is the angular radius of the cluster core (Section 2.7), $T_g(0)$ is the gas temperature at the cluster center, and the coefficient of proportionality depends on the distribution of gas in the cluster and the X-ray detector response (Cavaliere *et al.*, 1979). In fact, any assumptions about the gas distribution can be avoided by mapping the variation of both the X-ray surface brightness I_x and the microwave diminution as a function of the angle away from the cluster center θ. Silk and White (1978) find

$$D_A = \frac{F[T_g(0)]}{\pi(1+z)^4} \frac{\left\{\int_0^\infty [\Delta T_r(0) - \Delta T_r(\theta)] \frac{d\theta}{\theta^2}\right\}^2}{\int_0^\infty [I_x(0) - I_x(\theta)] \frac{d\theta}{\theta^2}} \tag{3.8}$$

where F is a known function of gas temperature, which for a hot solar abundance plasma contains only atomic constants. This determination of the distance of the cluster is independent of the distribution of the gas as long as it is spherically symmetric. Applying this method to nearby clusters and comparing the distances with the redshift could allow the determination of the Hubble constant H_o. Mapping high redshift clusters ($z \approx 1$) could give the cosmological deceleration parameter q_o; together, these two parameters determine the structure, dynamics, and age of the universe (Weinberg, 1972), yet remain very poorly determined after a half century of observational cosmology research.

Unfortunately, the difficulty of obtaining reliable microwave diminution measurements has made it impossible to apply this method at the present time (Birkinshaw, 1979; Boynton *et al.*, 1982). In general, cluster microwave diminutions have not been mapped with sufficient accuracy to allow the distance to be determined from equation (3.8). Even the optical data on clusters are not accurate enough to allow an accurate distance determination. In addition, the cause of false detections, such as A576, must be determined so that they can be weeded out of cluster samples. For example, if the detection of

A576 were taken seriously, it would imply a distance to this cluster at least an order of magnitude more than its redshift distance (White and Silk, 1980).

Gould and Rephaeli (1978) suggested that it might be easier to detect the Sunyaev–Zel'dovich effect unambiguously at high frequencies ($\lambda < \lambda_o$) at which ΔT_r is positive. Some observations have been attempted at $\lambda = 1$–3 mm (Meyer *et al.*, 1983), but no cluster diminutions were detected, and this wavelength range straddles λ_o. Observations at shorter wavelengths must be made from satellites.

Sunyaev and Zel'dovich (1981) point out that although their effect is often thought of as a small change in the cosmic microwave background, at $\lambda < \lambda_o$ the effect may also be considered as an enormous source of submillimeter luminosity for the cluster. Because the surface brightness of the submillimeter emission is proportional to $\tau_T \propto n_a l$, where $l \approx r$ is the path length through the cluster and r is the radius of the gas, and the surface area is proportional to r^2, the luminosity is proportional to $n_e r^3$ or the mass M_g of the gas in the cluster. The submillimeter luminosity at frequencies above the critical frequency ($\lambda < \lambda_o$) is given by

$$L^+ = 8.7 \times 10^{43} \left(\frac{T_g}{10^8 \text{ K}} \right) \left(\frac{T_r}{2.7 \text{ K}} \right)^4 \times \left(\frac{M_g}{10^{14} M_\odot} \right) (1+z)^4 \text{ erg/s} \tag{3.9}$$

where the factor $[T_r(1+z)]^4$ gives the cosmic radiation density at the redshift of the cluster. Thus clusters could be among the strongest sources of submillimeter radiation in the universe. Other strong submillimeter sources, such as quasars, would have different spectra, be more compact, and probably be variable.

The variation in the cosmic microwave intensity and polarization toward a cluster can also be used to determine the velocity of the cluster relative to the average of all material in that region of the universe (Sunyaev and Zel'dovich, 1980b). This velocity, measured relative to the local comoving cosmological reference frame, is as near as one can come to an absolute measure of motion in a relativistically invariant universe. In a sense, the cosmic background radiation acts as an 'ether'. Just as the thermal motion of electrons in a cluster changes the wavelength of the cosmic background radiation during scattering, their bulk motion has a similar effect. As long as $\tau_T \ll 1$, the variation in the brightness temperature due to cluster motion is independent of frequency and is given by

$$\frac{\Delta T_r}{T_r} = -\frac{v_r}{c} \tau_T, \tag{3.10}$$

where v_r is the radial component of the velocity. The tangential component of velocity (the component in the plane of the sky) can be detected if the polarization of the microwave background in the direction of clusters can be measured to very high accuracy. At low frequencies ($x \ll 1$), the polarization (which is in the direction of motion) is

$$p = \frac{1}{10}\beta_t^2 \tau_T + \frac{1}{40}\beta_t \tau_T^2 \tag{3.11}$$

to lowest order in τ_T and $\beta_t \equiv v_t/c$, where v_t is the tangential velocity.

The Sunyaev–Zel'dovich effect can also be used to determine the spectrum and angular distribution of the cosmic background radiation itself (Sunyaev and Zel'dovich, 1972; Fabbri *et al.*, 1978; Rephaeli, 1980, 1981; Sunyaev, 1981; Zel'dovich and Sunyaev, 1981). In principle, the different causes of variations in ΔT_r towards clusters (thermal motions of electrons, bulk motions, and variations in the cosmic background radiation itself) can be separated because they have different spectral variations (Sunyaev and Zel'dovich, 1981).

To summarize, the Sunyaev–Zel'dovich effect has a tremendous potential for providing information about the properties of hot gas in clusters and the nature of the universe as a whole. Unambiguous detections of this effect in clusters have proved elusive, but this situation may be improving.

3.6 Faraday Rotation

A plasma containing a magnetic field is birefringent; the speed of propagation of an electromagnetic wave depends on its circular polarization (Spitzer, 1978). While natural sources of circularly polarized radiation are rare, synchrotron emission in an ordered magnetic field produces linearly polarized radiation, and many radio galaxies and quasars produce radio emission that is somewhat linearly polarized. The plane of polarization of linearly polarized radiation is rotated during passage through a birefringent medium. For a magnetized plasma, the angle of rotation ϕ is

$$\phi = R_m \lambda_m^2, \tag{3.12}$$

where it is conventional to give the wavelength λ_m of the radiation in meters. The rotation measure R_m is given by

$$R_m = \frac{e^3}{2\pi m_e^2 c^4} \int n_e B_\parallel dl$$
$$= 8.12 \times 10^5 \int n_e B_\parallel dl_{pc} \, \mathrm{cm^3\, m^{-2}\, G^{-1}\, pc^{-1}} \tag{3.13}$$

where l is the path length through the medium, $B_\parallel$ is the component of the magnetic field parallel to the direction of propagation of the radiation, and n_e is the electron density. In the second line of equation (3.13), l is given in pc, n_a in cm^{-3}, and $B_\parallel$ in gauss (G). The wavelength dependence of the rotation is the feature that allows the observational separation of the initial polarization angle and the amount of rotation.

The rotation measures to radio sources lying within or behind clusters can be used to constrain the magnitude of the intracluster magnetic field and its geometry (Dennison, 1980a; Jaffe, 1980; Lawler and Dennison, 1982). One problem with these determinations is that the rotation measure due to other plasma along the line of sight to the radio source must be removed. If this can be done, and the electron density and path length through the gas are determined from X-ray observations of the cluster, the average value of $B_\parallel$ can be determined. Since this average of a single component of B must be less than the average magnitude of B, this gives a lower limit to the intracluster magnetic field. The Faraday rotation of the halo radio source in M87 in the Virgo cluster has been used to give a lower limit $B \gtrsim 2 \mu G$ ($\mu G \equiv 10^{-6}$ gauss) on the magnetic field in the gaseous halo around M87 (Andernach *et al.*, 1979; Dennison, 1980a). This limit implies that very little of the X-ray emission from M87 can be nonthermal.

Jaffe (1980) noted that the rotation measures to radio sources within or behind clusters are generally small $R_m \lesssim 100\ m^{-2}$, which for a cluster with $n_e \approx 3 \times 10^{-3}\ cm^{-3}$ and a path length of 500 kpc gives $B_\parallel \lesssim 0.1\ \mu G$. On the other hand, the halo radio emission observed in some clusters (Section 3.4) implies that the intracluster magnetic field strengths are $B \gtrsim 1\ \mu G$. Since it is unlikely that the fields are preferentially ordered perpendicular to our line-of-sight to each cluster, the difference in these two limits must be due to the cancellation of components of $B_\parallel$ along the path through the cluster. Jaffe argues that this implies that the field is tangled. If this tangled field can be thought of as consisting of cells of ordered field of size l_B randomly oriented along the path length l through the cluster, then statistically $\langle B_\parallel \rangle \approx B(l_B/l)^{1/2}$. Noting that only a portion of the rotation measure can be associated with the intracluster gas, Jaffe argued that the coherence length l_B of the intracluster magnetic field must be $l_B \lesssim 10$ kpc. Since tangled fields in a static medium can straighten and decay rapidly, he suggests that the fields are tangled by the turbulent wakes produced behind galaxies as they move through the intracluster gas.

With a larger sample of radio sources and clusters, Lawler and Dennison (1982) claimed that the sources seen through rich clusters had slightly larger rotation measures, although the two distributions only differ at the 80% confidence level. Attributing the difference to intracluster Faraday rotation, they derived an average rotation measure of $R_m \approx 130\ m^{-2}$ through the cluster center, which implies $l_B \gtrsim 20$ kpc; this is not inconsistent with the galactic wake model.

These limits on the strength and geometry of the intracluster *B* field are important to models for the X-ray emission for two reasons. First, the halo radio emission from a cluster depends on the product of the number of relativistic electrons and the magnetic field, and the relativistic electrons may heat the intracluster gas (Section 5.3.5). Second, the effectiveness of transport processes in the intracluster gas (such as thermal conduction) is determined by the geometry of the magnetic field because the gyroradius of electrons in even a very weak intracluster *B* field is very much less than the size of the cluster (Section 5.4.3).

3.7 21 cm Line Observations of Clusters

The 21 cm hyperfine line in hydrogen allows one to measure the mass of neutral hydrogen in galaxies or clusters of galaxies. There are an enormous number of 21 cm observations of galaxies, which would require an entire review to discuss. The main results for clusters are negative, and I shall briefly summarize these. The H I observations of clusters and superclusters have been reviewed recently by Chincarini (1984).

The 21 cm line observations have shown that the galaxies that make up clusters are deficient in neutral hydrogen gas. First, many observations have shown that elliptical and S0 galaxies in clusters have very little neutral hydrogen gas (Section 2.10.2; Davies and Lewis, 1973; Krumm and Salpeter, 1976; Bieging and Biermann, 1977). Similarly, spiral galaxies in clusters generally have less neutral hydrogen than spirals found in the field (Huchtmeier *et al.*, 1976; Sullivan and Johnson, 1978; Helou *et al.*, 1979; Krumm and Salpeter, 1979; Chamaraux *et al.*, 1980; Giovanelli *et al.*, 1981; Sullivan *et al.*, 1981), and the deficiency is stronger near the cluster center (Giovanelli *et al.*, 1982; Giovanardi *et al.*, 1983). As discussed in Section 2.10.2, these observations have been used to argue that stripping of gas from galaxies through ram pressure ablation or some other mechanism is an important process in clusters, and that S0 galaxies may be produced by this process.

Observations indicate that cD galaxies, including those accreting large amounts of intracluster gas (Sections 4.3.3 and 5.7.2), contain very little neutral hydrogen (less than $10^9 M_\odot$; Burns *et al.*, 1981a; Valentijn and Giovanelli, 1982). This indicates that the accreted gas is not all stored as neutral hydrogen gas.

Finally, 21 cm observations of clusters as a whole indicate that the missing mass component (Section 2.8) is not neutral hydrogen gas (Goldstein, 1966; Haynes *et al.*, 1978; Peterson, 1978; Baan *et al.*, 1978; Shostak *et al.*, 1980).

4

X-RAY OBSERVATIONS

4.1 Detections and Identifications

The first extragalactic object to be detected as an X-ray source was M87 in the Virgo cluster (Byram *et al.*, 1966; Bradt *et al.*, 1967). Sources associated with the Perseus cluster (Fritz *et al.*, 1971; Gursky *et al.*, 1971a) and the Coma cluster (Meekins *et al.*, 1971; Gursky *et al.*, 1971b) were detected next. The idea that extragalactic X-ray sources were generally associated with groups or clusters of galaxies was suggested by Cavaliere *et al.* (1971). While the early detections were made with balloon- or rocket-borne detectors, a major advance in the study of X-ray clusters (and all X-ray astronomy) came with the launch of the *Uhuru* X-ray satellite, which permitted more extended observations of individual sources and a complete survey of the sky in X-rays, the *Uhuru* Catalog (Giacconi *et al.*, 1972, 1974; Forman *et al.*, 1978a).

The early *Uhuru* observations established a number of properties of the X-ray sources associated with clusters. First, clusters of galaxies are the most common bright extragalactic X-ray sources. Second, clusters are extremely luminous in their X-ray emission, with luminosities $\approx 10^{43-45}$ erg/s, and they have a wide range of luminosities. This makes clusters as a class the most luminous X-ray sources in the universe, with the exception of quasars. Third, the X-ray sources associated with clusters are extended (Kellogg *et al.*, 1972; Forman *et al.*, 1972); the sizes found from the *Uhuru* data range from about 200 to 3000 kpc. Fourth, the clusters have X-ray spectra that show no strong evidence for low energy photoabsorption, unlike the spectra of the compact sources associated with discrete sources either in the nuclei of galaxies or stellar sources within our own galaxy. Fifth, the X-ray emission from clusters is not time variable, as is the emission from many point sources of X-rays in our galaxy or in the nuclei of other galaxies (Elvis, 1976). These last three results suggest that the emission is truly diffuse, and not the result of one or many compact sources. Many of the early *Uhuru* results are presented in several review papers by Kellogg (1973, 1974, 1975).

The original identifications of clusters with *Uhuru* sources were made by Gursky *et al.* (1972) and Kellogg *et al.* (1971, 1973). Other identifications of clusters as *Uhuru* sources were made by Bahcall (1974c), Disney (1974), Rowan-Robinson and Fabian (1975), Elvis *et al.* (1975), Melnick and Quintana (1975), Vidal (1975a, b), Bahcall *et al.* (1976), Ives and Sanford (1976), Pye and Cooke (1976), Lugger (1978), and Johnston *et al.* (1981). The situation on the identifications of clusters was summarized by Bahcall and Bahcall (1975), who argued that the many unidentified *Uhuru* sources at high galactic latitude were probably also clusters of galaxies. A systematic search for Abell cluster identifications in the *Uhuru* catalog was made by Kellogg *et al.* (1973). Cluster X-ray sources were also identified in surveys made with the *Ariel 5* satellite (Elvis *et al.*, 1975; Cooke and Maccagni, 1976; Maccacaro *et al.*, 1977; Mitchell *et al.*, 1977; McHardy, 1978a; Ricketts, 1978; McHardy *et al.*, 1981) and SAS-C satellite (Markert *et al.*, 1976, 1979). A combined sample of X-ray identification from the *Uhuru* and *Ariel* surveys was given by Jones and Forman (1978), and the pre-HEAO cluster identifications have been reviewed by Gursky and Schwartz (1977).

The next major advance in sensitivity came with the launch of the HEAO-1 X-ray observatory with proportional counters with a very large collecting area. Identifications of clusters with hard X-ray sources detected in sky surveys with the HEAO-1 A-2 instrument have been given by Marshall *et al.* (1979) and by Piccinotti *et al.* (1982) for high galactic latitude X-ray sources. Searches for X-ray emission from the richest Abell clusters were made by Pravdo *et al.* (1979). A soft X-ray survey of a few clusters was made by Reichert *et al.* (1981), and a complete soft X-ray catalog is given by Nugent *et al.* (1983). A statistically complete sample of Abell clusters was surveyed with the HEAO-1 A-2 instrument by McKee *et al.* (1980). A survey of a large sample (≈ 1900) of Abell clusters with the HEAO-1 A-1 was made by Ulmer *et al.* (1981) and Johnson *et al.* (1983), while a survey of the most distant Abell clusters with the HEAO-1 A-1 detector (Ulmer *et al.*, 1980b) detected 11 such clusters, suggesting that many of them are extremely luminous. The southern cluster catalog of Duus and Newell (1977) was surveyed for X-ray emission using the HEAO-1 A-1 detector by Kowalski *et al.* (1984); this paper includes a compilation of all HEAO-1 A-1 cluster detections and limits. Wood *et al.* (1984) is the complete HEAO-1 A-1 X-ray source catalog.

X-ray astronomy made a quantum leap forward with the launch of the *Einstein* observatory. This was the first satellite with focusing optics for extrasolar X-ray observing. Because of its focusing capability, the sensitivity of this instrument to small sources was orders of magnitude higher than that for any previous X-ray detector. The two major *Einstein* surveys of X-ray emission from clusters are Abramopoulos and Ku (1983) and Jones and

Forman (1984) (see Table 2 below). Other X-ray cluster detections with *Einstein* include those of Jones *et al.* (1979), Henry *et al.* (1979, 1982), Burns *et al.* (1981c), Forman *et al.* (1981), Maccagni and Tarenghi (1981), Perrenod and Henry (1981), White *et al.* (1981a, b, 1987), Bechtold *et al.* (1983), Soltan and Henry (1983), and Henry and Lavery (1984).

Compact, poor clusters were detected as X-ray sources by Schwartz *et al.* (1980a, b) and Kriss *et al.* (1980, 1981, 1983), and will be discussed in more detail in Section 4.7.

Murray *et al.* (1978) suggested that a number of sources in the 4U *Uhuru* catalog were associated intrinsically with superclusters (that is, there was more emission than could be accounted for simply by the sum of the emissions from the number of X-ray clusters one would have expected to find in the supercluster). Kellogg (1978) presented evidence that X-ray clusters were located in superclusters, but did not suggest that there was any intrinsic emission associated with the supercluster itself. Subsequent observations have not confirmed the detections of superclusters as a distinct class of X-ray sources (Ricketts, 1978; Pravdo *et al.*, 1979). Forman *et al.* (1978b) claimed to have detected very large and luminous haloes of X-ray emission about clusters of galaxies, and argued that the extra emission seen by Murray *et al.* from superclusters might really be the sum of the haloes of the clusters in the superclusters. In general, subsequent observations have not confirmed the presence of these luminous and extensive haloes (Pravdo *et al.*, 1979; Nulsen *et al.*, 1979; Ulmer *et al.*, 1980a; Nulsen and Fabian, 1980).

4.2 X-Ray Luminosities and Luminosity Functions

The number of clusters per unit volume with X-ray luminosities in the range L_x to $L_x + dL_x$ is defined as $f(L_x)dL_x$, where $f(L_x)$ is the X-ray luminosity function. In general, the luminosity function will depend on the method used to select the clusters. One can begin with a statistically complete catalog of optically detected clusters (such as the 'statistical sample' of Abell clusters; see Section 2.1), which is surveyed for X-ray emission. Alternatively, a complete catalog of X-ray sources can be examined optically to determine which sources are associated with clusters of galaxies. In addition to reproducing the observed statistics of X-ray cluster identifications, the X-ray luminosity function is subject to the additional constraints that the total number of X-ray clusters not exceed the total number of all clusters, and the total emissivity of X-ray clusters not produce a larger X-ray background than is observed. Most data on cluster luminosities have been fit to either an exponential or a power-law form of the luminosity function. Thus we define

$$f(L_x) = A_e h_{50}^5 \exp(-L_x/L_{xo})(10^{44}\,\mathrm{erg/s})^{-1}\,\mathrm{Mpc}^{-3}, \tag{4.1}$$

$$f(L_x) = A_p h_{50}^5 \left(\frac{L_x}{10^{44} h_{50}^{-2}\ \mathrm{erg/s}} \right)^{-p} (10^{44}\ \mathrm{erg/s})^{-1}\ \mathrm{Mpc}^{-3}. \tag{4.2}$$

All luminosities in this section are given for the photon energy range of 2–10 keV. It is convenient to define $L_{44} \equiv L_x / 10^{44}$ erg/s.

Schwartz (1978) derived an estimate of the luminosity function for a sample of 14 Abell clusters in distance class 3 or less, which were detected with the *Uhuru*, *Ariel* 5, or SAS-C satellites. More distant clusters are used only to give an upper limit to the luminosity function at high luminosities. This sample is only expected to be complete for $1 \leqslant L_{44} \leqslant 10$. Schwartz found that the best fit exponential luminosity function has $A_e = 4.5 \times 10^{-7}$ and $L_{xo} = 2.0 \times 10^{44} h_{50}^{-2}$ erg/s, although a power-law with $A_p = 7.9 \times 10^{-7}$ and $p \approx 2.45$ would also fit the data if suitably truncated at high and low luminosities.

McHardy (1978a) derived an X-ray luminosity function from the *Ariel* 5 fluxes for Abell clusters of distance class 3 or less; he argued that the *Uhuru* fluxes are unreliable for weak sources. While he did not fit his numerical luminosity function to any analytic expression, a suitable fit is given by a power law with $A_p = 2.5 \times 10^{-7}$ and $p \approx 2$ for $0.2 \leqslant L_{44} \leqslant 20$.

The HEAO-1 satellite provided a much more extensive data base for determining the luminosity function of clusters. Both the A-1 and A-2 experiments were used to survey the Abell clusters. A luminosity function was derived for a significant portion of the statistical sample of Abell clusters (Section 2.1), using the A-1 data by Ulmer *et al.* (1981). Exponential and power-law fits to these data gave $A_e = 0.49 \times 10^{-7}$, $L_{xo} = 2.9 \times 10^{44} h_{50}^{-2}$ erg/s, and $A_p = 1.1 \times 10^{-8}$, $p = 1.7$, respectively. When the sample was extended to all Abell clusters and luminosities, the normalization A_e and the characteristic luminosity L_{xo} both were roughly doubled.

The HEAO-1 A-2 data were used to derive a luminosity function both by surveying the Abell clusters (McKee *et al.*, 1980; Hintzen *et al.*, 1980) and by identifying a complete sample of X-ray sources in a flux-limited survey at high galactic latitude (Piccinotti *et al.*, 1982). The Abell cluster survey included all richness classes and all distance classes less than five. The luminosity function could be fit adequately with either an exponential or power-law form, and the coefficients in equations (4.1) and (4.2) were $A_e \approx 2.5 \times 10^{-7}$, $L_{xo} \approx 1.8 \times 10^{44} h_{50}^{-2}$ erg/s, $A_p \approx 3.8 \times 10^{-7}$, and $p \approx 2.2$. The cluster luminosity from the high latitude survey was not well represented by an exponential; a power-law fit gave $A_p \approx 3.6 \times 10^{-7}$ and $p \approx 2.15$, which agrees well with the A-2 Abell survey result.

Bahcall (1979b) has attempted to predict the X-ray luminosity function of clusters from their optical luminosity function by assuming a one-to-one correspondence between the optical and X-ray luminosity of clusters. She predicts that clusters have a luminosity function that can be represented by two intersecting power laws with $p \approx 2.5$ for $L_{44} \gtrsim 1$ and $p \approx 1.3$ for $L_{44} \lesssim 1$. While the HEAO-1 data do not show any clear evidence for a change in the slope of the luminosity function at $L_{44} \approx 1$, they are probably not inconsistent with such a change because they do not extend much below this luminosity.

One important application of cluster luminosity functions is in determining the contribution of clusters to the hard X-ray background (see Field, 1980, for a review of its properties). From the estimates of the luminosity function of clusters discussed above, it appears that clusters probably provide only about 3 to 10% of the X-ray background in the 2–10 keV photon energy band, assuming they do not evolve rapidly with time (Rowan-Robinson and Fabian, 1975; Gursky and Schwartz, 1977; McKee *et al.*, 1980; Hintzel *et al.*, 1980; Piccinotti *et al.*, 1982; Ulmer *et al.*, 1981).

4.3 X-Ray Spectra

Observations of the X-ray spectra of clusters of galaxies have played a critical role in establishing the primary emission mechanism (thermal emission from diffuse hot intracluster gas) and in testing models for the origin of this gas. Models in which the emission comes from diffuse thermal gas predict (1) that the spectrum will be roughly exponential (the intensity I_ν (erg/cm^2 s Hz) varies as $\approx \exp(-h\nu/kT_g)$ where T_g is the gas temperature); (2) that the gas temperature will be such that the thermal velocity of protons in the gas $\approx \sqrt{kT_g/m_p}$ be comparable to the velocity of the galaxies in the cluster, as both are bound by the same gravitational potential; (3) that there will be no strong low energy photoabsorption; and (4) that emission lines will be present if the gas contains a significant contamination of heavy elements like iron. Alternatively, models in which the emission is due to relativistic nonthermal electrons predict a power-law spectrum $I_\nu \propto \nu^{-\alpha}$, which implies an excess at low and high energies when compared to an exponential spectrum; no line emission would be expected for a nonthermal emission process. As another possibility, the emission might be thermal emission from a number of compact sources, such as galactic nuclei or the binary stellar X-ray sources which dominate the X-ray sky within our own galaxy; however, such sources are generally optically thick at low X-ray energies ($\lesssim 1$ keV). The theories for each of these classes of emission processes and the basis for these predictions are discussed in Section 5.1.

The first three of the predictions given above concern the broad-band form of the spectrum (the continuum), while the last prediction concerns lines. Accordingly, the properties of the continuum spectra will first be reviewed, and then those of the line spectra. Reviews devoted primarily to the observations of the X-ray spectra of clusters have been given recently by Canizares (1981) and Mushotzky (1980, 1984, 1985), while Holt and McCray (1982) review all of X-ray spectroscopy.

4.3.1 Continuum Features in the Spectrum

If the X-ray emission from clusters is due to a diffuse plasma of either thermal or nonthermal electrons, the optical depth of the gas should be quite low. On the other hand, compact X-ray sources (such as galactic nuclei or binary stellar X-ray sources) often contain significant quantities of relatively cool neutral gas, which absorbs soft X-rays through photoionization. Because the fluorescent yield of the light elements is low, the absorbed X-rays are not reemitted and are lost from the spectrum. This low energy photoabsorption occurs in a series of edges which correspond to the absorption edges of cosmically abundant elements. The opacity of a solar abundance, low density, cold neutral gas has been calculated, for example, by Brown and Gould (1970). It is conventional to parametrize the absorption observed in an X-ray spectrum by the column density of hydrogen N_H in a gas with assumed solar abundances required to produce the observed absorption. Typically, compact sources have $N_H \gtrsim 10^{22}\,\mathrm{cm}^{-2}$. Even the earliest X-ray spectra of clusters suggested that they had rather weak low energy absorption (Catura *et al.*, 1972; Kellogg, 1973; Davidsen *et al.*, 1975; Kellogg *et al.*, 1975; Margon *et al.*, 1975; Avni, 1976), with column densities $N_H \lesssim 10^{22}\,\mathrm{cm}^{-2}$, which were generally consistent with the amount of neutral hydrogen in our own galaxy along the line of sight to the cluster. This indicated that the emission from clusters comes from a diffuse, ionized plasma.

Initially, there were two competing models for the nature of this ionized plasma (see Section 5.1). It could be a hot, thermal plasma with a temperature $T_g \approx 10^8$ K, or it could be a relativistic, nonthermal plasma with a power-law electron energy distribution, such as the plasma responsible for the radio emission observed in clusters (see Section 3.1). In the first case, the X-ray continuum would be primarily due to thermal bremsstrahlung (see Section 5.1.3), with a spectrum given by equation (5.11). If the frequency variation of the Gaunt factor $g_{ff}(\nu, kT_g)$ is ignored and the gas is all at a single temperature, the spectrum is exponential $I_\nu \propto \exp(-h\nu/kT_g)$. In the second case, the emission is primarily due to the inverse Compton process (the scattering of low energy photons to X-ray energies by the relativistic electrons; see Section 5.1.1), and the expected spectrum is a power-law $I_\nu \propto \nu^{-\alpha}$.

Unfortunately, proportional counters have rather poor spectral resolution, and it is therefore difficult to distinguish between thermal and nonthermal spectra. Moreover, any sufficiently smooth and monotonic spectrum can be produced by the combination of the thermal spectra with varying temperatures, or nonthermal spectra with varying spectral indices α; thus the distinction between thermal and nonthermal spectra cannot be made unambiguously. It is not surprising, therefore, that the early proportional counter spectra of clusters could be fit consistently by either thermal (exponential) of nonthermal (power-law) spectra (for example, Kellogg *et al.*, 1975). However, spectra over a large energy range were better fit by the thermal model (Davidsen *et al.*, 1975; Scheepmaker *et al.*, 1976).

The first large surveys of cluster spectra came from observations with OSO-8 and *Ariel 5*. These satellites observed individual clusters for longer periods of time than had been possible with previous sky survey instruments, and had detectors that were optimized for spectral resolution. The spectra of clusters observed with OSO-8 and *Ariel 5* were significantly better fit by the thermal bremsstrahlung model than by the nonthermal model (Mushotzky *et al.*, 1978; Mitchell *et al.*, 1979). The required temperatures for the cluster gas were found to range from about 2×10^7 to 2×10^8 K from cluster to cluster, and some of the clusters required gas at several temperatures to fit the spectrum. Recently, a more extensive survey of X-ray cluster spectra was made with the A-2 experiment on the HEAO-1 satellite (Mushotzky, 1980, 1984, 1985; Henriksen, 1985; Henriksen and Mushotzky, 1985, 1986).

The two properties that can be derived most easily from the continuum X-ray spectrum are the gas temperature T_g and the emission integral

$$EI \equiv \int n_p n_e dV, \tag{4.3}$$

where n_p is the proton density, n_e is the electron density, and V is the volume of the gas in the cluster. The X-ray luminosity of a cluster is proportional to EI (equation (4.11)). The X-ray luminosity (or EI) and gas temperature are found to be strongly correlated (Mitchell *et al.*, 1977, 1979; Mushotzky *et al.*, 1978). The HEAO-1 A-2 sample (Figure 11) gives $L_x \propto T_g^3$ (Mushotzky, 1984). The OSO-8, *Ariel 5*, and HEAO-1 A-2 spectral surveys established a number of correlations between these X-ray spectral parameters and the optical properties of X-ray clusters, which are discussed in Section 4.6 below.

There was also some evidence from the OSO-8 survey that the gas in clusters was isothermal; that is, the range of temperatures within the gas in a single cluster was relatively small. However, in many cases the OSO-8 and *Ariel 5* temperatures were not in agreement within the errors; if these differences are real, they suggest that there are multiple temperature components to the emission. If that were the case, the OSO-8 and *Ariel 5* detectors, which have different spectral and spatial sensitivities, might give different weights to the different components, and produce different average temperatures.

The HEAO-1 A-2 detector has provided much more data on the spectra of clusters in the photon energy range 2–60 keV (Mushotzky, 1980, 1984, 1985; Henriksen, 1985; Henriksen and Mushotzky, 1985, 1986). There is now evidence that most clusters contain a range of gas temperatures, with typical values between $T_g \approx 2 \times 10^7$ and 8×10^7 K. These multiple temperature components appeared to be most significant in clusters with low X-ray luminosities, although it is possible that similar low-luminosity cool components might remain undetected if hidden in the spectrum of clusters with high-luminosity high-temperature emission. The information on the spatial distribution of the X-ray emission in clusters (Section 4.4) suggests two locations for this cool gas. First, in low luminosity clusters, the X-ray emission is often inhomogeneous, with clumps of emission being associated, in some cases, with individual galaxies. These clumps may contain cooler gas. Second, in some clusters there are enhancements in the X-ray surface brightness at the position of the cD or other centrally located dominant galaxy in the cluster. X-ray line observations suggest that these are regions at which the hot intracluster gas is cooling and being accreted by the central dominant galaxy (Section 5.7).

Fig. 11. The correlation between the gas temperatures derived from X-ray spectra with HEAO-1 A-2 and the cluster X-ray luminosities, from Mushotzky (1984).

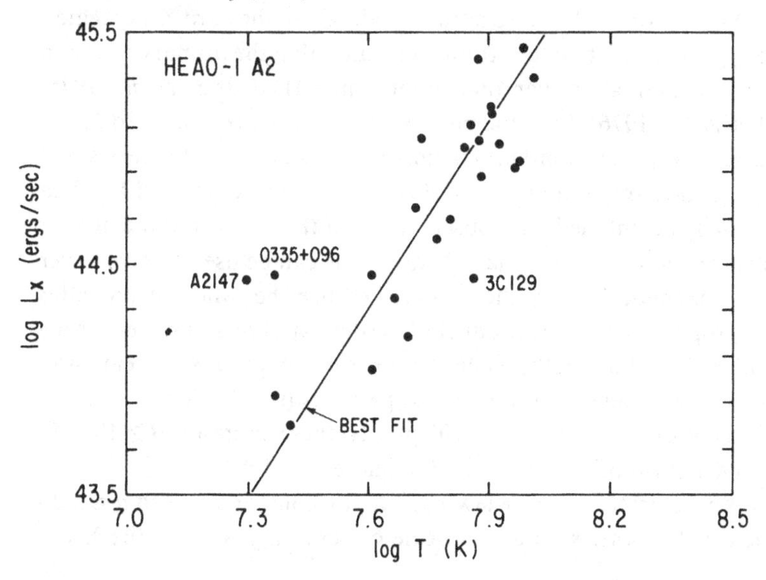

The *Einstein* X-ray observatory had two instruments capable of providing information on the continuum spectra of clusters. First, there was the Imaging Proportional Counter (IPC), which provided low resolution spatial and spectral information. Initially there were problems with the calibration of the energy scale of the spectra due to gain variations. These problems have now apparently been resolved, and a few cluster spectra from this instrument are available at the present time (Fabricant *et al.*, 1980; Perrenod and Henry, 1981; Fabricant and Gorenstein, 1983; White *et al.*, 1987). The second instrument was the Solid State Spectrometer (SSS), which had considerably better spectral resolution, but had no spatial resolution and less sensitivity than the IPC. Because of its small field of view (6 arc min), it could only observe a small portion of nearby clusters. Thus it was used primarily to determine spectra for the central regions of nearby clusters. It provided strong evidence for the presence of cool gas at the centers of a number of clusters (Mushotzky, 1980, 1984, 1985; Mushotzky *et al.*, 1981; Lea *et al.*, 1982); these observations are discussed further below. One problem with *Einstein* as an instrument for X-ray cluster spectroscopy is that the telescope was only sensitive to photons with energies of about 0.1–4.0 keV. With the typical temperatures of the gas in clusters being $kT_g \approx 8$ keV, observations with *Einstein* could not determine the thermal structure in this hot gas. However, the *Einstein* detectors were very sensitive to the presence of low temperature components of the emission.

Detections or limits on the hard X-ray spectrum and flux of clusters have been useful in limiting the contribution of nonthermal processes to their luminosity. As mentioned above, spectra extending into the hard X-ray region ($h\nu > 20$ keV) gave the first direct, strong indication that the primary emission mechanism was thermal, rather than nonthermal (Davidsen *et al.*, 1975; Scheepmaker *et al.*, 1976). Subsequently, stronger limits on the hard X-ray emission have shown that nonthermal emission makes at most a very small contribution to the X-ray luminosity of clusters (Mushotzky *et al.*, 1977; Lea *et al.*, 1981). When combined with observations of the diffuse radio emission in the cluster (Section 4.4), these hard X-ray limits can be used to give lower limits on the magnetic field in the cluster, because the synchrotron radio emissivity is proportional to the product of the density of relativistic electrons and the magnetic field strength, while the inverse Compton X-ray emission depends only on the density of relativistic particles (see Section 5.1.1 for a more detailed discussion of this point). Typically, these limits are $B \gtrsim 10^{-7}$ G (Lea *et al.*, 1981; Primini *et al.*, 1981; Bazzano *et al.*, 1984).

In the Perseus cluster, a power-law hard X-ray component with $\alpha \approx 2.25$ has been detected; it varies on a time scale of about a year, and the X-ray variations are correlated with variations in the radio flux of the compact radio source at the nucleus of the galaxy NGC 1275 (Primini *et al.*, 1981; Rothschild *et al.*, 1981). Much weaker power-law sources may also have been detected in the M87/Virgo, A2142, and 3C129 clusters (Lea *et al.*, 1981; Bazzano *et al.*, 1984).

4.3.2 *Line Features – The 7 keV Iron Line*

In many ways, the most significant observational discovery concerning X-ray clusters (following their identification as X-ray sources) was the detection of line emission due to highly ionized iron as a strong feature in their X-ray spectra. Immediately, this discovery established that the primary emission mechanism in X-ray clusters was thermal, and that the hot intracluster gas contained at least a significant portion of processed gas, which had at some point been ejected from stars. This line feature was first detected in the spectrum of the Perseus cluster by Mitchell *et al.* (1976) and shortly thereafter in the spectra of the Coma, Perseus, and Virgo clusters by Serlemitsos *et al.* (1977). It has subsequently been detected in the spectra of a total of about thirty clusters (Mitchell and Culhane, 1977; Mushotzky *et al.*, 1978; Malina *et al.*, 1978; Berthelsdorf and Culhane, 1979; Mitchell *et al.*, 1979; Mushotzky, 1980, 1984, 1985; Henriksen, 1985; Henriksen and Mushotzky, 1985, 1986). Figure 12 gives the HEAO-1 A-2 spectra of the Coma and Perseus clusters, showing these line features.

The line feature that was detected is actually a blend of lines from iron ions (mainly Fe^{+24} and Fe^{+25}) and weaker lines from nickel ions (see Section 5.2.3). These lines are mainly at photon energies between 6.5 and 7.0 keV; for convenience, this blend will be referred to as the '7 keV Fe line'. The resolution of this component structure and the measurement of the relative intensities of the various components can provide a wealth of diagnostic information on the physical state and environment of the X-ray emitting gas (Sarazin and Bahcall, 1977; Bahcall and Sarazin, 1978). Unfortunately, while the *Einstein* observatory contained a number of high resolution spectrometers, none were sensitive to the 7 keV Fe lines because the mirror in *Einstein* was ineffective for photon energies greater than about 4 keV. The application of the *Einstein* spectrometers to lower energy lines is discussed below. However, the proportional counters on the HEAO-1 A-2 experiment had sufficient spectral resolution to resolve the Kβ line[3] of Fe^{+24} from the Kα line in the Centaurus

[3] This notation gives the principal quantum number n of the lower level of the transition and the change in the principal quantum number $\Delta n \equiv n' - n$, where n' is the principal quantum number of the upper level of the transition. K indicates that the lower level is in the K-shell ($n = 1$), L indicates the lower level is in the L-shell ($n = 2$), and so on, while α indicates that $\Delta n = 1$, β indicates that $\Delta n = 2$, etc.

and Perseus clusters (Mitchell and Mushotzky, 1980; Mushotzky, 1980; see Figure 12). The observation of this line proves that the emission is thermal in nature, and not the result of the fluorescence of cold gas through photoionization by an X-ray continuum, because the fluorescent yield for the Kβ line is rather small. In fact, the observed Kβ lines are so strong that they require that the X-ray emission arises from a non-isothermal gas, with both cool and hot temperature components.

The line strengths are often given as 'equivalent widths' or *EW*. The equivalent width of any line feature is defined as

$$EW \equiv \int \left(\frac{I_\nu - I_\nu^o}{I_\nu^o} \right) d(h\nu), \tag{4.4}$$

Fig. 12. HEAO-1 A-2 low resolution X-ray spectra of clusters, showing the Fe K line at about 7 keV. The plots give the number flux of X-ray photons per cm^2-sec-keV versus photon energy in keV, for the (a) Coma (Henriksen and Mushotzky, 1986) and (b) Perseus (Henriksen, 1985) clusters.

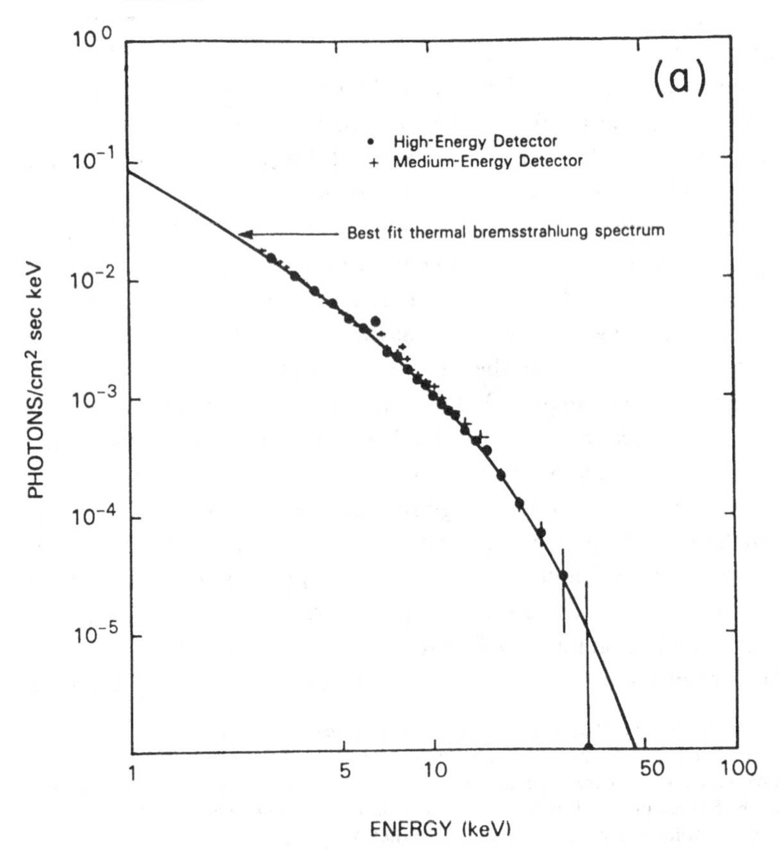

where I_ν is the observed intensity including the line as a function of frequency ν and I_ν^o is the continuum intensity without the line. The details of the emission processes for this feature are discussed in Section 5.2; here we note that the emissivity of the line is proportional to the square of the density and to the abundance of iron, and depends significantly on the electron temperature. Because the thermal bremsstrahlung emissivity also is proportional to the square of the density (Section 5.1.3), the equivalent width EW of the line is independent of density as long as the iron is well mixed in the gas (see Section 5.4.5 for a discussion of this point). If the shape of the X-ray continuum spectrum of a cluster is used to derive a temperature or a range of temperatures for the gas in the cluster, then the equivalent width of the 7 keV Fe line gives a measure of the abundance of iron in the gas (Section 5.2.3; Figure 35 below). The abundances by number of atoms determined from the

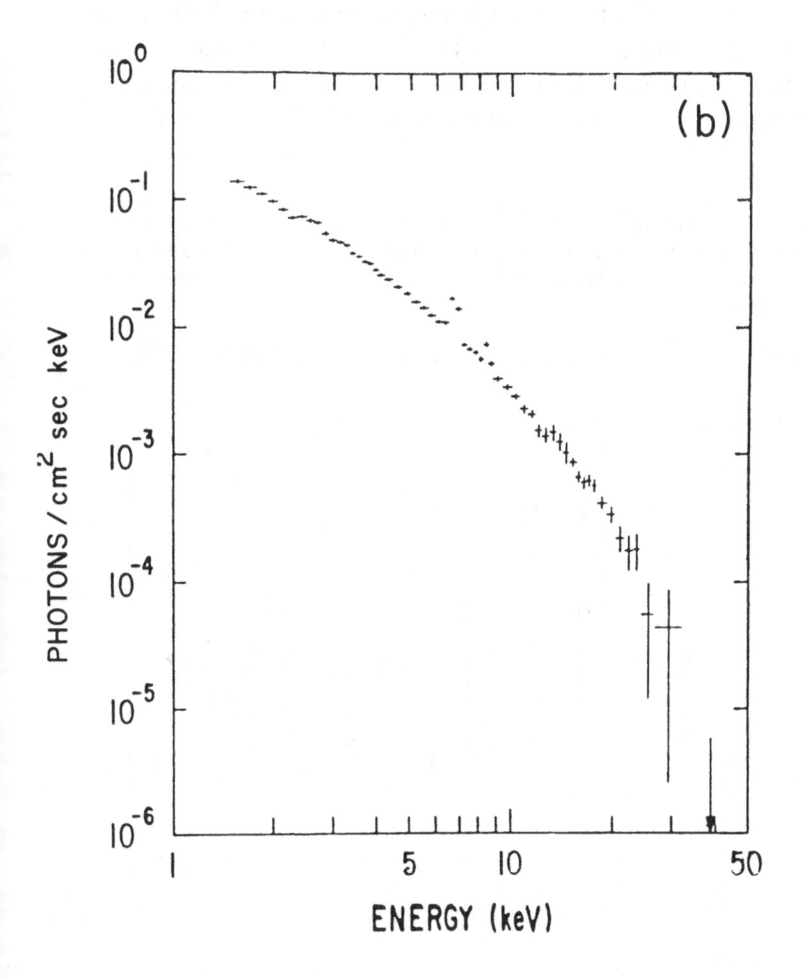

observations of clusters are all roughly Fe/H $\approx 2 \times 10^{-5}$, which is about one-half of the solar value (see the detection references above, as well as Bahcall and Sarazin, 1977). The upper limits on iron abundances in clusters without line detections are also generally consistent with this value. Figure 13 gives the derived iron abundances for clusters from the HEAO-1 A-2 sample, plotted as a function of the cluster X-ray luminosity. The uniformity of the iron abundance suggests that the intracluster gas in clusters has a similar origin in all clusters, regardless of the dynamical state of the cluster.

The strong 7 keV Fe line emission observed from clusters is very difficult to reconcile with any model for the origin of the X-ray emission except the thermal intracluster gas model (Section 5.1.3; Mitchell *et al.*, 1976; Serlemitsos *et al.*, 1977; Bahcall and Sarazin, 1977). This line emission occurs naturally in the intracluster gas model if the abundances of heavy elements are roughly solar (Section 5.2.2). However, it is not at all expected in the IC model (Section 5.1.1) and unlikely in the individual stellar source model because such sources are generally optically thick (Section 5.1.2). Only a small portion of the X-ray luminosity of a typical cluster is emitted in these lines, so it is

Fig. 13. The iron abundance of the gas in clusters as derived from their X-ray spectra, plotted versus their X-ray luminosity, from Mushotzky (1984). The iron abundance is relative to solar, and the X-ray luminosity is in erg/s.

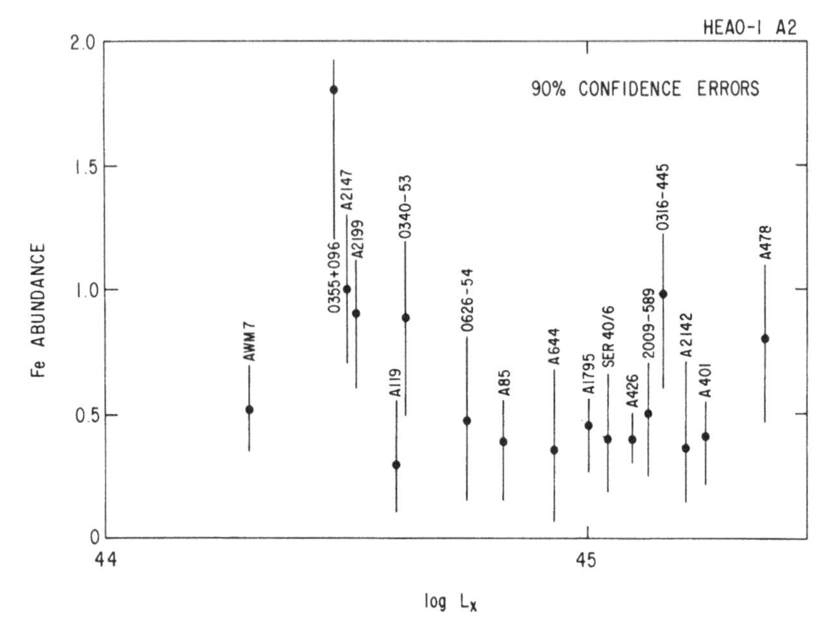

possible, in principal, that the lines might come from a different source than the majority of the X-ray emission. Several considerations show that this is extremely unlikely. First of all, the required abundances in the intracluster gas are nearly solar, and thus very high abundances would be needed if the lines were to come from gas that provided only a small fraction of the continuum X-ray emission. Second, the abundances derived for all the observed clusters are essentially the same within the errors (Section 4.3.2), although the clusters span a wide range in optical and X-ray properties. A very odd coincidence would be required to produce the appearance of constant abundances in such varied clusters, if the X-ray lines and continuum came from two distinct sources.

As discussed in Sections 5.1.3 and 5.10, the nearly solar iron abundance in the intracluster gas suggests that a significant portion of this gas has been processed in and ejected from stars.

The 7 keV Fe line can be used to determine the redshift of a cluster with moderate accuracy from even low resolution X-ray spectra (Boldt, 1976; Bahcall and Sarazin, 1978). Over a wide range of temperatures, the equivalent width of the 7 keV Fe line is at least an order of magnitude larger than that of any other feature in the spectrum of a hot plasma. Thus the detection of a strong feature at photon energies below 7 keV, coupled with the failure to detect a feature at higher energies up to 7 keV, permits the immediate identification of the feature with the 7 keV Fe line and the determination of a single-line redshift. This application of the 7 keV line has not been useful thus far because the *Einstein* X-ray observatory could not detect hard X-rays and had too little sensitivity to measure spectra from high redshift clusters. However, it promises to be of great importance in the future (see Chapter 6).

4.3.3 Lower Energy Lines

In addition to the 7 keV Fe line complex, the X-ray spectrum of a solar abundance low density plasma contains a large number of lower energy lines (Sarazin and Bahcall, 1977; Figure 34 below). These include the K lines of the common elements lighter than iron, such as C, N, O, Ne, Mg, Si, S, Ar, and Ca, as well as the L lines of Fe and Ni. Although the Fe L line complex was tentatively identified in proportional counter spectra of the M87/Virgo cluster by Fabricant *et al.* (1978) and Lea *et al.* (1979), the capability for detecting these lines was greatly increased with the launch of the *Einstein* X-ray observatory with its moderate resolution Solid State Spectrometer (SSS) and its high resolution Focal Plane Crystal Spectrometer (FPCS); see Giacconi *et al.* (1979) for a description of the satellite and its capabilities.

The SSS has detected the K-lines from Mg, Si, and S and the L-lines from Fe in the spectra of M87/Virgo, Perseus, A496, and A576 (Mushotzky, 1980, 1984; Mushotzky *et al.*, 1981; Lea *et al.*, 1982; Nulsen *et al.*, 1982; Rothenflug *et al.*, 1984). Figure 14 shows the SSS spectrum from Virgo. In general, line emission from both the helium-like and hydrogenic ions of Si and S is seen, indicating that the emission occurs at relatively low temperatures $T_g \approx 2 \times 10^7$ K. The observations in M87/Virgo are consistent with nearly solar abundances of Si, S, and Mg, while the observation in A576 may require lower abundances. Observations of SSS spectra away from the center of M87/Virgo show that the heavy element abundances are roughly constant throughout the gas (Lea *et al.*, 1982).

In Perseus, the SSS observations of the Fe L lines imply that the iron abundance is about one-half of the solar value (Mushotzky *et al.*, 1981), which agrees with the abundance derived from the 7 keV Fe line in proportional counter spectra. However, the SSS observations were made with a very small

Fig. 14. The moderate resolution X-ray spectrum of the M87/Virgo cluster taken with the Solid State Spectrometer on the *Einstein* satellite (Lea *et al.*, 1982). The spectral lines of Mg, Si, S, and the lower energy (L-shell) lines of Fe are marked.

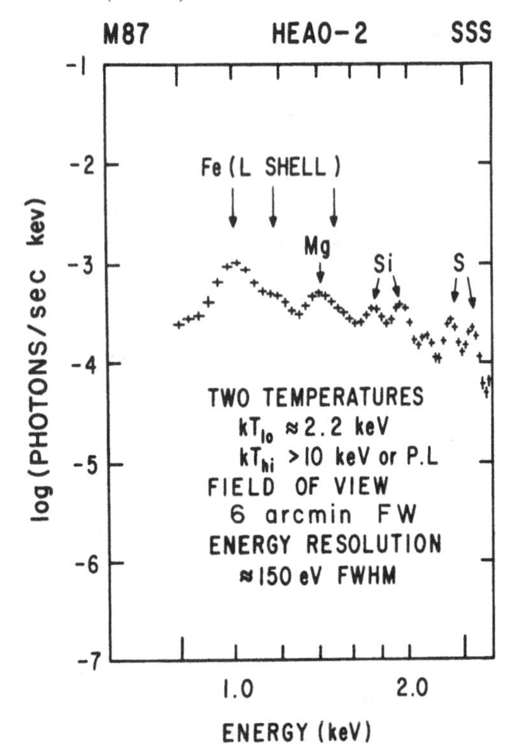

($\approx$6 arc min) aperture centered on the cluster center, while the proportional counter observations determine the spectrum of the entire cluster. The approximate agreement of the two abundances suggests that the iron is well mixed throughout the cluster, and not just concentrated in the cluster core.

In general, the SSS spectral line observations seem to imply that many clusters contain cooler gas than was required to explain their continuum spectra. Because the SSS has a small field of view and was centered on the cluster center in these observations, this cool gas must be concentrated at the center of the cluster; in the case of the Virgo and Perseus clusters, the center coincides with the central dominant galaxies M87 and NGC1275. In Perseus, this cool gas had a cooling time (see Section 5.3.1) of less than 2×10^9 yr, which is considerably less than the probable age of the cluster. It therefore seems likely that the cool gas observed is part of a steady-state cooling flow (see Section 5.7 for a discussion of the theory of such flows). From the observed line intensities, Mushotzky *et al.* (1981) determined that about $300 M_\odot$ per year of gas must currently be cooling onto NGC1275 in the Perseus cluster, and Nulsen *et al.* (1982) found that about $200 M_\odot$ per year must be accreting onto the cD in A496. These rates assume that the gas is not being heated.

The FPCS has also provided strong evidence for the cooling and accretion of gas onto M87 in the Virgo cluster and NGC1275 in the Perseus cluster (Canizares *et al.*, 1979, 1982; Canizares, 1981). In M87, the FPCS has detected the O^{+7} Kα line, as well as blends of the $Fe^{+(16-23)}$ L lines and the Ne^{+9} Kα line. Figure 15 shows the FPCS detection of the O^{+7} Kα line in M87. The ratio of the abundance of oxygen to iron is apparently 3–5 times higher than the solar ratio. The relative strengths of the various Fe L line blends cannot result from gas at any single temperature. Apparently, a range of temperatures is necessary, with the X-ray luminosity originating from gas in any range of temperature dT_g being roughly proportional to dT_g. This is just what is predicted if the cool gas results from the cooling and accretion of hotter gas onto the center of M87 (Cowie, 1981). Canizares *et al.* (1979, 1982) and Canizares (1981) show that the spectra are consistent with radiative accretion at a rate of ≈ 3–$10 M_\odot$ per year. Similar results were found for the Perseus cluster, except that the required accretion rate is very large $\approx 300 M_\odot$ per year (Canizares, 1981), in agreement with the results from the SSS. The SSS spectra of about a half dozen other clusters also show evidence for such accretion flow, with rates between those of the M87/Virgo cluster and the Perseus cluster (Fabian *et al.*, 1981b; Mushotzky, 1984).

Thus the two primary observational results of the *Einstein* spectrometers are these: first, the intracluster gas contains the heavy elements oxygen, magnesium, silicon, and sulfur, as well as iron; second, gas is cooling and being accreted onto central dominant galaxies in many clusters at rather high rates (3–400$M_{\odot}$/yr). Such accretion had been predicted by Cowie and Binney (1977), Fabian and Nulsen (1977), and Mathews and Bregman (1978); models for these cooling flows are discussed in Section 5.7. The rates of cooling are so high that if they have persisted for the age of the cluster, the entire mass of the inner portions of the central dominant galaxies might be due to accretion. Models for central dominant galaxies based on this idea have been given by Fabian *et al.* (1982a) and Sarazin and O'Connell (1983), who argue that the majority of the accreted gas is converted into low mass stars.

X-ray line observations have established that the primary emission mechanism of X-ray clusters is thermal emission from hot, diffuse intracluster gas. They have also shown that at least part of that gas has been ejected from stars and presumably from galaxies. Apparently, some of this intracluster gas

Fig. 15. The very high resolution X-ray spectrum of the M87/Virgo cluster, showing the O VIII K line, from Canizares *et al.* (1979) using the Focal Plane Crystal Spectrometer on the *Einstein* satellite.

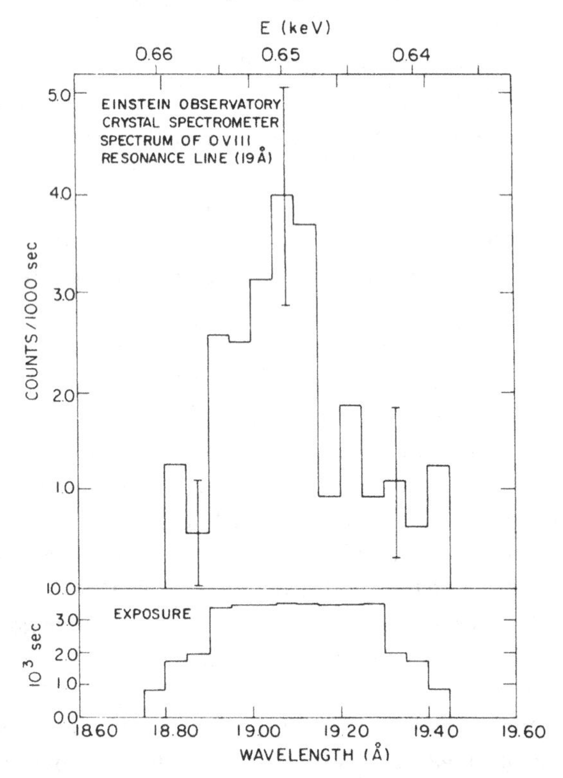

is now completing the cycle and returning to the central galaxies, and possibly being formed into stars!

4.4 The Spatial Distribution of X-Ray Emission

4.4.1 X-Ray Centers, Sizes, and Masses

Early *Uhuru* observations indicated that the X-ray sources were centered on the cluster center, as determined from the galaxy distribution, or on an active galaxy in the cluster (Bahcall, 1977a). In many of the X-ray clusters, the cluster center also corresponds with the position of a cD or other central dominant galaxy.

The earliest X-ray observations showed that the X-ray sources in clusters were extended. With the nonimaging proportional counters used in these observations, the spatial resolution of the detectors was usually determined by mechanical collimators in front of the detectors and was therefore relatively crude (generally an angular resolution of $\approx 1°$). Since this is comparable to the sizes of the X-ray emission regions in the nearest clusters, the distribution of emission could not generally be observed in any detail. Only estimates of the size could be determined by convolving a model for the distribution of the emission with the resolution of the detector and comparing the result to the observations.

Lea *et al.* (1973), Kellogg and Murray (1974), and Abramopoulos and Ku (1983) derived sizes for cluster X-ray sources, assuming that the gas had the density distribution given by the King approximation to a self-gravitating isothermal sphere (equations (2.9), (2.13))

$$\rho_g \propto \left[1 + \left(\frac{r}{r_x}\right)^2\right]^{-3/2} . \tag{4.5}$$

Here, ρ_g is the gas density and r_x is the X-ray core radius. While this model is not physically consistent (see Section 5.5.1), it does provide a convenient fitting form for comparison to the galaxy distribution, which is often fit by the same function. Values of r_x were derived from the X-ray distribution in nine clusters by Lea *et al.* (1973) and Kellogg and Murray (1974). More recently, equation (4.5) has been fit to the distribution of X-ray emission from 53 clusters detected in an extensive survey of clusters with the *Einstein* X-ray observatory (Abramopoulos and Ku, 1983; Table 2). In general, these studies found that the X-ray core radii were significantly larger than the core radii of the galaxies (Section 2.7). These results were fairly uncertain because of the difficulties in determining either of these radii accurately.

More physically consistent hydrostatic models for the intracluster gas have also been used to fit the observed X-ray distributions (Lea, 1975; Gull and Northover, 1975; Cavaliere and Fusco-Femiano, 1976; Bahcall and Sarazin, 1977; Cavaliere, 1980). One model that has been used extensively is the hydrostatic isothermal model for the intracluster gas (Cavaliere and Fusco-Femiano, 1976, 1981; Bahcall and Sarazin, 1977, 1978; Sarazin and Bahcall, 1977; Gorenstein *et al.*, 1978; Jones and Forman, 1984; Section 5.5.1). In this model, both the galaxies and the intracluster gas are assumed to be isothermal, bound to the cluster, and in equilibrium. The galaxies are assumed to have an isotropic velocity dispersion. However, the gas and galaxies are not assumed to have the same velocity dispersion; the square of the ratio of the galaxy-to-gas velocity dispersions is

$$\beta \equiv \frac{\mu m_p \sigma_r^2}{k T_g}, \tag{4.6}$$

where μ is the mean molecular weight in amu, m_p is the mass of the proton, σ_r is the one-dimensional velocity dispersion, and T_g is the gas temperature. Then, the gas and galaxy densities vary as $\rho_g \propto \rho_{gal}^{\beta}$. If the galaxy distribution is taken to be a King analytical form for the isothermal sphere (equation (2.13)), then the X-ray surface brightness $I_x(b)$ at a projected radius b varies as

$$I_x(b) \propto \left[1 + \left(\frac{b}{r_c}\right)^2\right]^{-3\beta + 1/2}, \tag{4.7}$$

where r_c is the galaxy core radius (Section 2.7). The self-gravitating isothermal model (equation (4.5)) has the same surface brightness distribution if one makes the replacements $\beta = 1$ and $r_c = r_x$. These and other models for the distribution of the intracluster gas will be discussed in detail in Section 5.5.

The earliest determinations of the extent of the intracluster gas used low spatial resolution proportional counters. Observations with somewhat higher spatial resolution were made using detectors with modulation collimators (Schwarz *et al.*, 1979) or one-dimensional imaging detectors (Gorenstein *et al.*, 1973). Some clusters were found to contain compact X-ray sources associated with the central dominant galaxies in the cluster (Schnopper *et al.*, 1977), such as NGC1275 in the Perseus cluster (Wolff *et al.*, 1974, 1975, 1976; Cash *et al.*, 1976; Malina *et al.*, 1976; Helmken *et al.*, 1978). In Perseus, this point source contributes $\approx$20–25% of the X-ray luminosity at moderate photon energies ($\approx$keV); at very high energies $\gtrsim$20 keV the point source is dominant (Primini *et al.*, 1981; Rothschild *et al.*, 1981). The X-ray surface brightness of the Perseus cluster was shown to be extended in the east–west direction (Wolff *et al.*, 1974, 1975, 1976; Malina *et al.*, 1976; Cash *et al.*, 1976); this is the same direction as the line of bright galaxies seen optically in this L cluster. Later high resolution observations of the center of the cluster by Branduardi-

Raymont *et al.* (1981) show a smaller elongation. A similar but smaller elongation was observed in the X-ray emission from the Coma cluster (Johnson *et al.*, 1979; Gorenstein *et al.*, 1979).

The *Einstein* X-ray telescope has allowed much more accurate determinations of the distribution of the X-ray emitting gas in clusters, using the moderate resolution (≈ 1 arc min) Imaging Proportional Counter (IPC) or the High Resolution Imager (HRI, ≈ 8 arc sec). Moderate resolution *Einstein* IPC images of the X-ray emission in 46 clusters have been fit by equation (4.7) (Jones and Forman, 1984). Because galaxy core radii, velocity dispersions, and X-ray temperatures are poorly determined for most of these clusters, both the core radius and β were derived from the observed X-ray surface brightness. Table 2 gives the X-ray luminosities (L_x), X-ray core radii r_x, and temperature parameters β derived by Jones and Forman (the entries without an *). Note that the X-ray luminosity is only for gas within a projected radius of 0.5 Mpc of the cluster center. Abramopoulos and Ku (1983) fit IPC observations of 53 clusters to equation (4.5); that is, they assumed $\beta = 1$. Table 2 also contains the results of their fits (the entries marked with an *) for the clusters with X-ray detections which were not studied by Jones and Forman. The values of the total X-ray luminosities of Abramopoulos and Ku were converted to luminosities within 0.5 Mpc using their best fit X-ray distribution. Of course, the value of β for all the Abramopoulos and Ku fits is listed as one. It is worth noting that in nearly every case where both Abramopoulos and Ku, and Jones and Forman studied the same cluster, Jones and Forman found best fit values of β which were much smaller than the $\beta = 1$ assumed by Abramopoulos and Ku, and Jones and Forman could rule out the value $\beta = 1$ with high statistical confidence. Thus the validity of the results of Abramopoulos and Ku is questionable because of the value of β assumed.

Figure 16 shows the X-ray surface brightness as a function of radius from the center of the X-ray emission (in minutes of arc) for A400, A2063, and A1795 from Jones and Forman (1984). The solid histogram is the data; the dots are the best fit using equation (4.6). About two-thirds of the clusters were fit quite well with this formula; however, one-third appeared to have an excess of emission near the cluster center. For those clusters with a central excess, the central surface brightness points were excluded for the fits given in Table 2. In Figure 16, A400 and A2063 were reasonably fit by equation (4.6), while A1795 was not. The second panel on A1795 shows the fit to equation (4.6) if the central several minutes of arc are ignored. Jones and Forman suggest that this excess central X-ray emission correlates with the radio luminosity of the cluster, and that the excess emission may be due to a cooling flow in the cluster core (Section 5.7). A1795 does in fact have a very large cooling flow (Table 4 below).

Table 2. *The results of fits to the X-ray surface brightness profiles of X-ray clusters using the IPC detector on the Einstein X-ray observatory*

Cluster	L_x (0.5–3.0 keV) (10^{43} erg/s)	r_x (Mpc)	β	M_g ($10^{14} M_\odot$)
A56*	19.51 ± 9.17	0.33†	1.00	0.25
A76*	1.90 ± 0.21	0.75†	1.00	0.22
A85	41.77 ± 0.41	0.19–0.26	0.60–0.65	2.30
A119*	6.58 ± 0.15	0.71–0.83	1.00	0.42
A133*	15.24 ± 0.64	0.24–0.28	1.00	0.17
A136*	3.93 ± 0.95	0.22–1.49	1.00	0.28
A154	3.57 ± 0.20	0.05–0.30	0.40–0.70	1.00
A168	1.94 ± 0.08	0.34–0.88	0.50–1.00	1.50
A194	0.27 ± 0.02	0.09–0.38	0.40–1.00	0.29
A262	3.07 ± 0.06	0.07–0.12	0.50–0.60	0.36
A348*	8.61 ± 1.48	0.67†	1.00	0.40
A358*	1.46 ± 0.39	0.51–0.83	1.00	0.16
A376*	6.28 ± 0.81	0.33–0.47	1.00	0.18
A399	14.79 ± 0.31	0.18–0.25	0.47–0.57	1.20
A400	1.84 ± 0.04	0.13–0.20	0.50–0.65	0.41
A401*	35.69 ± 0.55	0.53–0.57	1.00	0.64
A407*	3.22 ± 1.18	0.36–0.48	1.00	0.13
A426	46.10 ± 0.15	0.23–0.34	0.55–0.60	3.00
A496*	15.33 ± 0.27	0.22–0.23	1.00	0.14
A501*	6.24 ± 1.40	0.00–0.53	1.00	0.04
A514*	1.45 ± 0.20	0.19–0.29	1.00	0.04
A539*	2.70 ± 0.13	0.27–0.29	1.00	0.08
A568*	2.96†	0.41†	1.00	0.13
A569*	0.78 ± 0.14	0.01–0.05	1.00	0.00
A576	6.55 ± 0.10	0.09–0.14	0.47–0.52	0.47
A592	3.89 ± 0.13	0.16–0.37	0.63–1.00	1.20
A644*	43.59 ± 1.42	0.36–0.42	1.00	0.46
A646*	22.38 ± 2.98	0.14–0.28	1.00	0.14
A671	3.83 ± 0.26	0.09–0.34	0.55–1.00	1.40
A754*	18.09 ± 0.45	0.68–0.77	1.00	0.63
A854*	35.50 ± 2.75	0.23–0.68	1.00	0.44
A882*	3.77†	0.46†	1.00	0.16
A910*	11.04 ± 1.33	0.86–1.24	1.00	0.80
A957	2.66 ± 0.14	0.05–0.23	0.40–0.63	0.58
A1060	2.37 ± 0.02	0.09–0.11	0.60–0.75	0.40
A1142	1.03 ± 0.17	0.10–0.41	0.43–1.00	0.64
A1185	1.60 ± 0.06	0.07–0.41	0.40–0.80	0.81
A1213*	1.03†	0.31†	1.00	0.05
A1291	2.92 ± 0.16	0.05–0.24	0.50–1.00	1.00
A1314	1.58 ± 0.07	0.21–0.50	0.60–1.00	1.00
A1367	4.49 ± 0.04	0.26–0.60	0.40–0.65	1.30
A1377	1.98 ± 0.11	0.14–0.62	0.43–1.00	1.30
A1569*	7.41 ± 1.18	0.40–0.97	1.00	0.35
A1631*	1.03 ± 0.21	0.11–0.23	1.00	0.03
A1656*	23.55 ± 0.14	0.48–0.52	1.00	0.46

Table 2 *continued*

Cluster	L_x (0.5–3.0 keV) (10^{43} erg/s)	r_x (Mpc)	β	M_g ($10^{14} M_\odot$)
A1677*	24.91 ± 2.84	0.37–0.57	1.00	0.44
A1767*	13.59 ± 0.84	0.49–0.58	1.00	0.37
A1775	10.83 ± 0.40	0.12–0.25	0.52–0.80	1.20
A1795	51.26 ± 0.53	0.20–0.40	0.65–0.80	4.20
A1809	6.93 ± 0.47	0.13–0.49	0.50–1.00	2.00
A1890	3.01 ± 0.21	0.17–0.59	0.50–1.00	1.50
A1904*	1.98 ± 0.48	0.81–1.42	1.00	0.34
A1913	1.73 ± 0.08	0.36–1.10	0.45–1.00	1.70
A1983	2.12 ± 0.44	0.05–0.25	0.50–1.00	0.90
A1991	8.47 ± 0.25	0.03–0.10	0.50–0.63	0.62
A2029	68.18 ± 1.76	0.08–0.30	0.63–0.83	5.30
A2040	1.91 ± 0.08	0.08–0.25	0.50–1.00	0.63
A2052*	10.88 ± 0.40	0.20–0.24	1.00	0.12
A2063	9.65 ± 0.19	0.15–0.20	0.58–0.67	1.10
A2065*	20.57 ± 0.47	0.56–0.62	1.00	0.53
A2079*	2.87 ± 0.54	0.61–1.05	1.00	0.29
A2107	7.08 ± 0.26	0.13–0.19	0.60–1.00	0.89
A2124	5.61 ± 0.26	<0.04–0.12	0.45–0.53	0.51
A2125*	6.89†	0.56†	1.00	0.29
A2142*	69.08 ± 0.90	0.49–0.53	1.00	0.80
A2151*	1.51 ± 0.10	0.41–0.43	1.00	0.09
A2152*	2.02 ± 0.20	0.20–0.29	1.00	0.06
A2162*	0.58 ± 0.32	0.05–0.79	1.00	0.03
A2165*	3.39 ± 2.26	0.00–0.32	1.00	0.02
A2197*	0.66 ± 0.11	0.31–0.53	1.00	0.05
A2199	20.68 ± 0.21	0.12–0.16	0.63–0.73	1.50
A2255	13.85 ± 0.24	0.53–0.65	0.70–0.83	3.45
A2256	27.34 ± 0.31	0.43–0.47	0.68–0.77	3.65
A2271	2.70 ± 0.19	0.04–0.23	0.40–0.73	0.70
A2312	6.60 ± 0.35	0.05–0.15	0.50–0.65	0.73
A2319	41.74 ± 0.39	0.36–0.46	0.46–0.63	3.90
A2410	2.42 ± 0.14	0.08–0.63	0.40–0.65	1.60
A2415	9.37 ± 0.57	<0.04–0.13	0.45–0.57	0.72
A2424*	13.76†	0.42†	1.00	0.28
A2440*	3.36 ± 2.35	0.22–1.66	1.00	0.12
A2521*	7.44 ± 1.33	0.27–0.75	1.00	0.21
A2580*	31.11 ± 3.01	0.20–0.36	1.00	0.26
A2593	4.34 ± 1.41	0.15–0.29	0.50–0.70	0.85
A2626	8.43 ± 0.34	0.08–0.31	0.57–1.00	2.10
A2634	3.67 ± 0.58	0.45–0.85	0.61–1.10	2.20
A2657	7.29 ± 0.11	0.10–0.19	0.50–0.57	0.66
A2670	9.90 ± 0.51	0.05–0.18	0.50–0.67	1.00
SC0107-46	1.83 ± 0.05	0.07–0.52	0.40–1.00	1.40
SC0559-40	5.44 ± 0.01	0.60–0.95	0.65–1.00	0.33

See top of next page for Notes to Table 2

Note: The X-ray luminosities scale as $L_x \propto h_{50}^{-2}$, the X-ray core radii scale as $r_x \propto h_{50}^{-1}$, and the gas masses scale as $M_g \propto h_{50}^{-5/2}$. The data are from Jones and Forman (1984) and Abramopoupos and Ku (1983); the entries from Abramopoulos and Ku are marked with an *. The value of L_x is only for gas within 0.5 Mpc of the cluster center, while the gas mass is for gas within 3.0 Mpc of the center. The values of r_x and β are from fits of the X-ray surface brightness to equation (3.7). In the data from Abramopoulos and Ku, β was assumed to be unity, and was not derived from the data. The data from Abramopoulos and Ku were adjusted for the definitions of L_x and M_g from Jones and Forman, using the fits given by Abramopoulos and Ku.
† Very uncertain value.

A wide range of core radii (0.07–0.9h_{50}^{-1} Mpc) were derived by Jones and Forman. They found a strong anticorrelation between the presence of a dominant cluster galaxy in the core and the size of the X-ray core radius. (Unfortunately, Abramopoulos and Ku (1983) found the opposite effect with their sample of clusters observed with *Einstein*.) The average value of β found by Jones and Forman from the X-ray distributions is $\beta = 0.65$, which implies that the gas is considerably more extensively distributed than the mass in the cluster. Unfortunately, the average value of β determined by applying equation (4.6) to those clusters with measured velocity dispersions and X-ray temperatures is $\beta = 1.1$. It is not clear whether this discrepancy results from errors in the measured cluster properties, velocity anisotropies, or a failure of the isothermal model (see Section 5.5.1).

While in most clusters the X-ray emission was found to be as broadly distributed as the galaxy distribution, an exception was Virgo/M87. Here the soft X-ray emission comes from a small region around M87, while the galaxy core radius of this irregular cluster, although hard to define, is certainly much larger. While early observations suggested the existence of weaker hard X-ray emission originating from a larger region of the cluster (Davison, 1978; Lawrence, 1978), recent observations indicate that this emission actually is due to the nucleus of M87 (Lea *et al.*, 1981, 1982).

The thermal X-ray emission from intracluster gas is proportional to the emission integral *EI* (equation (4.3)), while the mass of intracluster gas is proportional to $\int n_p dV$. All other things being equal, the mass of the intracluster gas is then $M_g \propto (L_x r_x^3)^{1/2}$. Thus, if the size and distribution of the X-ray emitting gas can be determined from the X-ray surface brightness, the mass in intracluster gas can be estimated. Unfortunately, the estimate is very uncertain because the X-ray emission falls off rapidly as the density decreases in the outer parts of the cluster, where a large fraction of the mass of the intracluster gas may be located. Early estimates of the mass of the intracluster gas based on the self-gravitating isothermal model (equation (4.7)) were given in Lea *et al.* (1973) and Kellogg and Murray (1974). They found that the total

Fig. 16. The X-ray surface brightness of several clusters, as determined by the IPC on the *Einstein* satellite by Jones and Forman (1984). The surface brightness is normalized to its central value and is given as a function of the angular distance from the cluster center. The solid curves give the observed surface brightness, and the dots are the best fit using equation (4.7). The bottom two panels show the A1795 cluster with the inner eight data points either included in the fit (left) or removed (right). The improvement in the fit in the outer points when the inner regions are removed suggests excess emission in the center, due to a cooling flow.

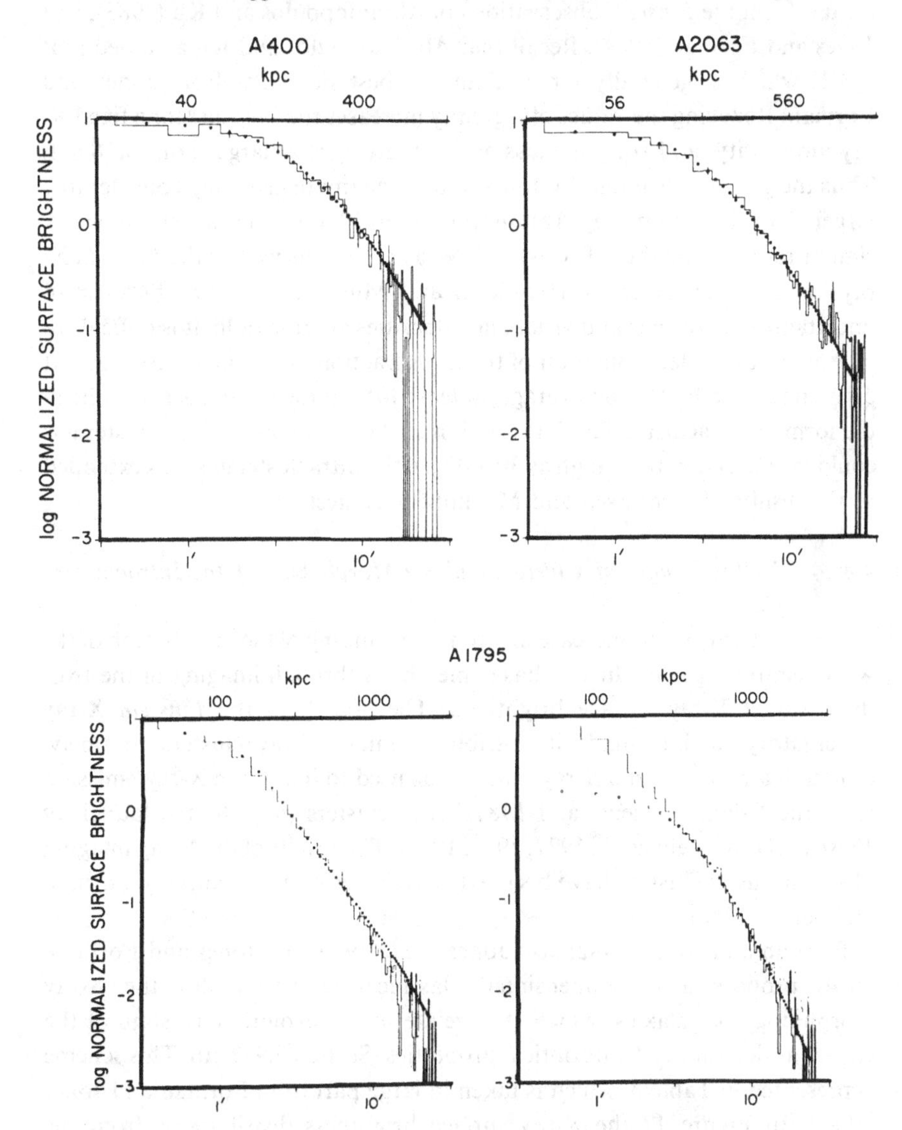

mass of intracluster gas determined from the X-ray observations is similar to or somewhat larger than the total mass of the galaxies in the cluster estimated from their total luminosity and a typical galaxy mass-to-light ratio. However, based on these mass estimates, the intracluster gas is still only 5–15% of the total virial mass of the cluster, and thus the discovery of the intracluster gas has not resolved the missing mass problem in clusters (Section 2.8). Table 2 gives the intracluster gas masses (here within 3.0 Mpc radius of the cluster center) from the *Einstein* observations of Abramopoulos and Ku (1983), and Jones and Forman (1984). Recall that Abramopoulos and Ku assumed that $\beta = 1$, which is generally larger than the best fit values from Jones and Forman. Reducing the values of β greatly increases the gas mass for a fixed X-ray luminosity, and the gas mass actually diverges at large radii for $\beta \leqslant 1$. Thus the gas masses found by Jones and Forman are generally considerably larger than those found by Abramopoulos and Ku or the earlier studies. A similar result, using the HEAO-1 A-2 X-ray spectra as well as the *Einstein* X-ray images, was found by Henriksen and Mushotzky (1985). Because of uncertainties in the gas and virial mass densities at large radii, it is difficult to give an accurate determination of the mass fraction of the intracluster gas. It does appear likely that, on average, at least 10% of the virial mass must be in the form of intracluster gas. The fraction of the total mass in intracluster gas could in some cases be as high as 30–60%, if the intracluster gas is as extended as the results of Henriksen and Mushotzky suggest.

4.4.2 *X-Ray Images of Clusters and the Morphology of the Intracluster Gas*

A tremendous increase in our understanding of the distribution of the X-ray emitting gas in clusters has come about through imaging of the two-dimensional X-ray surface brightness. The launch of the *Einstein* X-ray observatory satellite made it possible to image X-ray clusters routinely, although a rocket-borne X-ray mirror was used to image the X-ray emission from the Coma, Perseus, and M87/Virgo clusters prior to the launch of *Einstein* (Gorenstein *et al.*, 1977, 1978, 1979). The results of the X-ray imaging observations of clusters have been extensively reviewed recently by Forman and Jones (1982).

Forman and Jones (1982; also Jones *et al.*, 1979, and Jones and Forman, 1984) propose a two-dimensional classification scheme for the X-ray morphology of galaxies, which they relate to the evolutionary state of the cluster as determined by its optical properties (Section 2.9, 2.10). This scheme is presented in Table 3, which is taken in large part from Forman and Jones (1982). In Figure 17 the X-ray surface brightness distributions from the

Imaging Proportional Counter (IPC) on the *Einstein* X-ray observatory are shown for representative examples of clusters of each morphological type. The X-ray brightness distribution is shown as a contour plot, in which the lines are loci of constant X-ray surface brightness. The contours are superimposed on optical photographs of the cluster for comparison.

First of all, Forman and Jones classify the clusters as being irregular ('early') or regular ('evolved'), based on their overall X-ray distribution. The early clusters have irregular X-ray surface brightnesses, and often show small peaks in the X-ray surface brightness, many of which are associated with individual galaxies in the cluster. The X-ray luminosities L_x and X-ray spectral temperatures T_g are low. Optically, these clusters tend to be irregular clusters (Table 1). That is, they have irregular galaxy distributions with subclustering and with low central concentration. They are not generally very rich, and tend to be Bautz–Morgan types II to III and Rood–Sastry types F and I. They generally have low velocity dispersions σ_r. They are often, but not always, spiral rich.

On the other hand, the evolved clusters have regular, centrally condensed X-ray structures (Forman and Jones, 1982). The X-ray distribution is smooth, and X-ray emission peaks are not found associated with individual galaxies, except possibly with a central dominant galaxy at the cluster center. The evolved clusters have high X-ray luminosities and gas temperatures.

Table 3. *Morphological classification of X-ray clusters, adapted from Forman and Jones (1982)*

	No X-ray Dominant Galaxy (nXD)	X-ray Dominant Galaxy (XD)
Irregular	Low $L_x \lesssim 10^{44}$ erg/s	Low $L_x \lesssim 10^{44}$ erg/s
	Cool gas $T_g = 1$–4 keV	Cool gas $T_g = 1$–4 keV
	X-ray emission around many galaxies	Central galaxy X-ray halo
	Irregular X-ray distribution	Irregular X-ray distribution
	High spiral fraction >40%	High spiral fraction >40%
	Low central galaxy density	Low central galaxy density
	Prototype: A1367	Prototype: Virgo/M87
Regular	High $L_x \gtrsim 10^{44}$ erg/s	High $L_x \gtrsim 3 \times 10^{44}$ erg/s
	Hot gas $T_g \gtrsim 6$ keV	Hot gas $T_g \gtrsim 6$ keV
	No cooling flow	Cooling flow onto central galaxy
	Smooth X-ray distribution	Compact, smooth X-ray distribution
	Low spiral fraction $\lesssim 20$%	Low spiral fraction $\lesssim 20$%
	High central galaxy density	High central galaxy density
	Prototype: Coma (A1656)	Prototype: Perseus (A426)

Optically, these are regular, symmetric clusters, generally of Bautz–Morgan type I and II and Rood–Sastry types cD, B, L, or C. They are rich and have a high central concentration of galaxies. They have high velocity dispersions and are spiral poor in their galaxy composition.

The second determinant of the X-ray morphology of clusters is the presence or absence of a central, dominant galaxy in the cluster. The X-ray emission from a cluster tends to peak at the position of such a galaxy. Clusters

Fig. 17. The X-ray morphology of several clusters of galaxies from Jones and Forman (1984). Contours of constant X-ray surface brightness are shown superimposed on optical images of the clusters. (Top left), the prototypical irregular nXD cluster A1367. (Top right), the irregular XD cluster A262. (Bottom left), the regular nXD cluster A2256. (Bottom right), the regular XD cluster A85, showing the X-ray emission centered on the cD galaxy.

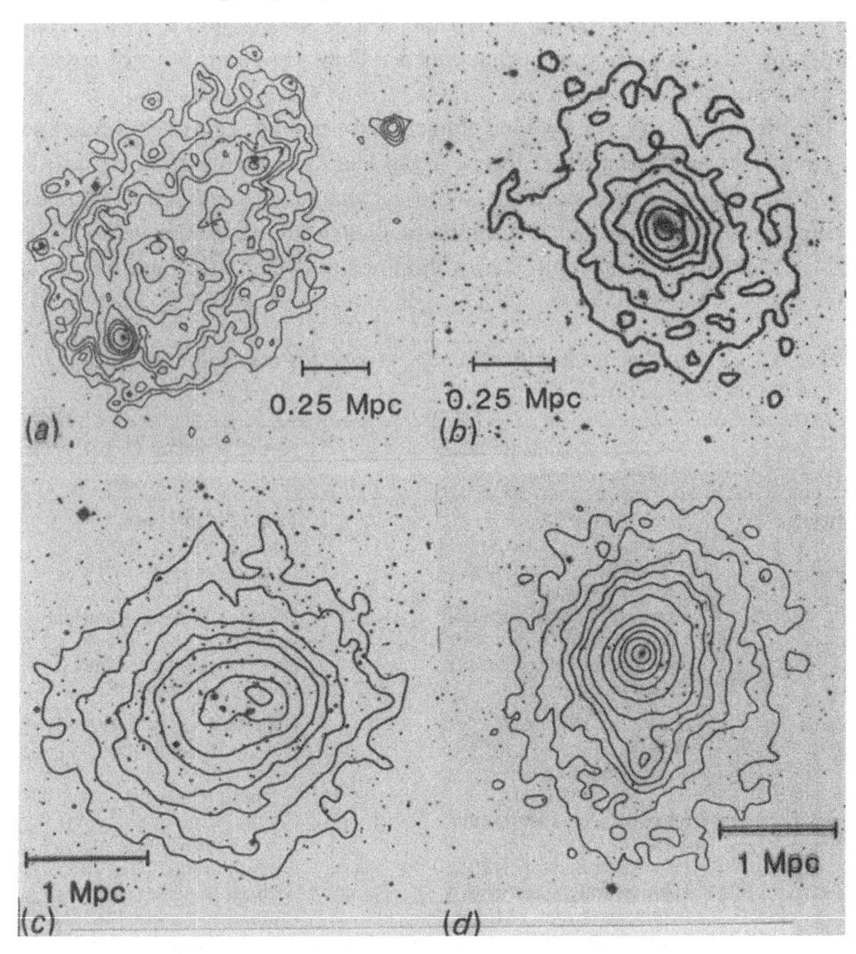

containing such central dominant galaxies are classified by Forman and Jones (1982) as X-ray dominant (XD); those without such a galaxy are classified as non-X-ray dominant (nXD). The nXD clusters have larger X-ray core radii $r_x \approx 500/h_{50}$ kpc. There is no strong X-ray emission associated with any individual galaxy in these systems.

The XD clusters have small X-ray core radii $r_x \approx 250/h_{50}$ kpc. The X-ray emission is strongly peaked on the central dominant galaxy. In many cases, spectral observations indicate that gas is cooling in this central peak and being accreted onto the central galaxy (Sections 4.3.3 and 5.7). The data of Jones and Forman (1984) suggest that *all* cooling flows are centered on the dominant galaxies in XD clusters. Regular XD clusters have, on average, the highest X-ray luminosities of any clusters (Forman and Jones, 1982; Jones and Forman, 1984).

Examples of irregular nXD clusters include A1367 (Jones *et al.*, 1979; Bechtold *et al.*, 1983), A194 (Forman and Jones, 1982), A566 (Harris *et al.*, 1982), and A2069 (Gioia *et al.*, 1982). A1367, which is shown in Figure 17a, is the prototype of this class.

Virgo/M87 is the prototype of the irregular XD clusters, of which A262 (Forman and Jones, 1982), A347 (Maccagni and Tarenghi, 1981), A2147 (Jones *et al.*, 1979), and A2384 (Ulmer and Cruddace, 1982) are also members. Figure 17b shows the X-ray image of A262.

Regular nXD clusters probably include Coma (A1656; Abramopoulos *et al.*, 1981; see Figure 19 below), A576 (White and Silk, 1980), A1763 (Vallee, 1981), A2255, A2256 (Jones *et al.*, 1979; Forman and Jones, 1982), and CA0340-538 (Ku *et al.*, 1983). Figure 17c shows the X-ray image of A2256. Although the statistics are poor, it appears that clusters with radio halo emission (see Section 3.4) are primarily of this class.

The regular XD clusters include Perseus (A426, Branduardi-Raymont *et al.*, 1981; Fabian *et al.*, 1981a), A85, A1413, A1795 (Jones *et al.*, 1979), A478 (Schnopper *et al.*, 1977), A496 (Nulsen *et al.*, 1982), A2029 (Johnson *et al.*, 1980), A2218 (Boynton *et al.*, 1982), and possibly A2319 (White and Silk, 1980). Figure 17d shows the X-ray image of the cD cluster A85.

The prototypes of each of these classes (A1367, Virgo/M87, Coma, and Perseus) are discussed individually in Section 4.5 below.

Forman and Jones (1982) and Jones *et al.* (1979) argue that this classification scheme represents a sequence of cluster evolution, just as the sequence of cluster optical morphology (Section 2.5) was related to the dynamical evolution of the cluster in Section 2.9. Specifically, the evolution of the overall cluster distribution may be due to violent relaxation during the collapse of the cluster. This occurs on the dynamical time scale of the cluster, which depends only on its density (equation (2.32)). On the other hand, the formation of central dominant galaxies may be due to mergers of galaxies; as demonstrated in Section 2.10.1, this mechanism favors compact but poor regions. Thus, while both a regular overall distribution of a cluster and the presence of a central dominant galaxy may indicate that the cluster has undergone dynamical relaxation, the processes are different and depend in different ways on the size and mass of the region. This provides a possible qualitative explanation for this two-dimensional classification system of X-ray morphology.

Double peaked X-ray emission has been seen in a number of clusters, of which the best studied example is A98 (Henry *et al.*, 1981; Forman *et al.*, 1981). Other possible examples include A115, A1750, SC0627-54 (Forman *et al.*, 1981), A1560, A2355–A2356 (Ulmer and Cruddace, 1982), A982, A2241 (Bijleveld and Valentijn, 1982), CA0329-527 (Ku *et al.*, 1983), and possibly A399/401 (Ulmer *et al.*, 1979; Ulmer and Cruddace, 1981) and A2204-2210 (Ulmer *et al.*, 1985). In Figure 18, the X-ray distributions are shown for four double clusters observed by Forman *et al.* (1981). In most of these systems, the galaxy distribution is also double peaked (Dressler, 1978c; Henry *et al.*, 1981; Beers *et al.*, 1983), and the two peaks correspond to slightly different radial velocities (Faber and Dressler, 1977; Beers *et al.*, 1982). The small velocity differences indicate that the two subclusters are bound, and would be expected to collapse together and merge in a time of typically a few billion years.

These double systems may represent an intermediate stage in the evolution of clusters from irregular to regular distributions (Tables 1 and 3). In fact, numerical N-body simulations of cluster formation often show an intermediate bimodal stage to the galaxy distributions (see Figure 5c; White, 1976c; Ikeuchi and Hirayama, 1979; Carnevali *et al.*, 1981). With the current (rather poor) statistics, the fraction of clusters detected in this double phase ($\approx 10\%$) is consistent with the relatively short lifetime of the phase.

In the irregular and double clusters, the gas distribution is correlated with the galaxy distribution. The gas is probably roughly in hydrostatic equilibrium with the cluster gravitational potential (Section 5.5), which is primarily influenced by the distribution of the dynamically dominant missing mass component (Section 2.8). The fact that the X-ray surface brightness and galaxy distribution correlate suggests that the galaxies and the missing mass have a similar distribution in clusters; other evidence favoring this viewpoint was given in Section 2.8.

4.5 Individual Clusters

4.5.1 *Coma*

Coma is the prototype of the nXD regular cluster, and is often used as a general prototype for comparing models for relaxed clusters. Its optical image is given in Figure 1b. It is worthwhile noting that clusters as rich and compact as Coma are actually rather uncommon. Coma is also unusual in having the most prominent radio halo observed among clusters (Section 3.4). Coma and A1367 (see below) appear to be portions of a large supercluster system (Gregory and Thompson, 1978).

The X-ray image of Coma, which is shown in Figure 19, shows that the X-ray emission is somewhat elongated (Johnson *et al.*, 1979; Gorenstein *et al.*, 1979; Helfand *et al.*, 1980) in the same direction as the galaxies in the cluster (Section 2.7). The X-ray emission shows a central uniform core (Helfand *et al.*, 1980; Abramopoulos *et al.*, 1981) and falls off with radius more slowly than the galaxy distribution ($\beta \approx 0.8$ in equation (4.7)). Coma has an unusually high X-ray temperature for its galaxy velocity dispersion (Mushotzky *et al.*, 1978), although this is consistent with its small value for β. The X-ray surface brightness from Coma is very smooth, and there is no apparent excess emission associated with the two central galaxies (Forman and Jones, 1982; Bechtold *et al.*, 1983).

4.5.2 *Perseus*

The Perseus cluster is one of the most luminous X-ray clusters known, and has an unusually high radio luminosity as well (Gisler and Miley, 1979). It may represent an extreme in the evolution of the gas component in clusters.

Optically, Perseus is an L cluster (Bahcall, 1974a); the brightest galaxies form a chain oriented roughly east–west (see Figure 1c). The very prominent and unusual galaxy NGC1275 (Figure 20) is located near the east end of this chain. Together, Perseus, the clusters A262 and A347, and some smaller groups from the Perseus supercluster, which is also elongated in an east–west direction (Gregory *et al.*, 1981). On a moderately large scale, the X-ray emission from Perseus also appears to be elongated in the same direction as the galaxy distribution (Wolff *et al.*, 1974, 1976; Cash *et al.*, 1976; Malina *et al.*, 1976; Branduardi-Raymont *et al.*, 1981). The X-ray emission is peaked on NGC1275 and becomes more spherically symmetric near this galaxy (see Figures 21 and 22). Oddly, the center of the extended X-ray emission in Perseus appears to lie slightly to the east of NGC1275, while the galaxy distribution is centered to the west (Figure 21).

Fig. 18. The X-ray surface brightness in four double clusters, from Forman *et al.* (1981). Contours of constant X-ray surface brightness are shown superimposed on optical images of the clusters.

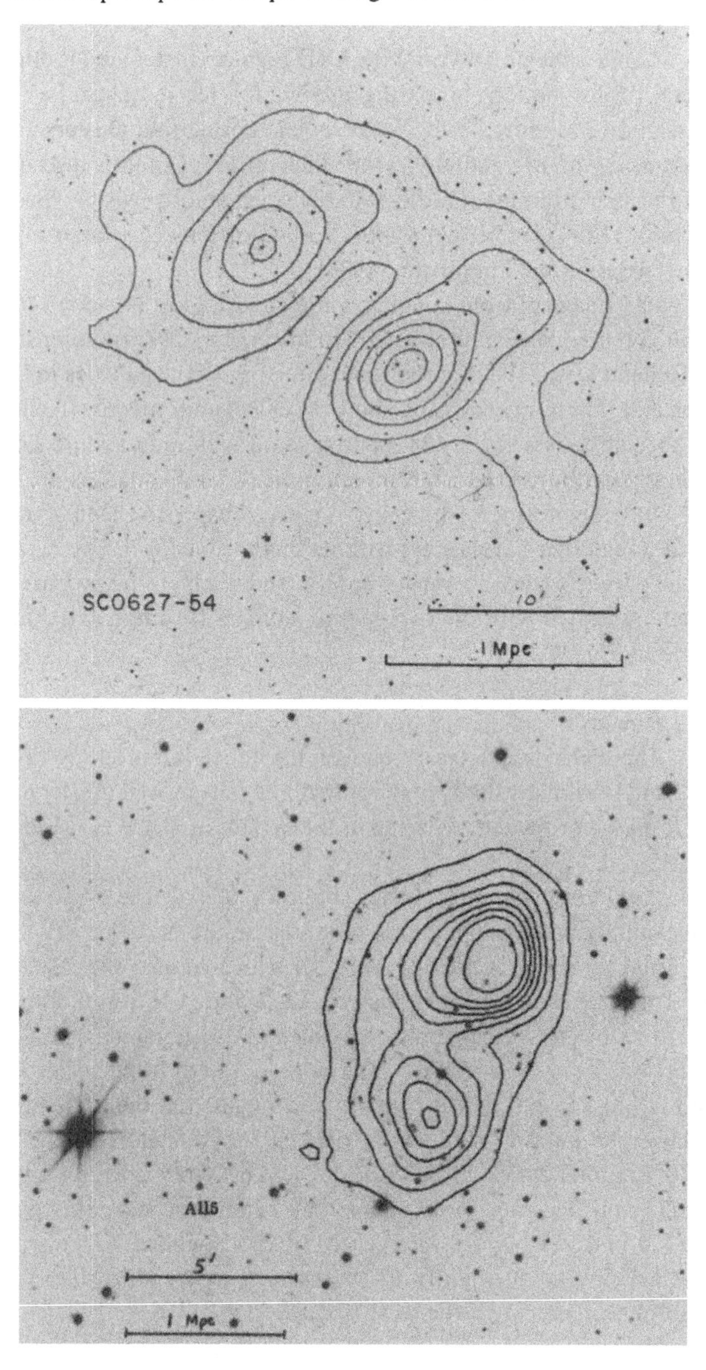

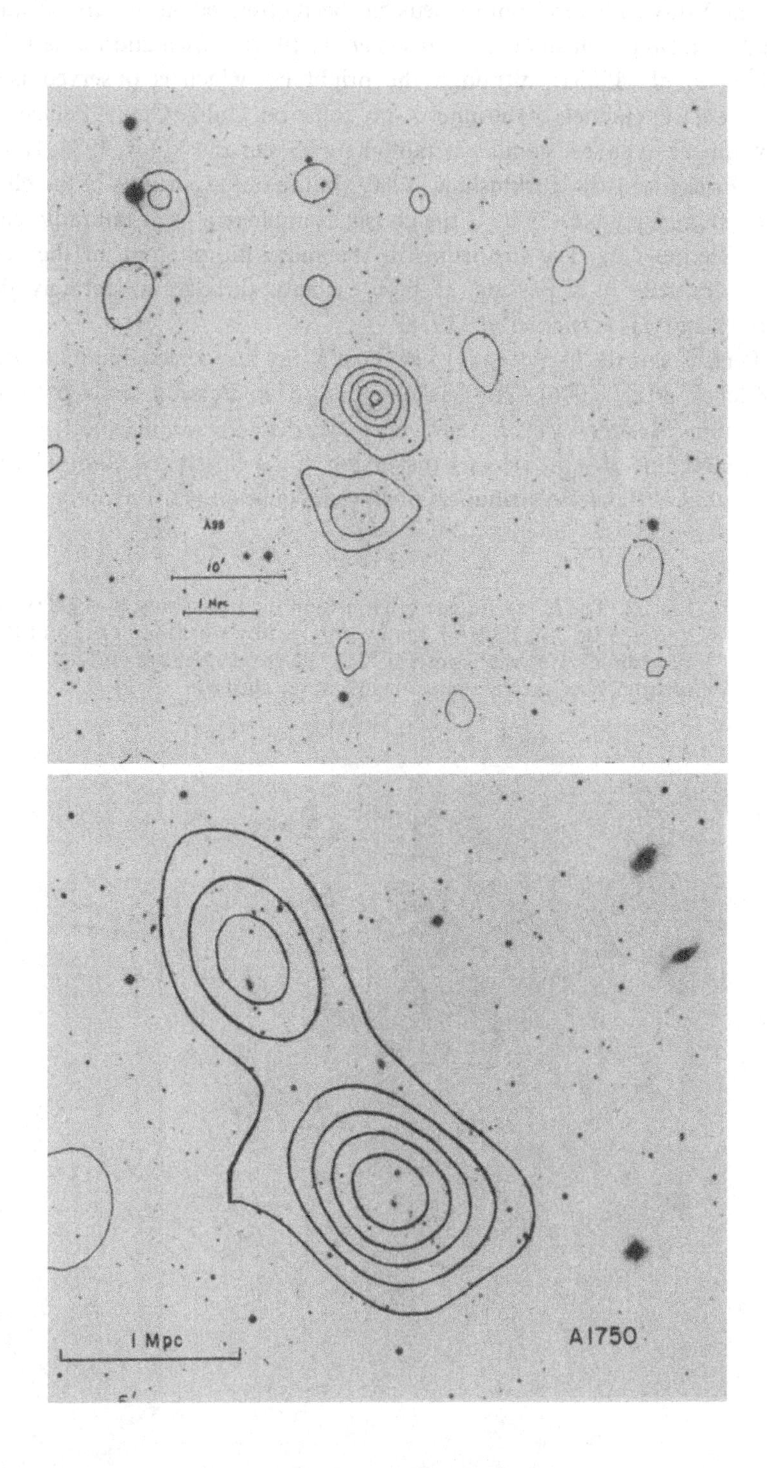
A98
10'
1 Mpc
1 Mpc
A1750

The X-ray emission from Perseus has been observed out to large distances ($\approx 2.5°$) from the cluster core (Nulsen *et al.*, 1979; Nulsen and Fabian, 1980; Ulmer *et al.*, 1980a), although the brightness which is observed is that expected from models of the inner X-ray emission. Unlike Coma, Perseus does not appear to have a significant radio halo (Gisler and Miley, 1979; Hanisch and Erickson, 1980; Birkinshaw, 1980). In the outer portions of the cluster, the radio galaxy NGC1265 is the classic example of a head-tail radio galaxy (see Section 4.3). The distortion in the radio morphology of the source indicates that it is passing at high velocity through moderately dense intracluster gas (Owen *et al.*, 1978).

Perseus was the first cluster to have an X-ray line detected in its spectrum (Mitchell *et al.*, 1976). The gas temperature in Perseus, derived from its spectrum (Mushotzky *et al.*, 1981), is smaller than one would expect, given the very large line-of-sight velocity dispersion in the cluster (equation (4.10)) or the extent of the gas distribution observed (equation (4.6)). However, recent

Fig. 19. The X-ray surface brightness of the Coma cluster of galaxies from the IPC on *Einstein*, kindly provided by Christine Jones and Bill Forman. Contours of constant X-ray surface brightness are shown superimposed on the optical image of the cluster.

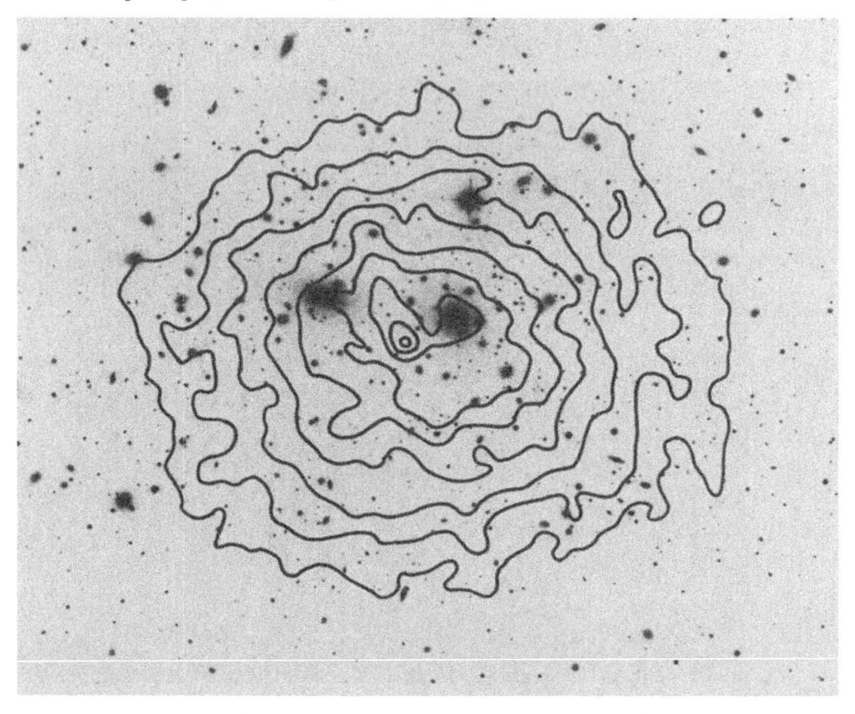

work on the galaxy spatial and velocity distribution by Kent and Sargent (1983) has reduced the discrepancy considerably.

The galaxy NGC1275 in the core of the Perseus cluster is one of the most unusual objects in the sky; it occupies a role in extragalactic astronomy

Fig. 20. An optical photograph of the spectacular galaxy NGC1275 in the Perseus cluster from Lynds (1970), copyright 1970 AURA, Inc., the National Optical Astronomy Observatories, Kitt Peak. The photograph was taken with a filter sensitive to the Hα emission line, and shows the prominent optical emission line filaments around this galaxy.

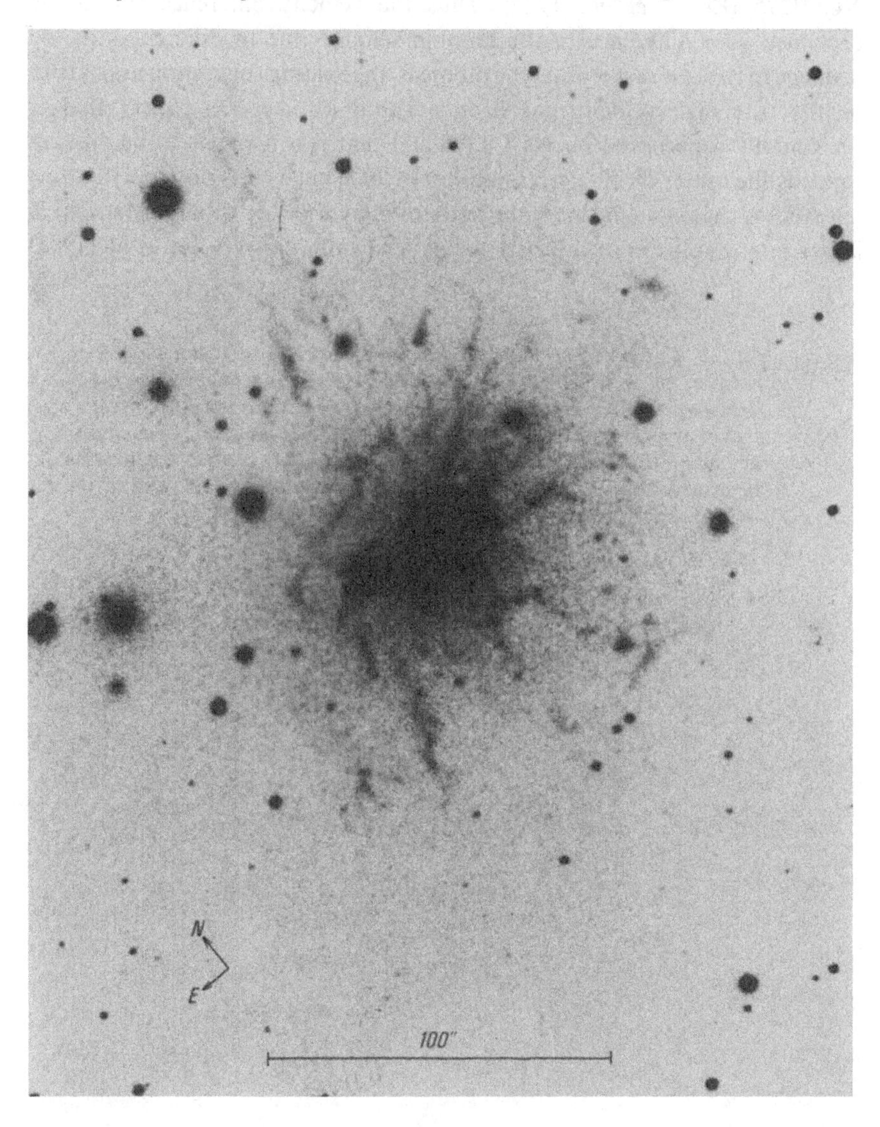

similar to the position of the Crab Nebula (which it visually resembles) in galactic astronomy. The visual appearance of the galaxy (Figure 20) is dominated by a complex network of filaments, which show an emission line spectrum (Kent and Sargent, 1979). These filaments form two distinct velocity systems (Rubin *et al.*, 1977, 1978): a 'low velocity' system with the same velocity as the stars in NGC1275, and a 'high velocity' system with a radial velocity about 3000 km/s higher. Strangely enough, 21 cm absorption line measurements indicate that the high velocity system lies between us and NGC1275 (Rubin *et al.*, 1977). Thus the velocity difference cannot be explained as a difference in the Hubble velocity due to differences in the distance to the two systems. At the moment, the leading suggestion appears to be that the high velocity gas is in a spiral or irregular galaxy that is accidentally superposed on NGC1275 and that just happens to be moving towards the center of the Perseus cluster at 3000 km/s. It is possible that this intervening galaxy is in the outer parts of the cluster or supercluster and is falling into the cluster on a nearly radial orbit. Alternatively, Hu *et al.* (1983)

Fig. 21. The X-ray surface brightness of the Perseus cluster of galaxies, observed by Branduardi-Raymont *et al.* (1981) with the IPC on the *Einstein* satellite. Contours of constant X-ray surface brightness are shown superimposed on the optical image of the cluster. The center of the galaxy distribution in the cluster is shown as a dashed circle, while the centroid of the extended X-ray emission is the ×. The peak in the X-ray surface brightness is centered on the galaxy NGC1275.

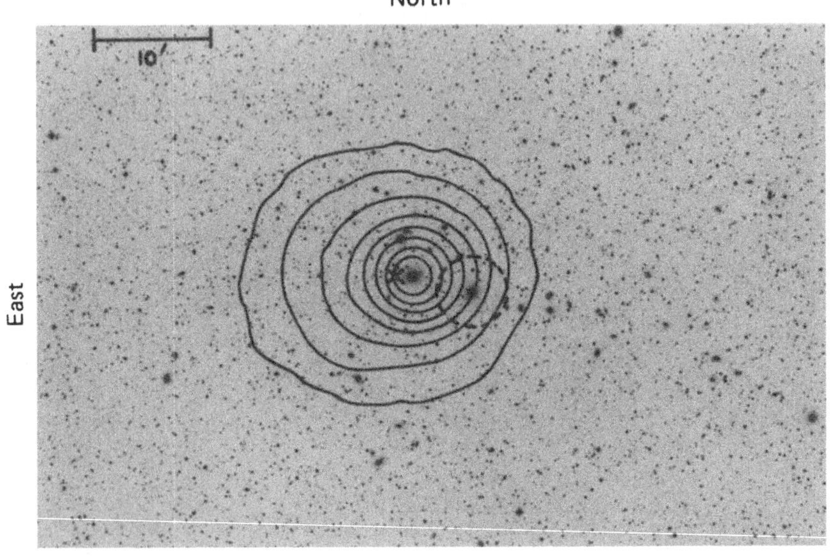

suggest that the high velocity system is a spiral galaxy that is actually colliding with the X-ray emitting gas around NGC1275 (Section 5.7.3). Unfortunately, no stellar component of this possible intervening galaxy has ever been convincingly detected (Kent and Sargent, 1979; but see Adams, 1977), and the galaxy seems to have very little neutral hydrogen, given its strong line emission (van Gorkom and Ekers, 1983).

Although the spatial distribution of stellar light from NGC1275 is that of a giant elliptical galaxy (Oemler, 1976), the galaxy has very blue colors and an A-star stellar spectrum (Kent and Sargent, 1979), suggesting ongoing or recent star formation. The nucleus of the galaxy contains a very compact, highly variable, powerful nonthermal source of radiation with a spectrum that extends from radio to hard X-rays. Because of its line emission, filaments, blue color, and nuclear emission, NGC1275 was classified as a Seyfert galaxy, although Seyferts are generally spiral galaxies and are not usually located in compact cluster cores.

The diffuse cluster X-ray emission from Perseus is very strongly peaked on the position of NGC1275 (Fabian *et al.*, 1981a; Branduardi-Raymont *et al.*, 1981; Figures 21 and 22). This is in addition to the highly variable, very hard X-ray point source associated with the nucleus of NGC1275 (Primini *et al.*, 1981; Rothschild *et al.*, 1981). The gas around NGC1275 has a positive temperature gradient $dT_g/dr > 0$ (Ulmer and Jernigan, 1978), which is not expected if the gas is hydrostatic and not cooling. High resolution X-ray spectra of the region about NGC1275 show the presence of significant quantities of cool gas in this region (Canizares *et al.*, 1979; Mushotzky *et al.*, 1981). Together, the X-ray surface brightness, X-ray spectra, and temperature gradients are best explained if gas is cooling onto NGC1275 at a rate of $\dot{M} \approx 400 M_\odot/\text{yr}$ (Section 5.7).

This cooling flow can provide an explanation of all the unusual properties of NGC1275 except the high velocity filaments. As the gas continues to cool through lower temperatures, it could produce the optical line emission seen in the low velocity filaments. The line emission becomes filamentary because the cooling is thermally unstable (Fabian and Nulsen, 1977; Mathews and Bregman, 1978; Cowie *et al.*, 1980; Section 5.7.3). If a small fraction of the accreted mass reaches the nucleus, it could power the nonthermal nuclear emission. Over the age of the cluster of $\approx 10^{10}$ yr, roughly $3 \times 10^{12} M_\odot$ would have been accreted. There is no reasonable reservoir for this much mass except in low mass star formation, which may be favored under the physical conditions in accretion flows (Fabian *et al.*, 1982b; Sarazin and O'Connell, 1983). Ongoing star formation would explain the blue color and A-star spectrum of NGC1275. Moreover, if accretion has been going on for the lifetime of the cluster, the entire luminous mass of NGC1275 might be due to accretion (Fabian *et al.*, 1982b; Sarazin and O'Connell, 1983; Wirth *et al.*, 1983).

4.5.3 *M87/Virgo*

Virgo is the nearest rich cluster to our galaxy. The cluster is quite irregular and spiral rich (de Vaucouleurs, 1961; Abell, 1975; Bautz and Morgan, 1970), although the X-ray emission comes from an elliptical-rich core surrounding the galaxy M87 (Figure 1a).

M87 is classified as a peculiar elliptical galaxy. It has fairly extended optical emission (de Vaucouleurs and Nieto, 1978), but is not a cD. This galaxy is one of the brightest radio sources in the sky. Nonthermal radio emission is observed from the nucleus and from the prominent optical jet (see Figure 23), and both the nucleus and jet also produce nonthermal X-ray emission (Schreier *et al.*, 1982; Figure 25). There is also a larger scale radio halo source

Fig. 22. A higher resolution X-ray image of the Perseus cluster emission around the galaxy NGC1275, from Branduardi-Raymont *et al.* (1981) and Fabian *et al.* (1984), using the HRI on the *Einstein* satellite. Contours of constant X-ray surface brightness are shown superimposed on the optical image of the galaxy NGC1275.

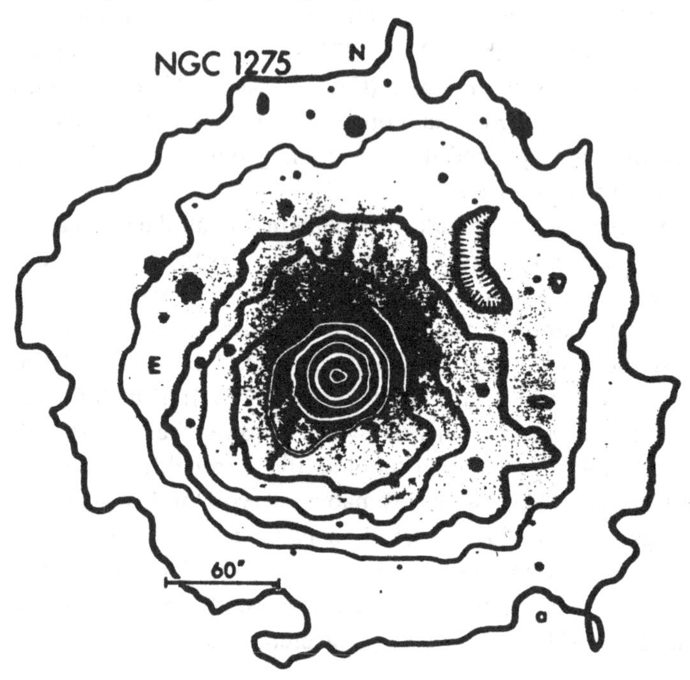

with a size comparable to that of the entire galaxy (Andernach *et al.*, 1979; Hanisch and Erickson, 1980).

The X-ray emission from this cluster is strongly concentrated to the region around M87, and is much less broadly distributed than the galaxies in the cluster (Malina *et al.*, 1976; Gorenstein *et al.*, 1977; Fabricant and Gorenstein, 1983). Figure 24 shows the *Einstein* IPC X-ray image of this cluster; Figure 25 is the HRI image. The top panel shows the outer contours, while the bottom panel is an expanded view of the center, showing X-ray emission from the jet. The emission is also considerably cooler ($T_g \approx 2.5$ keV) than is usually associated with cluster emission (Lea *et al.*, 1982). Nearly all of the emission comes from a 1° region around M87, in which more than 95% of the optical emission comes from M87. As a result, it seems reasonable to assume that the bulk of the emission is associated with M87 itself, rather than the cluster as a whole.

Similar arguments led Bahcall and Sarazin (1977) and Mathews (1978b) to suggest that the X-ray emitting gas is gravitationally bound to M87 itself.

Fig. 23. An optical photograph of the elliptical galaxy M87 in the Virgo cluster, showing the jet in the interior of the galaxy (Arp and Lorre, 1976).

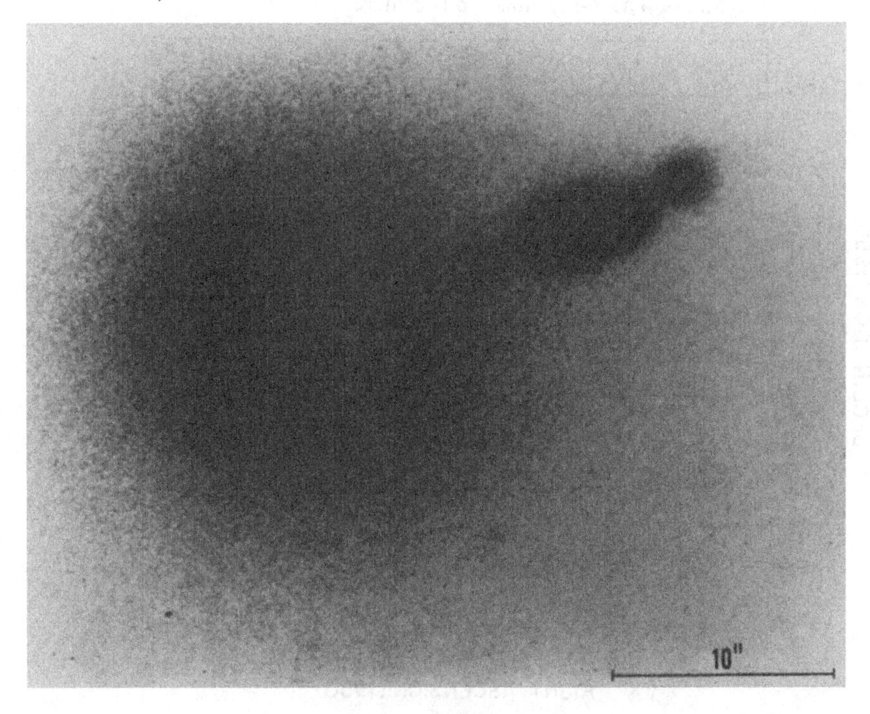

Bahcall and Sarazin showed that M87 could only bind the gas if it had a very massive halo, with a total mass of 1–6 × $10^{13}M_{\odot}$. This estimate was based on early low resolution X-ray observations. More recent *Einstein* observations (Fabricant *et al.*, 1980; Fabricant and Gorenstein, 1983; Stewart *et al.*, 1984a) appear to support this model and provide a more accurate estimate of the mass of 3–6 × $10^{13}M_{\odot}$. These observations measure, at least approximately, the temperature gradient in M87 and allow a direct determination of the mass profile of the galaxy. The massive halo must extent out to roughly 1° from M87. The optical surface brightness of the galaxy is very low in this region, and thus the mass-to-light ratio of the material making up this halo must be rather large. These observations suggest that at least this one elliptical galaxy possesses a massive, dark missing mass halo of the type discussed in Section 2.8. Whether this is typical of giant ellipticals or whether it is a consequence of M87's position at the center of the Virgo cluster is uncertain.

Alternatively, Binney and Cowie (1981) have suggested that the mass of M87 might actually be rather small. They argue that the cooler ($T_g \approx 2.5$ keV)

Fig. 24. The X-ray surface brightness of the M87/Virgo cluster of galaxies as observed by Fabricant and Gorenstein (1983) with the IPC on the *Einstein* satellite. The lines are contours of constant X-ray surface brightness. The X-ray emission is centered on M87.

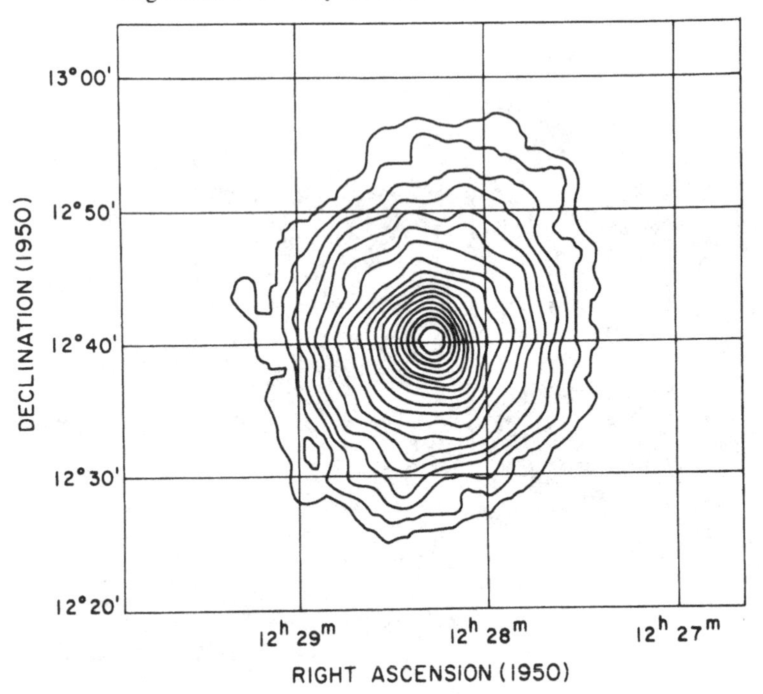

Fig. 25. A higher resolution X-ray image of the M87/Virgo cluster emission centered on the galaxy M87, from Schreier *et al.* (1982). The lines are contours of constant X-ray surface brightness. (Upper), a lower spatial resolution version, showing the peaking of the emission on the center of M87, and its asymmetrical structure. (Lower), a blowup of the center of the galaxy, showing X-ray emission along the optical jet.

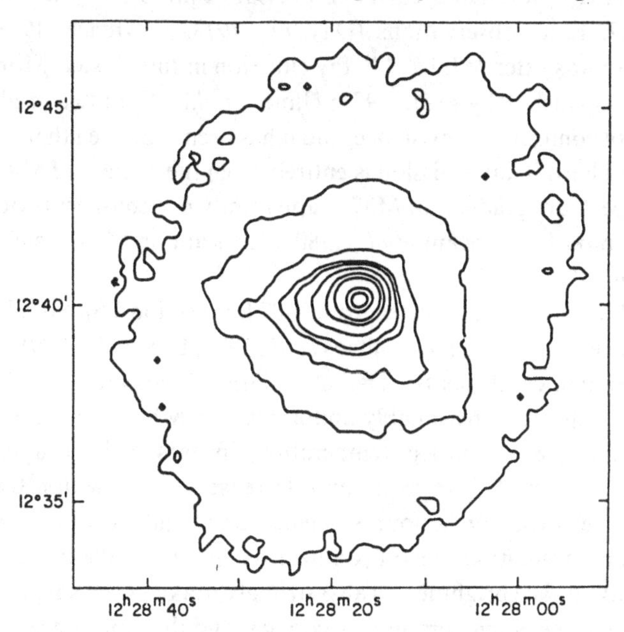

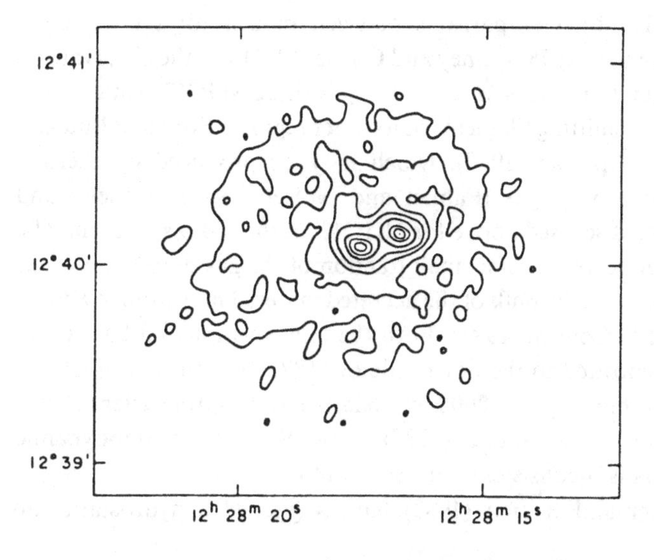

gas providing the bulk of the X-ray emission is confined by the pressure of a hotter ($T_g \approx 8$ keV), lower density, intracluster medium.

A key production of the Binney–Cowie model is the existence of a significant amount of hot gas ($T_g \approx 8$ keV) surrounding M87. The density of this gas is determined by the requirement of pressure equilibrium with the cooler gas in M87. Early observations (Davison, 1978; Lawrence, 1978) suggested that there was extended, hard X-ray emission in this cluster. More recent observations (Mushotzky *et al.*, 1977; Ulmer *et al.*, 1980a; Lea *et al.*, 1981, 1982) have not confirmed its existence, and it has been suggested that the previously observed hard X-ray emission is entirely from the nucleus of M87. The observed temperature gradient in M87 is apparently not consistent with the Binney–Cowie model (Fabricant *et al.*, 1980; Fabricant and Gorenstein, 1983; Stewart *et al.*, 1984a).

A host of X-ray lines have been detected from M87 (see Section 4.3). Both Fe L and K lines have been observed (Serlemitsos *et al.*, 1977; Lea *et al.*, 1979), as well as lines from lighter elements. The observations indicate that the abundance of these elements is reasonably uniform within M87 (Fabricant *et al.*, 1978; Lea *et al.*, 1982). The gas temperature appears to be roughly constant at projected radii of $\gtrsim 5$ arcmin, and decreases rapidly within this radius. In the inner regions, large amounts of emission by quite cool gas are observed in the higher resolution spectra (Canizares *et al.*, 1979, 1982; Lea *et al.*, 1982). The X-ray surface brightness is also strongly peaked in this region.

The presence of a range of temperatures of cool gas and the central peak in the X-ray surface brightness both suggest that the gas in M87 is radiatively cooling and being accreted onto the center of the galaxy (Gorenstein *et al.*, 1977; Mathews, 1978b). Comparisons between models for the accretion (Mathews and Bregman, 1978; Binney and Cowie, 1981) and the observations suggest that the accretion rate is $\dot{M} \approx 3\text{–}10 M_\odot$/yr. Like NGC1275 in Perseus, M87 has optical line emitting filaments in its inner regions (Ford and Butcher, 1979; Stauffer and Spinrad, 1979), which may be produced by thermal instabilities in the cooling gas (Fabian and Nulsen, 1977; Mathews and Bregman, 1978). As discussed above for NGC1275, the radio source may be powered by further accretion of a small fraction of the gas onto the nucleus (Mathews, 1978a), while the bulk of the accreted material may form low mass stars (Sarazin and O'Connell, 1983). Both the line emission and halo radio emission are concentrated to the north side of M87. De Young *et al.* (1980) suggest that M87 is moving at ≈ 200 km/s relative to the intracluster gas it is accreting, but Dones and White (1985) show that the thermodynamic structure of the gas is inconsistent with this motion.

Recently, Tucker and Rosner (1983) have suggested a hydrostatic (no accretion) model for M87. The gas in the outer portions of the galaxy is heated by the nonthermal electrons that produce the halo radio emission. Heat is conducted from these hotter outer regions into the cooler central regions, where it is radiated away; unlike Binney and Cowie (1981), Tucker and Rosner assume full thermal conductivity. The heating by nonthermal electrons balances the cooling, and the gas is in thermal equilibrium and hydrostatic. The radio source is itself powered by accretion; thus they argue that the behavior of the system is episodic, with alternating periods of accretion and nonthermal heating.

Many other galaxies in the central regions of the Virgo cluster have been detected as X-ray sources with luminosities in the range $L_x \approx 10^{39-41}$ erg/s (Forman *et al.*, 1979). Although no spectra are available for these galaxies, it is likely that the X-rays are due to thermal emission from hot gas. Figure 26 shows the X-ray emission from the very optically luminous galaxy M84 and

Fig. 26. The X-ray emission from the Virgo cluster galaxies M86 (east or left) and M84 (west or right) from Forman and Jones (1982). Contours of constant X-ray surface brightness are shown superimposed on an optical image of the galaxies. Note the plume of X-ray emission extending to the north of M86, which may indicate that this galaxy is currently being stripped of its gas.

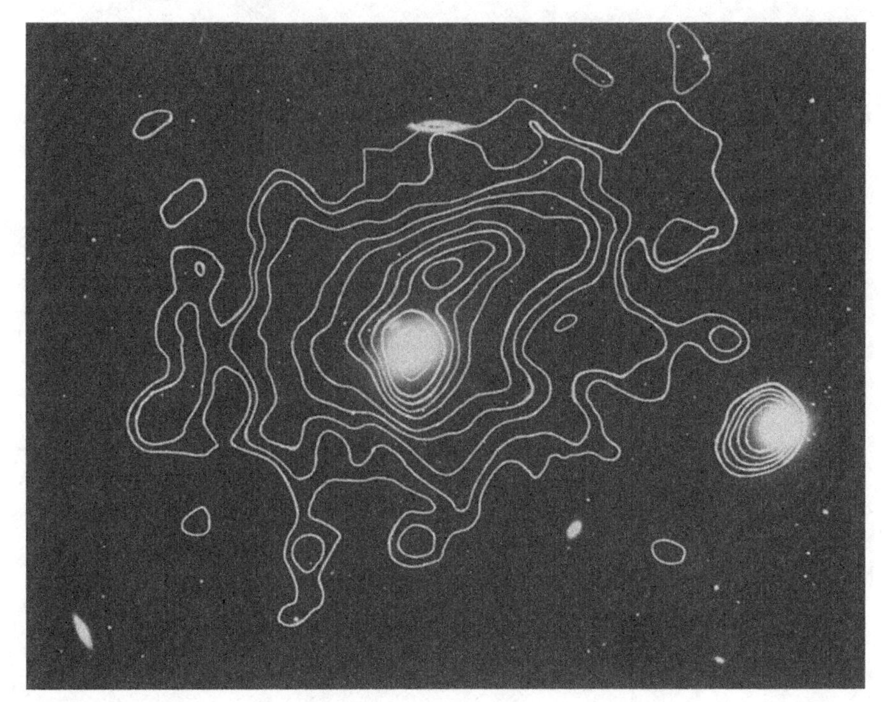

the galaxy M86, which has an X-ray plume extending away from M87. Fabian *et al.* (1980) suggest that M86 contains hot gas, which is currently being stripped by ram pressure as it moves into the core of the cluster (see Sections 2.10.2 and 5.9). The interpretation of the X-ray emission from M87 and the other galaxies in the Virgo cluster is discussed in more detail in Section 5.8.

4.5.4 A1367

A1367 is an irregular, BM type II–III cluster (Bautz and Morgan, 1970; Carter and Metcalfe, 1980) with a moderately high spiral fraction (Bahcall, 1977c). It has a low X-ray luminosity $L_x \approx 4 \times 10^{43} h_{50}^{-2}$ erg/s and a temperature $T_g \approx 3 \times 10^7$ K (Mushotzky, 1984). The X-ray emission is extended and elongated (Figure 27a); Bechtold *et al.* (1983) found an X-ray core radius (equation (4.7)) of $r_x = (0.8 \times 0.4) h_{50}^{-1}$ Mpc for the semimajor and

Fig. 27. The X-ray surface brightness of the A1367 cluster of galaxies, observed by Bechtold *et al.* (1983) with the *Einstein* satellite. Contours of constant X-ray surface brightness are shown superimposed on the optical image of the cluster. (a) The lower resolution IPC image, showing the irregular cluster emission. (b) The higher resolution HRI image, showing many discrete sources within the cluster, many of which are associated with individual cluster galaxies.

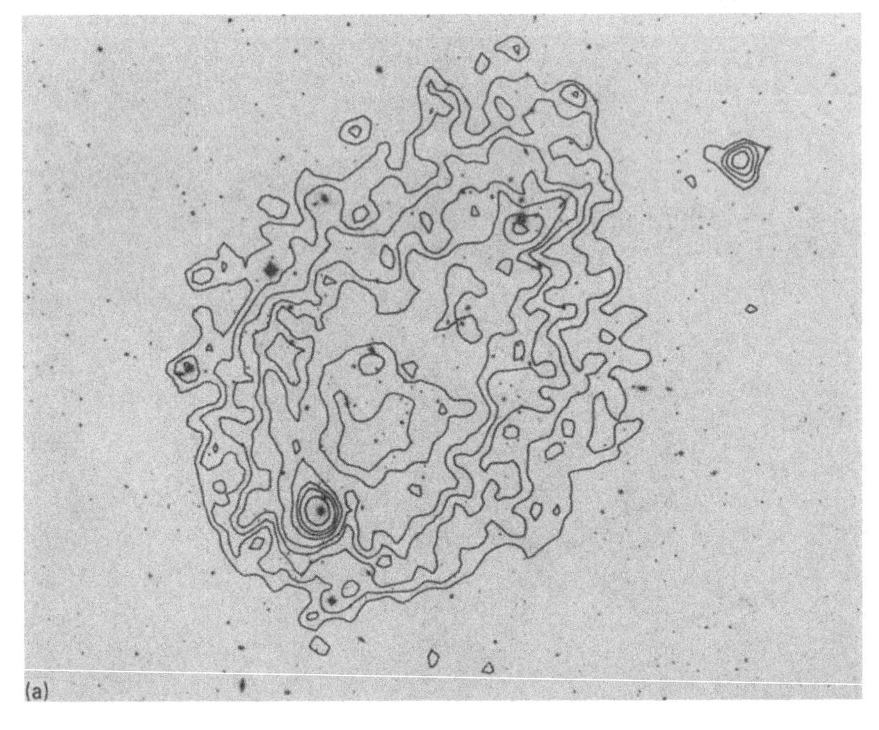

semiminor axes of the distribution. In addition, 8 point sources and 13 resolved peaks in the X-ray emission (with typical sizes of an arc minute) were observed at higher spatial resolution (Figure 27b), 11 of which were associated with cluster galaxies. These galaxies have X-ray luminosities in the range $10^{40-42} h_{50}^{-2}$ erg/s. 21 cm observations indicate that some of these galaxies also contain neutral hydrogen (Chincarini *et al.*, 1983). A1367 is the only irregular cluster that appears to have an extended radio halo (Hanisch, 1980).

4.6 X-Ray–Optical Correlations

A number of correlations between the optical and X-ray properties of clusters have been found (Bahcall, 1977a). The optical cluster properties that have been used to study X-ray clusters include the richness (Section 2.3), morphology (RS or BM type; Section 2.5), the galactic content (cD galaxies and spiral fractions; Section 2.10), the core radius r_c (or other radii; Section 2.7), the velocity dispersion σ_r (Section 2.6), and the central galaxy density $\bar{N}_o$ of Bahcall (1977b; Section 2.7). The largest X-ray surveys (Section 4.2) provide only X-ray fluxes (and thus luminosities L_x) for a given X-ray photon energy range. In addition, there are now smaller samples of clusters with X-ray

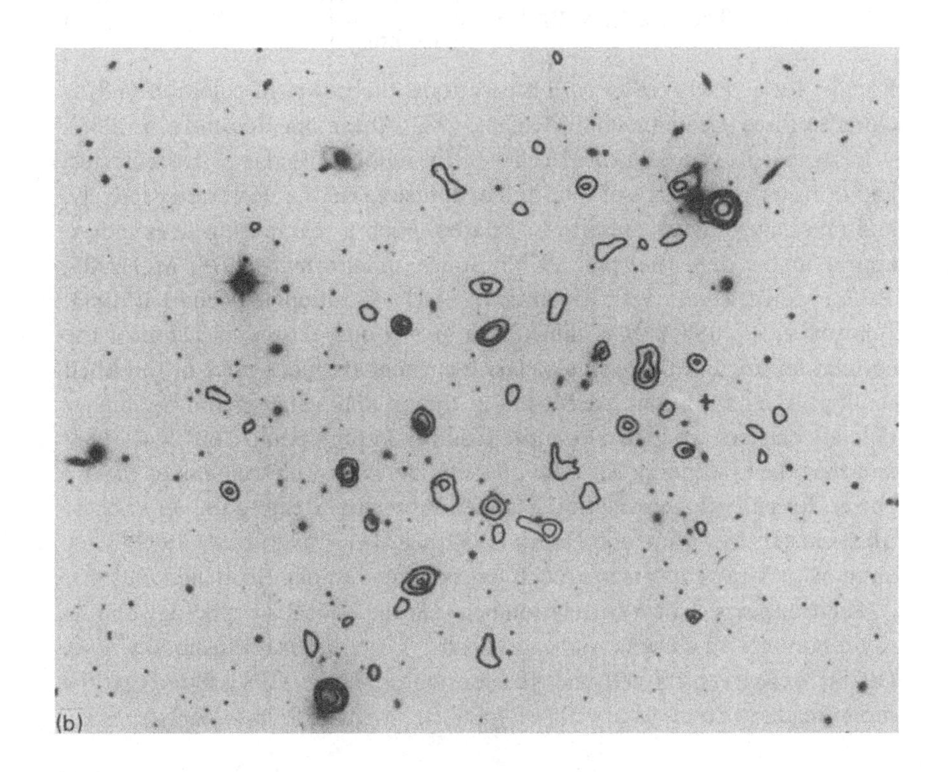

surface brightness determinations, giving r_x or β (Section 4.4), and samples with X-ray spectra, yielding T_g and EI (Section 4.3).

Solinger and Tucker (1972) first suggested that the X-ray luminosity of a cluster correlates with its velocity dispersion $L_x \propto \sigma_r^4$, based on the small sample of known X-ray clusters at that time. This sample included M87/Virgo, in which the X-ray emission comes from the galaxy M87 rather than the entire cluster, and several clusters have multiple velocity components, which probably cause the velocity dispersion to be overestimated. Solinger and Tucker gave a simple model to explain the correlation based on the assumption that the emission comes from intracluster gas. This correlation has been reexamined a number of times and other explanations of its physical significance have been given (Yahil and Ostriker, 1973; Katz, 1976; Silk, 1976). As the sample of X-ray clusters has grown, the correlation has become less convincing (McHardy, 1978a), and certainly there is a great deal of scatter about any such correlation. However, Quintana and Melnick (1982) find basically the same relationships for larger data samples from the HEAO-1 and *Einstein* observatories. Their relationship for the HEAO-1 data is shown as Figure 28; roughly, the correlation is

$$L_x(2\text{–}10\,\text{keV}) \approx 4.2 \times 10^{44}\,\text{erg/s}\left[\frac{\sigma_r}{10^3\,\text{km/s}}\right]^4. \tag{4.8}$$

For the lower energy *Einstein* observations, the power in equation (4.8) is closer to three (Quintana and Melnick, 1982; Abramopoulos and Ku, 1983).

In some sense, the richness of a cluster (the number of galaxies in the cluster) and L_x measure the mass of stars and of diffuse gas in the cluster, respectively, and one might expect these to be related. Such a relationship does indeed appear in the data (Bahcall, 1974b; Jones and Forman, 1978; McHardy, 1978a; McKee *et al.*, 1980; Ulmer *et al.*, 1981; Abramopoulos and Ku, 1983; Johnson *et al.*, 1983). It is difficult to give a quantitative measure of the correlation because the richness class 0 clusters are incomplete in the Abell catalog, Abell richnesses are binned in the original catalog, and the higher richness clusters are generally more distant. Abramopoulos and Ku (1983) find that the low energy X-ray luminosity increases with richness to the 1.2 power. Recently this correlation has been shown to extend to the very richest Abell clusters by Soltan and Henry (1983; see also Pravdo *et al.*, 1979). Low luminosity X-ray clusters are much less common among the richest clusters.

There appears to be a correlation between the optical morphology and X-ray luminosity of clusters. Bahcall (1974b), Owen (1974), Mushotzky *et al.* (1978), McKee *et al.* (1980), and Abramopoulos and Ku (1983) found that the more regular Rood–Sastry types (cD, B) in general have higher X-ray luminosities than the less regular clusters. The reality of this correlation has been disputed by Jones and Forman (1978) and Lugger (1978), although it does appear in the larger HEAO-1 and *Einstein* data samples. Bautz–Morgan type I clusters are found to be more luminous than the less regular clusters, although the correlation does not seem to continue to less regular BM types (McHardy, 1978a; McKee *et al.*, 1980; Ulmer *et al.*, 1981; Abramopoulos and Ku, 1983; Johnson *et al.*, 1983). Clusters that contain optically dominant galaxies near the cluster center tend to be stronger X-ray sources than those that do not (Bahcall, 1974b; Jones and Forman, 1984). (In the Jones–Forman classification scheme these are XD clusters; see Section 4.4).

Jones and Forman (1978) have argued that all of these X-ray–optical correlations are due to the correlation of X-ray luminosity and richness *plus* the tendency of rich clusters to be more regular (lower BM type, etc.). As the sample of X-ray clusters has enlarged, it has become possible to test this

Fig. 28. The correlation between the X-ray luminosity of clusters observed with HEAO-1 A-2 and their line-of-sight velocity dispersion, from Quintana and Melnick (1982).

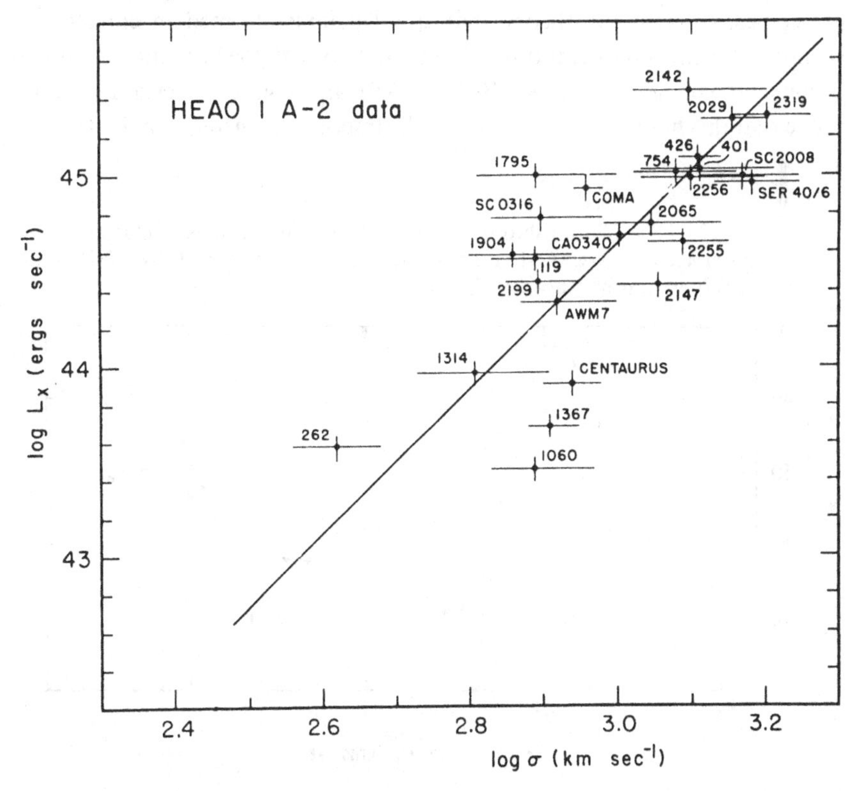

possibility by considering subsamples of fixed richness, and correlations with optical morphology appear in these samples as well (e.g. McHardy, 1978a).

Bahcall (1977b) showed that the X-ray luminosity of clusters was well-correlated with the projected central galaxy density parameter $\bar{N}_o$ (Section 2.7). This correlation is tighter than the richness correlation; an explanation of this may be that thermal X-ray emission depends on the square of the density of the gas, and thus is most sensitive to the deepest portion of the cluster potential. This correlation has been confirmed by Mushotzky *et al.* (1978), Mitchell *et al.* (1979), Abramopoulos and Ku (1983), and Mushotzky (1984). The correlation in the HEAO-1 A-2 sample (Figure 29) is consistent with $L_x \propto (\bar{N}_o)^{3.5}$, and is one power weaker in the lower energy *Einstein* sample.

Bahcall (1977c) showed that the X-ray luminosity of a cluster correlated with its galactic content, in that luminous X-ray clusters generally have a small proportion of spiral galaxies. This correlation has been confirmed by McHardy (1978a), Mushotzky *et al.* (1978), Tytler and Vidal (1979), and Abramopoulos and Ku (1983). Melnick and Sargent (1977) showed that the spiral fraction increased with radius in X-ray clusters, and that the spiral fraction was inversely correlated to the velocity dispersion. These correlations are consistent with the theory that spiral galaxies formed in clusters are stripped of their gas content through interactions with the intracluster gas and become S0 galaxies (Sections 2.10.2 and 5.9), although they certainly do not prove that this has occurred. Let f_{Sp} be the fraction of cluster galaxies that are

Fig. 29. The relationship between the X-ray luminosities of clusters observed with HEAO-1 A-2 and the central galaxy density N_o of Bahcall (1977b), from Mushotzky (1984).

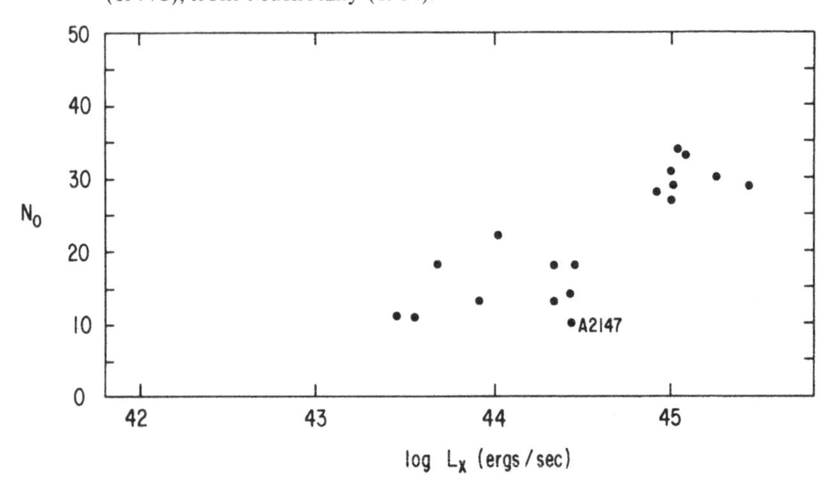

spirals. This spiral fraction is plotted against the X-ray luminosity of clusters in Figure 30. Bahcall (1977c) showed that the correlation could be represented as

$$f_{Sp} \approx 0.37 - 0.26 \log\left(\frac{L_x}{10^{44}\ \mathrm{erg/s}}\right), \tag{4.9}$$

which she derived based on a simple model for stripping of spirals in a cluster. An important counterexample to this correlation is A194, which has a low spiral fraction $f_{Sp} \approx 0.27$ (Oemler, 1974; Dressler, 1980a). Yet this cluster has a very low X-ray luminosity (Jones and Forman, 1984) and low velocity dispersion, which make it unlikely that the spirals were stripped by the ram pressure of intracluster gas.

Kellogg and Murray (1974) suggested a correlation between the sizes of the X-ray sources in clusters and the galaxy core radii, $r_x \approx 2r_c$, but very few clusters have accurately determined galaxy core radii (Section 2.7). Ulmer *et al.* (1981) found that the X-ray luminosity of clusters in the HEAO-1 A-1 survey correlated with the cluster radius R_{LV} as given by Leir and van den Bergh (1977; Section 2.7), with $L_x \propto R_{LV}^2$. This correlation was confirmed in the *Einstein* data by Abramopoulos and Ku (1983).

The OSO-8, *Ariel 5*, and HEAO-1 A-1 spiral surveys established a number of correlations between the X-ray spectral parameters and the optical

Fig. 30. The correlation between the spiral fraction in clusters and their X-ray luminosity, from Bahcall (1977c). The line shows the relation from equation (4.9).

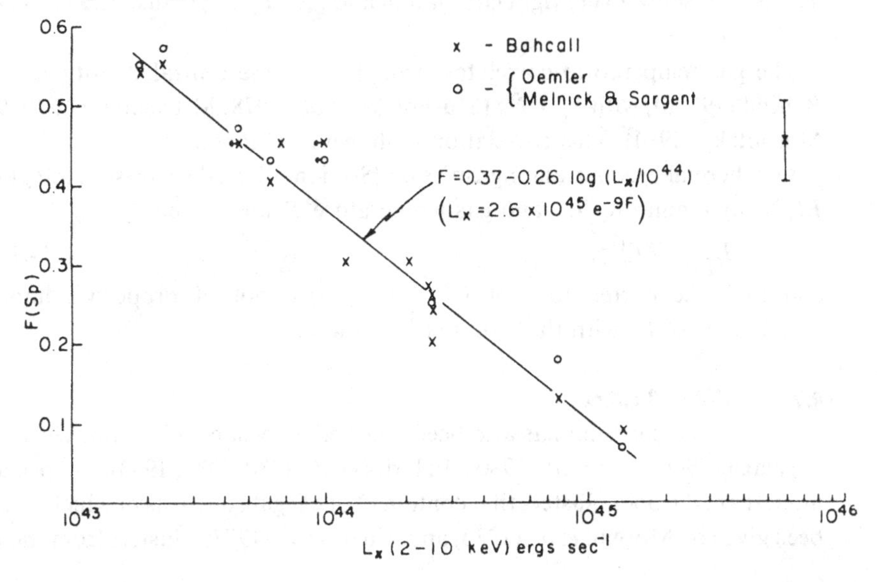

properties of X-ray clusters (Mitchell *et al.*, 1977, 1979; Mushotzky *et al.*, 1978; Smith *et al.*, 1979a; Mushotzky, 1980, 1984). The two properties that can be derived most easily from a continuum X-ray spectrum are the gas temperature T_g and the emission integral *EI* (equation (4.3)). The X-ray luminosity of a cluster is proportional to *EI* (equation (4.11)).

An important correlation was found between the gas temperatures T_g, determined from the X-ray spectra, and the velocity dispersion of the galaxies in the cluster (see Section 2.6). The older surveys (Mushotzky *et al.*, 1978; Mitchell *et al.*, 1979; Smith *et al.*, 1979a) had found that, when cD clusters were excluded, the temperatures varied roughly as

$$T_g \approx 6 \times 10^7 \,\mathrm{K}\left(\frac{\sigma_r}{10^3 \,\mathrm{km/s}}\right)^2 . \tag{4.10}$$

As will be demonstrated in Section 5.5, such a correlation would be expected if the gas were gravitationally bound to the cluster, because the velocity dispersion measures the depth of the cluster potential (equation (2.24)). Moreover, this correlation between T_g and σ_r^2 supported the view that clusters must contain large masses of unseen material (the missing mass problem, Section 2.8). The fact that the galaxy and gas velocity dispersions were similar and proportional to one another suggested that both gas and galaxies were bound by the same gravitational potential, which required a very large mass for the cluster. The cD clusters showed less of a correlation, which may indicate that in these systems there is a significant amount of gas bound to the cD itself, rather than to the entire cluster. Unfortunately, the HEAO 1 A-2 data do not show a very tight correlation and give $T_g \propto \sigma_r$ (Mushotzky, 1984; Figure 31).

The gas temperature correlates strongly with the central density $\bar{N}_o$ of Bahcall (1977b), with $T_g \propto \bar{N}_o$ (Mushotzky *et al.*, 1978; Mitchell *et al.*, 1979; Mushotzky, 1984). This correlation is shown in Figure 32.

For thermal bremsstrahlung emission (Section 5.1.3), the emission integral *EI*, X-ray luminosity L_x, and gas temperature T_g are related by

$$L_x \propto EI\, T_g^{1/2}, \tag{4.11}$$

and thus the correlations of L_x and T_g with optical properties imply correlations of *EI* with these properties as well.

4.7 Poor Clusters

X-ray emission has also been detected associated with poor clusters of galaxies (Schwartz *et al.*, 1980a, b; Kriss *et al.*, 1980, 1981, 1983). Of special interest are the poor clusters that contain D or cD galaxies, lists of which have been given by Morgan *et al.* (1975) and Albert *et al.* (1977); clusters from these two lists are identified as MKW and AWM clusters, respectively. The optical properties of these cD galaxies were discussed in Section 2.10.1, where it was concluded that the cDs in poor clusters were similar to cDs in rich clusters, except that they lacked the very extended, low surface brightness envelopes seen in rich cluster cDs (Thuan and Romanishin, 1981; Oemler, 1976; van den Bergh, 1977a). This suggests that the main bodies of cDs are formed by a process, such as mergers, which is not strongly dependent on richness, while the halos are formed by a process, such as tidal interactions, which does depend strongly on richness.

In general, the optical properties of the poor clusters appear to be simple extensions of the properties of rich clusters to lower richness (Stauffer and Spinrad, 1978; Thomas and Batchelor, 1978; Schild and Davis, 1979; Bahcall, 1980; Malumuth and Kriss, 1986). The radio properties of the poor cluster cDs are also very similar to those of rich clusters (White and Burns, 1980; Burns *et al.*, 1980, 1981b).

Fig. 31. The relationship between gas temperatures derived from X-ray spectral observations with HEAO-1 A-2 and the central velocity dispersion ($\Delta V_C = \sigma_r$ at the cluster center), from Mushotzky (1984). The lines show the predictions of polytropic models with various indices (Section 5.5.2). The dots are the data for non-cD clusters; the open circles are cD clusters.

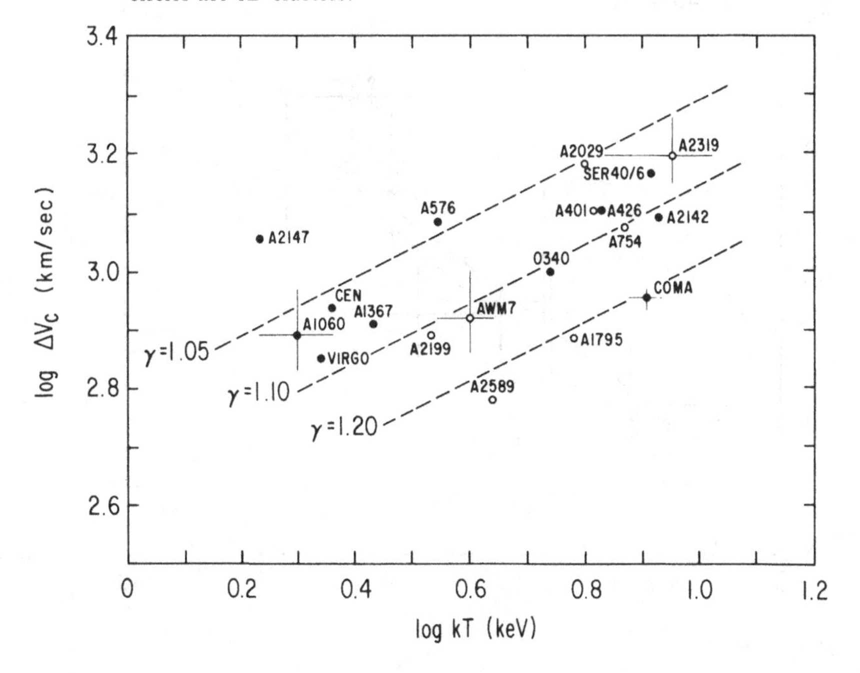

The poor clusters containing cD galaxies are generally observed to be X-ray sources with luminosities of $\approx 10^{42-44}$ erg/s (Schwartz *et al.*, 1980a, b; Kriss *et al.*, 1980, 1981, 1983; Malumuth and Kriss, 1986). Kriss *et al.* (1983) surveyed 16 MKW and AWM clusters, and detected X-ray emission from 12. In the brighter clusters, the X-ray emission was found to be smoothly distributed, relatively symmetrical, and fairly extended (out to radii of ≈ 1 Mpc). Kriss *et al.* found that the X-ray temperatures in these poor clusters were fairly low $T_g \approx 1$–5 keV, in keeping with their low velocity dispersions (see equation (4.10)). The X-ray emission is strongly peaked on the position of the cD galaxy (Canizares *et al.*, 1983; Malumuth and Kriss, 1986; see the image of AWM4 in Figure 33). This suggests that the cDs in poor clusters are located at the bottoms of cluster potential wells, as was shown to be the case for rich cluster cDs (Section 2.10.1). In many ways, the bright poor cluster X-ray sources resemble the regular XD clusters discussed above. The X-ray emission in a number of cases is elongated in the same direction as the long axis of the cD

Fig. 32. The correlation between the gas temperatures derived from X-ray spectral measurements with HEAO-1 A-2 and the central galaxy density of Bahcall (1977b), from Mushotzky (1984).

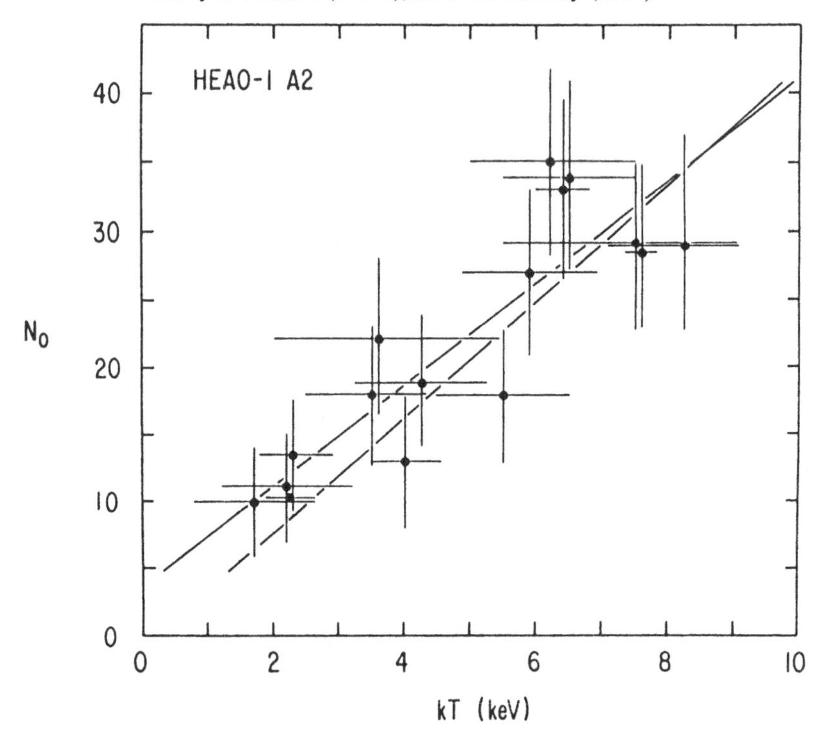

galaxy (Kriss *et al.*, 1983; a similar effect is seen in rich clusters, as discussed in Section 2.10.1 and 4.5).

From the distribution of the gas in the cluster, the cluster gravitational potential and mass can be derived (Kriss *et al.*, 1983; Malumuth and Kriss, 1986). When these are compared to the optical luminosity of the clusters, a considerable missing mass problem is found. The mass-to-light ratios in the inner parts of these clusters are found to be roughly $M/L_V \approx 100h_{50}M_{\odot}/L_{\odot}$. The X-ray emitting gas makes up roughly 15% of the total cluster mass in these inner regions. Because the cD galaxy contributes a large fraction of the luminosity in these poor clusters, this dark matter can also be thought of as a massive halo around the central cD, as in the M87/Virgo cluster.

Fig. 33. The X-ray emission from the AWM4 poor cluster of galaxies, from Kriss *et al.* (1983) with the IPC on the *Einstein* satellite. Contours of constant X-ray surface brightness are shown superimposed on the optical image of the cluster. The emission is centered on the cD galaxy.

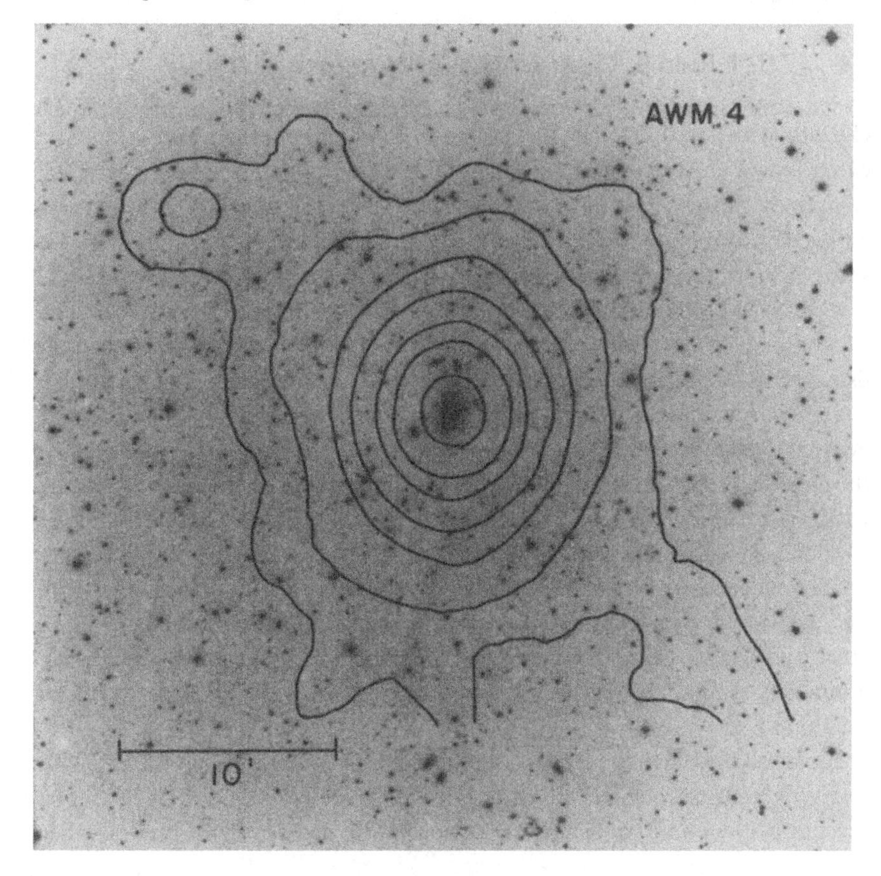

The X-ray emission is strongly peaked on the position of the cD in these poor clusters (Canizares *et al.*, 1983; Kriss *et al.*, 1983; Malumuth and Kriss, 1986), and the cooling times in the central portions of the gas are generally estimated to be less than the probable age of the cluster (the Hubble time). This suggests that the X-ray emitting gas forms a cooling accretion flow onto the cD (Canizares *et al.*, 1983), as has been observed in many rich clusters. The accretion rates in MKW4, MKW3s, AWM4, and AWM7 are estimated to be in the range of 5–100 $M_\odot$/yr (Canizares *et al.*, 1983; Malumuth and Kriss, 1986). It is possible that star formation from the accretion flow may provide a portion of the optical luminosity of the cD (Fabian *et al.*, 1982b; Sarazin and O'Connell, 1983). Very little cold gas (neutral hydrogen) is detected in these cDs (Burns *et al.*, 1981a; Valentijn and Giovanelli, 1982).

Poor clusters containing head-tail radio sources have also been detected in X-rays (Burns and Owen, 1979; Holman and McKee, 1981). Head-tail radio sources are believed to be produced by galaxy motions through intracluster gas (Section 4.3).

4.8 High Redshift Clusters and X-Ray Cluster Evolution

Observations of high redshift X-ray clusters (taken here to mean clusters with $z \geqslant 0.2$) offer the possibility of studying the formation and evolution of clusters. At the moment this study is hampered by the small sample of clusters that have been observed and the difficulty in obtaining detailed information (spectra and spatial distributions) for these distant sources. Samples of high redshift X-ray clusters have been given by Henry *et al.* (1979, 1982), Helfand *et al.* (1980), Perrenod and Henry (1981), White *et al.* (1981a, b, 1987), and Henry and Lavery (1984). These samples are based on lists of clusters detected initially through optical or radio emission.

One purpose of this study is to determine whether X-ray clusters have evolved in any detectable way over the cosmological time span represented by the redshift. Henry *et al.* (1979) compared a sample of six high redshift clusters with more nearby clusters, and concluded that they were similar in many ways. They found marginal evidence that X-ray luminosities increase with time in accordance with the models of Perrenod (1978b) and Sarazin (1979), but the sample was too small to make any strong statements. Henry *et al.* (1982) and Henry and Lavery (1984) studied the evolution of the X-ray luminosity function at redshifts $z < 0.5$. No significant evidence for evolution was found.

Perrenod and Henry (1981) estimated gas temperatures for a sample of seven high redshift clusters observed with the *Einstein* observatory. With one exception (the remarkable cluster 0016 + 16 discussed below), the six other clusters with $z > 0.3$ all had $T_g < 4$ keV. Nearby clusters of similar X-ray luminosity have average temperatures of $T_g \approx 7$ keV. Thus it appeared that gas temperatures in clusters might be increasing with time. Since the temperature of gas in a cluster reflects in part the depth of the cluster potential well (Section 5.5), this suggested that clusters might be growing through the merger of smaller clusters, in accord with models developed by Perrenod (1978b). White *et al.* (1981a) also detected a fairly distant cluster (SC2059-247) with a very high X-ray luminosity and a low X-ray temperature. However, White *et al.* (1987) observed a sample of 10 high redshift clusters; of the five detected cluster X-ray sources, all have reasonably high X-ray temperatures $T_g \geqslant 6$ keV. They suspect that the previous evidence for low X-ray temperatures by Perrenod and Henry was an artifact of the small size of the sample being discussed. They also point out that some high luminosity, low temperature clusters, such as SC2059-247, may be extreme examples of clusters with cooling accretion flows at their centers.

Several of the high redshift clusters that have been observed as X-ray sources (Henry *et al.*, 1979, 1982; Helfand *et al.*, 1980) are Butcher–Oemler clusters (Butcher and Oemler, 1978a). As discussed in Section 2.10.2, these are high redshift clusters that appear to contain blue galaxies with colors similar to those of spirals in nearby clusters. Prior to the X-ray observations, the most straightforward explanation of these clusters was that they did not contain enough gas to strip spiral galaxies effectively by ram pressure ablation (Norman and Silk, 1979; Sarazin, 1979). Obviously, the detection of large quantities of hot gas in these clusters has made that explanation untenable. Recently optical observations indicate that the blue galaxies in these clusters are not normal spirals and are unlike any class of present day galaxies (Dressler and Gunn, 1982).

The most spectacular high redshift cluster to be detected so far is probably the $z = 0.54$ cluster 0016 + 16 (White *et al.*, 1981b). Unfortunately, the cluster field also contains a foreground cluster at $z = 0.30$ (Ellis *et al.*, 1985), which probably affected early estimates of the richness of the $z = 0.54$ cluster and of the colors of its galaxies (Koo, 1981). Assuming the emission is entirely due to the higher redshift cluster, 0016 + 16 has one of the highest X-ray luminosities observed $L_x = 3 \times 10^{45} h_{50}^{-2}$ erg/s, and may have a high temperature, although this is very uncertain. The cluster is probably about as rich as Coma (Ellis *et al.*, 1985; Koo, 1981). The cluster 0016 + 16 may also be similar to nearby rich clusters in that it appears to have predominantly red galaxies (it does not show the Butcher–Oemler effect), although the foreground cluster makes this conclusion somewhat uncertain. A microwave decrement (Section 4.5) of -1.4 mK has been detected from the cluster (Birkinshaw *et al.*, 1981a). This effect has not been detected in many nearby clusters, and if the detection were confirmed it would demonstrate the extreme luminosity and temperature of 0016 + 16.

The tentative conclusion at this point is that the X-ray properties of clusters do not appear to evolve dramatically to redshifts of $z \approx 0.5$, and that evolution of the X-ray properties is probably not the explanation of the Butcher–Oemler effect.

5

THEORETICAL PROGRESS

5.1 Emission Mechanisms

When clusters of galaxies were found to be an important class of X-ray sources, there were a number of suggestions as to the primary X-ray emission mechanism. The three most prominent ideas were that the emission resulted from thermal bremsstrahlung from a hot diffuse intracluster gas (Felten *et al.*, 1966), or that the emission resulted from inverse Compton scattering of cosmic background photons up to X-ray energies by relativistic electrons within the cluster (Brecher and Burbidge, 1972; Bridle and Feldman, 1972; Costain *et al.*, 1972; Perola and Reinhardt, 1972; Harris and Romanishin, 1974; Rephaeli, 1977a, b), or that the emission was due to a population of individual stellar X-ray sources, like those found in our galaxy (Katz, 1976; Fabian *et al.*, 1976). Subsequent observations have provided a great deal of support for the thermal bremsstrahlung model, and have generally not supported the other two suggestions. These three emission mechanisms will be briefly reviewed, and a few of the arguments in favor of thermal bremsstrahlung and against the other two models will be cited.

5.1.1 Inverse Compton Emission

As noted in Section 3.2, many X-ray clusters also have strong radio emission, and cluster radio emission is distinguished by having a large spectral index α_r. This led to the idea that the X-ray emission might be produced by the same relativistic electrons that produce the synchrotron radio emission. Very high energy electrons, with very short lifetimes, would be required to produce the emission by synchrotron emission, but much lower energy electrons could scatter low energy background photons up to X-ray energies (Felten and Morrison, 1966). If the electrons are extremely relativistic, with energies $E_e = \gamma m_e c^2$ for $\gamma \gg 1$, and the initial frequency of the background photon is ν_b, then on average the frequency after the scattering will be

$$\nu_x = \frac{4\gamma^2 \nu_b}{3}. \tag{5.1}$$

Because the cluster radio spectra vary as power-laws, it is generally assumed that the relativistic electrons have a power-law energy distribution

$$N_e(\gamma)d\gamma = N\gamma^{-p}d\gamma \qquad \gamma_l < \gamma < \gamma_u, \tag{5.2}$$

where $N_e d\gamma$ is the number of electrons with energies between γ and $\gamma + d\gamma$. Then, the resulting inverse Compton (IC) radiation has a spectrum given by

$$\frac{dL_x}{d\nu_x} = 3\sigma_T ch 2^p N \frac{p^2 + 4p + 11}{(p+3)^2(p+1)(p+5)} \times \left[\int \nu_b^{\alpha_x} n_b(\nu_b) d\nu_b \right] \nu_x^{-\alpha_x}, \tag{5.3}$$

where L_x is the X-ray luminosity, n_b is the number density of background photons as a function of frequency ν_b, σ_T is the Thomson cross section, and the X-ray spectral index is

$$\alpha_x = \frac{p-1}{2}. \tag{5.4}$$

A source of low energy photons that is present everywhere and dominates the overall photon density in the universe is the 3 K cosmic background radiation. If the background is taken to be a blackbody radiation field at a temperature T_r, then

$$\frac{dL_x}{d\nu_x} = \frac{3\pi\sigma_T}{h^2 c^2} b(p) N (kT_r)^3 \left(\frac{kT_r}{h\nu_x} \right)^{\alpha_x} \tag{5.5}$$

for frequencies in the range $\gamma_l^2 \ll (h\nu_x/kT_r) \ll \gamma_u^2$. Here

$$b(p) = \frac{2^{p+3}(p^2 + 4p + 11)\Gamma[(p+5)/2]\zeta[(p+5)/2]}{(p+3)^2(p+1)(p+5)}, \tag{5.6}$$

and Γ and ζ are the gamma and Riemann zeta functions.

Now, the synchrotron radio luminosity L_r produced by the same electron population is given by

$$\frac{dL_r}{d\nu_r} = \frac{8\pi^2 e^2 \nu_B N}{c} a(p) \left(\frac{3\nu_B}{2\nu_r} \right)^{\alpha_r}, \tag{5.7}$$

where $\nu_B \equiv (eB/2\pi m_e c)$ is the gyrofrequency in the magnetic field B, and

$$a(p) = 2^{(p-7)/2} \left(\frac{3}{\pi} \right)^{1/2} \frac{\Gamma\left(\frac{3p-1}{12} \right) \Gamma\left(\frac{3p+19}{12} \right) \Gamma\left(\frac{p+5}{4} \right)}{(p+1)\Gamma\left(\frac{p+7}{4} \right)}. \tag{5.8}$$

Equation (5.8) assumes that the electron distribution is isotropic and that the frequency is in the range $\gamma_l^2 \ll (\nu_r/\nu_B) \ll \gamma_u^2$. The radio spectral index α_r is equal to the X-ray spectral index α_x.

Thus the synchrotron radio and IC X-ray emission from the same relativistic electrons will have the same spectral shape, and the luminosities in the two spectral regimes are also very simply related:

$$\frac{L_x}{L_r} = \frac{U_r}{U_B}, \tag{5.9}$$

where U_r and $U_B = (B^2/8\pi)$ are the energy densities of the background radiation and the magnetic field, respectively. Because fluxes (f_x, f_r) at single frequencies are easier to measure than integrated luminosities, it is useful to note that (in cgs units)

$$\left(\frac{f_x}{f_r}\right)\left(\frac{\nu_x}{\nu_r}\right)^{\alpha_x} = \frac{2.47 \times 10^{-19} T_r^3 b(p)}{Ba(p)} \left(\frac{4960 T_r}{B}\right)^{\alpha_x}. \tag{5.10}$$

Since the temperature of the cosmic background radiation is known, the ratio of radio and IC X-ray fluxes determines the magnetic field strength (Harris and Romanishin, 1974). If the X-ray emission from clusters were due to inverse Compton emission, the magnetic field would typically be $B \approx 0.1\ \mu$G. Fields at least ten times larger than this are expected if the relativistic electrons and the magnetic field have equal energy densities; the small magnetic fields required by the IC model would increase the energy requirements for cluster radio sources by about two orders of magnitude. On the other hand, if the X-rays are not due to IC emission, then this value is a lower limit to the magnetic field strength in the radio emitting region (Section 4.3.1).

For a magnetic field of $B \approx 1\ \mu$G, radio photons at a frequency of 1 GHz are produced by electrons with $\gamma \approx 2 \times 10^4$. Similarly, IC X-ray emission at a photon energy of 1 keV is produced by electrons with $\gamma \approx 10^3$. Thus the electrons that could produce the observed X-ray emission would produce very low frequency radio emission, typically at a frequency of 20 MHz if $B \approx 1\ \mu$G, and at an even lower frequency if B is lower as required in this model. Such low frequency radio emission generally cannot be detected from the Earth. However, one argument in favor of the IC model of cluster X-ray emission was that cluster radio sources have steep spectra (Section 3.1), indicating that they have many lower energy relativistic electrons, as required to explain their X-ray emission.

There is now considerable evidence against the IC model. In this model, one would expect a very strong correlation between the low frequency radio flux and the X-ray flux (equation (5.10)). The reality of any radio–X-ray correlation is questionable (Section 3.2), and recent larger samples of X-ray clusters from HEAO-1 and *Einstein* do not support such a strong correlation. In the IC model, the radio and X-ray emission would come from identical spatial regions and would therefore have identical distributions on the sky. However, while the X-ray emission is extended and diffuse (Section 4.4), the radio emission comes primarily from individual radio galaxies. Only a very small fraction of clusters appear to have significant diffuse radio halo emission (Section 3.4). The IC model predicts that clusters have power-law X-ray spectra, which is not consistent with the best X-ray spectral observations (Section 2.3.1); one clear distinction is that a power-law provides much more high energy X-ray emission than an exponential thermal spectrum (Section 5.1.3), and such emission is either not seen or very weak. The IC model would not produce any of the X-ray line emission that is seen universally in clusters that have been observed with reasonable sensitivity (Sections 4.3.2 and 4.3.3). The magnetic field is required to be much weaker than would be favored by radio observations. The distorted radio morphologies in clusters would not be explained in this model. Finally, most of the X-ray–optical correlations could not be accounted for naturally in the IC model.

5.1.2 Individual Stellar X-Ray Sources

Katz (1976) suggested that the X-ray emission from clusters was produced by a large number of individual cluster stellar X-ray sources, located either in galactic haloes or in intracluster space. He suggested that these sources might be similar to the binary X-ray sources found in our galaxy, and assumed that the X-ray luminosity to virial mass ratio would be the same for our galaxy and for clusters of galaxies. This predicts too little X-ray emission from clusters (see also Felten *et al.*, 1966); however, globular clusters within our galaxy have a much higher X-ray luminosity to mass ratio, and so they might be used as a model for cluster X-ray sources (Fabian *et al.*, 1976). In fact, M87 and some other central dominant galaxies are known to possess very extensive systems of globular star clusters.

Katz argued that the observed X-ray luminosities of clusters were consistent (in 1976) with a fixed X-ray luminosity to virial mass ratio; the current data would appear to rule out such a correlation (Section 4.6). It is difficult to see how this model could produce the variation in X-ray emission properties of clusters, or the optical–X-ray and radio–X-ray correlations. Of course, if the properties of the individual stellar X-ray sources are not constrained in any manner, any X-ray observations could be explained but the model has no predictive power. The granularity in the cluster X-ray emission produced by these individual sources should have been detected in M87/Virgo, but has not been seen (Schreier *et al.*, 1982). Finally, we note that luminous stellar X-ray sources are generally very optically thick. As a result, they do not show any X-ray line emission except for the fluorescence line of low ionization iron (which is probably produced by reprocessing of X-ray radiation in other parts of the binary star system). In particular, they do not show lines from highly ionized iron (Fe^{+24} and Fe^{+25}) or from lighter elements (for which the fluorescent yields are low). Clusters, on the other hand, do show strong lines from Fe^{+24} and Fe^{+25}, as well as lines from lighter elements (Sections 4.3.2 and 4.3.3).

In short, there is no real evidence in favor of the individual stellar source model, and considerable evidence against it.

5.1.3 *Thermal Bremsstrahlung from Intracluster Gas*

Felten *et al.* (1966) first suggested that the X-ray emission from clusters (the Coma cluster in particular) was due to diffuse intracluster gas at a temperature of $T_g \approx 10^8$ K and an atomic density of $n \approx 10^{-3}\,\text{cm}^{-3}$. (Unfortunately, the early X-ray detection of Coma that they sought to explain was spurious.) At such temperatures and densities, the primary emission process for a gas composed mainly of hydrogen is thermal bremsstrahlung (free–free) emission. The emissivity at a frequency ν of an ion of charge Z in a plasma with an electron temperature T_g is given by

$$e_\nu^{ff} = \frac{2^5 \pi e^6}{3 m_e c^3} \left(\frac{2\pi}{3 m_e k} \right)^{1/2} Z^2 n_e n_i g_{ff}(Z, T_g, \nu) \times T_g^{-1/2} \exp(-h\nu/kT_g) \qquad (5.11)$$

where n_i and n_e are the number density of ions and electrons, respectively. The emissivity is defined as the emitted energy per unit time, frequency, and volume V,

$$\varepsilon_\nu \equiv \frac{dL}{dV d\nu}. \qquad (5.12)$$

The Gaunt factor $g_{ff}(Z, T_g, \nu)$ corrects for quantum mechanical effects and for the effect of distant collisions, and is a slowly varying function of frequency and temperature given in Karzas and Latter (1961) and Kellogg *et al.* (1975). If the intracluster gas is mainly at a single temperature, then equation (5.11) indicates that the X-ray spectrum should be close to an exponential of the frequency. In fact, the observed X-ray spectra are generally fit fairly well by this equation (Section 4.3.1), with gas temperatures of 2×10^7 to 10^8 K. This equation predicts that the emission from clusters should fall off rapidly at high frequencies, as is observed.

As first noted by Felten *et al.* (1966), if the intracluster gas either came out of galaxies or fell into the cluster from intergalactic space, it would have a temperature such that the typical atomic velocity was similar to the velocity of galaxies in the cluster. That is,

$$\frac{kT_g}{\mu m_p} \approx \sigma_r^2, \tag{5.13}$$

where μ is the mean molecular weight in amu, m_p is the proton mass, and σ_r is the line-of-sight velocity dispersion of galaxies in the cluster. If the gas came out of galaxies, it would have the same energy per unit mass as the matter in galaxies. If it fell into clusters, it has roughly the same velocity dispersion because it responds to the same gravitational potential as the galaxies. Finally, if the gas were heated so that its initial energy per mass were much greater than that given by equation (5.13), it would not be bound to the cluster and would escape as a wind. Even in this situation, most of the extra energy would go into the kinetic energy of the wind (Section 5.6), and the gas temperature would stay within a factor of 3 or so of equation (5.13). Thus this model predicts

$$T_g \approx 7 \times 10^7 \, \mathrm{K} \left(\frac{\sigma_r}{1000 \, \mathrm{km/s}} \right)^2 \tag{5.14}$$

for a solar abundance plasma, in reasonable agreement with the temperatures required to explain the X-ray spectra (see equation (4.10)).

If the intracluster gas is in hydrostatic equilibrium in the cluster potential, and if the gas temperature is given roughly by equation (5.14), then the spatial distribution of the gas and galaxies will be similar. In fact, the observed X-ray surface brightness of clusters is similar to the projected distribution of galaxies, with the gas distribution being slightly more extended in the inner regions (Section 4.4). The thermal emission model thus explains the extent of the X-ray emission from clusters (more detailed models and comparisons are given in Section 5.5). Moreover, gas at these temperatures and densities has a very long cooling time, and the time for sound waves to cross a cluster is much less than its probable age. These conditions suggest that the gas distribution will be smooth (at least, the pressure will be smooth), as is observed (Jones and Forman, 1984).

Given the extent of the X-ray emission, the gas temperature derived from the spectrum, and the observed X-ray flux, equation (5.11) gives the atomic number density if the emission is assumed to come mainly from hydrogen. As discussed in Section 4.3, the required mass of gas is less than the total virial mass of the cluster, as required for consistency with this model for the emission. Moreover, the derived densities are very similar to those required to explain the distortions of radio sources in clusters of galaxies (Section 3.3) as the result of ram pressure by intracluster gas.

Whether the intracluster gas came out of galaxies or fell into the cluster, the mass of this gas should increase as the total mass of the cluster increases, and as a result the X-ray luminosity should increase with the virial mass. Solinger and Tucker (1972) first pointed out that the X-ray luminosity (L_x) does appear to increase with the velocity dispersion of clusters, in a way consistent with a constant fraction of the total cluster mass being intracluster gas. They found $L_x \propto \sigma_r^4$, and were able to predict X-ray emission successfully in a number of clusters based on this relationship. As discussed in Section 4.6, there are many other correlations observed between L_x or T_g and optical properties such as richness, velocity dispersion, and projected central galaxy density $\bar{N}_o$. These correlations are generally consistent with the hypothesis that the mass of intracluster gas increases with the virial mass of the cluster, and that the temperature of the gas increases with the velocity dispersion (depth of the cluster gravitational potential well). There is also an inverse corrrelation between L_x and the fraction of spirals in the cluster, which may reflect the stripping of spiral galaxies by the intracluster gas. All of these optical–X-ray correlations have natural explanations if the cluster X-ray emission is due to intracluster gas, but not if it is due to the other two mechanisms discussed above.

The clearest evidence in favor of the thermal bremsstrahlung model is the detection of strong X-ray line emission from clusters (Sections 4.3.2 and 4.3.3). The strong 7 keV Fe line emission observed from clusters is very difficult to reconcile with any nonthermal model for the origin of the X-ray emission (Mitchell *et al.*, 1976; Serlemitsos *et al.*, 1977; Bahcall and Sarazin, 1977). This emission occurs naturally in the intracluster gas model if the abundances of heavy elements are roughly solar (see Section 5.2.2 below), while nonthermal emission processes would not directly produce any significant line emission (Sections 5.1.1 and 5.1.2). It is possible, of course, that the line emission arises from some different source than the generally distributed cluster X-ray emission; the line observations that have so far been made do not have sufficient spatial resolution to determine the location of the line emission region (Ulmer and Jernigan, 1978). However, the fact that the required *abundance* of iron is roughly constant from cluster to cluster (Section 4.3.2 and Figure 13) argues strongly against such a possibility. The clusters that have been observed differ widely in their properties, ranging from regular to irregular clusters, high X-ray luminosity to low luminosity clusters, clusters with strong radio sources to clusters without such sources, clusters with hard X-ray spectra to clusters with softer spectra, and so on. In fact, the strengths of the lines also vary widely; it is only the required abundances that are nearly constant. It is very unlikely that two independent sources for the line and continuum would vary simultaneously from cluster to cluster so as to maintain the appearance of a constant Fe abundance. Moreover, if the 7 keV Fe line were from a source that did not contribute significantly to the continuum X-ray emission, the iron abundance in this source would have to be vastly greater than solar. Such large iron abundances are certainly not common in astrophysical systems. Thus it appears very unlikely that the line emission and continuum emission could come from different sources; since the line emission is thermal, the continuum emission must be as well.

The nearly solar iron abundance in the intracluster gas, which has been derived from the 7 keV iron line, suggests that a significant portion of this gas has been processed in and ejected from stars. The total mass of X-ray emitting gas is very large, probably at least as large as the total mass in galaxies in a typical X-ray cluster (see Section 4.4.1). Because of this large mass, the intracluster gas was initially believed to be primordial gas, which had never been bound in stars or even in galaxies (Gunn and Gott, 1972). During the formation of the universe (the big bang), hydrogen and helium were formed, but it is generally believed that no heavier elements could have been produced because of the lack of any stable isotopes with atomic weights of 5 or 8 (see, for example, Weinberg, 1972). The only sources which have been suggested for the formation of iron involve processing in stars. Moreover, the observed abundance in the intracluster gas is nearly the same as that in the solar system. The solar system is a second generation stellar system (the Sun is a Population I star). Thus it is possible that much of the intracluster gas has been processed through stars.

At the present epoch, there is no significant population of stars known that is not bound to galaxies. This may indicate that part of the intracluster gas may originally have been located within the galaxies. Alternatively, it is possible that there was a 'pregalactic' generation of stars (sometimes referred to as Population III stars; see, for example, Carr *et al.*, 1984). Now, the mass of intracluster gas implied by the X-ray observations is nearly as large as the total mass of the galaxies in the cluster (see Section 4.4.1). If the intracluster gas did originate in part in galaxies, there has been a considerable exchange of gas between stars, galaxies, and the intracluster medium. Thus we are led by the X-ray line observations to a much more complicated picture of galaxy formation and evolution than we might have envisaged without this crucial piece of information. Theories of the origin of the intracluster gas and the formation of galaxies and clusters are discussed in more detail in Section 5.10.

5.2 Ionization and X-Ray Emission from Hot, Diffuse Plasma

The ionization state and X-ray line and continuum emission from a low density ($n \approx 10^{-3}\,\mathrm{cm}^{-3}$), hot ($T \approx 10^8$ K) plasma will now be discussed.

Several simple assumptions will be made. First, the time scale for elastic Coulomb collisions between particles in the plasma is much shorter than the age or cooling time of the plasma, and thus the free particles will be assumed to have a Maxwell–Boltzmann distribution at the temperature T_g (Section 5.4.1). This is the kinetic temperature of electrons, and therefore determines the rates of all excitation and ionization processes. Second, at these low densities collisional excitation and de-excitation processes are much slower than radiative decays, and thus any ionization or excitation process will be assumed to be initiated from the ground state of an ion. Three-body (or more) collisional processes will be ignored because of the low density. Third, the radiation field in a cluster is sufficiently dilute that stimulated radiative transitions are not important, and the effect of the radiation field on the gas is insignificant. Fourth, at these low densities, the gas is optically thin and the transport of the radiation field can therefore be ignored. These assumptions together constitute the 'coronal limit'. Under these conditions, ionization and emission result primarily from collisions of ions with electrons, and collisions with ions can be ignored. Finally, the time scales for ionization and recombination are generally considerably less than the age of the cluster or any relevant hydrodynamic time scale, and the plasma will therefore be assumed to be in ionization equilibrium.

In nearly all astrophysical plasmas, hydrogen is the most common element and helium is the next commonest, with all the heavier elements being considerably less abundant. For example, this is true of the abundances of elements observed on the surface of the Sun. It is conventional to use these solar abundances as a standard of comparison when studying other astrophysical systems. Since most of the electrons originate in hydrogen and helium atoms, and they are fully ionized under the conditions considered here, the electron density is nearly independent of the state of ionization and is given by $n_e = 1.21 n_p$, where n_p is the density of hydrogen.

5.2.1 *Ionization Equilibrium*

In equilibrium, the ionization state is determined by the balance between processes that produce or destroy each ion:

$$[C(X^i, T_g) + \alpha(X^{i-1}, T_g)]n(X^i)n_e = C(X^{i-1}, T_g)n(X^{i-1})n_e + \alpha(X^i, T_g)n(X^{i+1})n_e. \quad (5.15)$$

Here $n(X^i)$ is the number density of the ion X^i (X is the element), T_g is the electron temperature, and $C(X^i, T_g)$ and $\alpha(X^i, T_g)$ are the rate coefficients for collisional ionization out of ion X^i and recombination into ion X^i, respectively.

The collisional ionization rate is the sum of two processes: direct collisional ionization and collisional excitation of inner shell electrons to autoionizing levels which decay to the continuum. This last process is often referred to as autoionization. Recombination is also the sum of two processes, radiative and dielectronic recombination. Recent compilations of ionization and recombination rates and discussions of their accuracy include Mewe and Gronenschild (1981), Shull and Van Steenberg (1982), and Hamilton *et al.* (1983).

The electron density dependence drops out of equation (5.15), and the equilibrium ionization state of a diffuse plasma depends only on the electron temperature. Tables of ionization fractions of various elements are given by Shull and Van Steenberg (1982). Generally, each ionization fraction reaches a maximum at a temperature that is some fraction of its ionization potential. At the temperatures which predominate in clusters, iron is mainly in the fully stripped, hydrogenic, or heliumlike stages.

5.2.2 *X-Ray Emission*

The X-ray continuum emission from a hot diffuse plasma is due primarily to three processes, thermal bremsstrahlung (free–free emission), recombination (free–bound) emission, and two-photon decay of metastable levels. The emissivity for thermal bremsstrahlung is given by equation (5.11) above. The radiative recombination (bound–free) continuum emissivity is usually calculated by applying the Milne relation for detailed balance to the photoionization cross sections, which gives (Osterbrock, 1974)

$$\varepsilon_\nu^{bf}(X^i)d\nu = n(X^{i+1})n_e \sum_l \frac{\omega_l(X^i)}{\omega_{gs}(X^{i+1})}\, a_\nu^l(X^i) \times \frac{h^4\nu^3}{c^2}\left[\frac{2}{\pi(m_e k T_g)^3}\right]^{1/2} \exp\left\{-\left[\frac{h\nu - \chi_l(X^i)}{kT_g}\right]\right\} d\nu. \quad (5.16)$$

Here, l sums over all of the energy levels of the ion X^i, gs refers to the ground state of the recombining ion X^{i+1}, ω are the statistical weights of the levels, a_ν^l is the photoionization cross section, and $\chi_l(X^i)$ is the ionization potential for each energy level in the ion.

The two-photon continuum comes from the metastable $2s$ states of hydrogenic and heliumlike ions. These levels are excited by the same processes, discussed below, that excite line emission from less forbidden transitions. For hydrogenic ions, the spectral distribution of two-photon emission is given by Spitzer and Greenstein (1951).

At the high temperatures which predominate in clusters (outside of accretion flows), thermal bremsstrahlung is the predominant X-ray emission process. For solar abundances, the emission is primarily from hydrogen and helium.

Processes that contribute to the X-ray line emission from a diffuse plasma include collisional excitation of valence or inner shell electrons, radiative and dielectronic recombination, inner shell collisional ionization, and radiative cascades following any of these processes. The emissivity due to a collisionally excited line is usually written (Osterbrock, 1974)

$$\int \varepsilon_\nu^{line} d\nu = n(X^i) n_e \frac{h^3 \nu \Omega(T_g) B}{4\omega_{gs}(X^i)} \left[\frac{2}{\pi^3 m_e^3 k T_g}\right]^{1/2} e^{-\Delta E/kT_g}, \tag{5.17}$$

where $h\nu$ is the energy of the transition, ΔE is the excitation energy above the ground state of the excited level, B is the branching ratio for the line (the probability that the upper state decays through this transition), and Ω is the 'collision strength', which is often a slowly varying function of temperature. Recent compilations of emissivities for X-ray lines and continua include Kato (1976), Raymond and Smith (1977), Mewe and Gronenschild (1981), Shull (1981), Hamilton *et al.* (1983), and Gaetz and Salpeter (1983). Lines and line ratios that are particularly suited for determining the temperature, ionization state, and elemental abundances in the intracluster gas are described in Bahcall and Sarazin (1978).

Shapiro and Bahcall (1980) and Basko *et al.* (1981) have suggested that X-ray absorption lines due to intracluster gas might be observed in the spectra of background quasars.

5.2.3 *Resulting Spectra*

All of these emission processes give emissivities that increase in proportion with the ion and electron densities, and otherwise depend only on the temperature, so that

$$\varepsilon_\nu = \sum_{X,i} \Lambda_\nu(X^i, T_g) n(X^i) n_e, \tag{5.18}$$

where Λ is the emission per ion at unit electron density. If $n(X)$ is the total density of the element X, then in equilibrium the ionization fractions $f(X^i) \equiv n(X^i)/n(X)$ depend only on the temperature, and equation (5.18) becomes

$$\varepsilon_\nu = n_p n_e \sum_{X,i} \frac{n(X)}{n(H)} [f(X^i, T_g) \Lambda_\nu(X^i, T_g)]. \tag{5.19}$$

As previously noted (equation (4.3)), it is useful to define the emission integral EI as

$$EI \equiv \int n_p n_e dV, \tag{5.20}$$

where V is the volume of the cluster. Then the shape of the spectrum depends only on the abundances of elements $n(X)/n(H)$ and the distribution of temperatures $d(EI)/dT_g$. The normalization of the spectrum (the overall level or luminosity) is set by EI.

Detailed calculations of the X-ray spectra predicted by different models of the intracluster gas have been given by Sarazin and Bahcall (1977) and Bahcall and Sarazin (1977, 1978). Figure 34 gives the X-ray spectrum for isothermal (T_g constant) models for the intracluster gas at a variety of different temperatures, showing the continuum and X-ray emission. The emission integral for these models was taken to be $6.3 \times 10^{-7}\,\mathrm{cm}^{-6}\,\mathrm{Mpc}^3$.

In these models most of the X-ray emission is thermal bremsstrahlung continuum, and the strongest lines (highest equivalent width) are in the 7 keV iron line complex. This line complex is a blend of K lines from many stages of ionization, although Fe^{+24} and Fe^{+25} predominate at typical cluster temperatures. As noted in Section 4.3.2, the 7 keV iron line is indeed the strongest line feature observed from clusters. Weaker lines at lower energies from lighter elements, such as oxygen, silicon, and sulfur, as well as from L shell transitions in less ionized iron were also predicted to be present in the spectra of clusters, particularly at lower temperatures. Such low energy lines have recently been detected (Section 4.3.3).

Because the line intensities depend on the abundances of heavy elements, while the continuum intensity is mainly due to hydrogen, the line-to-continuum ratio of a line is proportional to the abundance of the element responsible. This ratio is given by the equivalent width (equation (4.4)). Figure 35 gives the equivalent width of the iron 7 keV line complex as a function of temperature in a gas with solar abundances. Comparison of these models to the observed strengths of the lines from clusters leads to the determination that the iron abundances are roughly one-half of solar (Section 4.3.2).

The spectral observations of clusters also indicate that in a number of cases the low energy X-ray lines are stronger than would be expected based on these hydrostatic models. This indicates that gas is cooling at the cluster center (Section 4.3.3).

5.3 Heating and Cooling of the Intracluster Gas

In this section, processes that heat or cool the intracluster gas are reviewed. Only processes that affect the total energy of the gas are considered here, while processes (such as heat conduction or mixing) that redistribute the gas energy are discussed in Section 5.4.

5.3.1 Cooling

The primary cooling process for intracluster gas is the emission of radiation by the processes discussed in Section 5.2.2 above. At temperatures

Fig. 34. The predicted X-ray spectra of intracluster gas at various gas temperatures (shown at the upper right of each panel), from Sarazin and Bahcall (1977). The calculations assume the gas is isothermal, and the intensities are normalized to a sphere of gas with a proton number density of 0.001 cm^{-3} and a radius of 0.5 Mpc. E is the photon energy. The strongest line features are labeled; the lower curves, where present, show the bound-free emission.

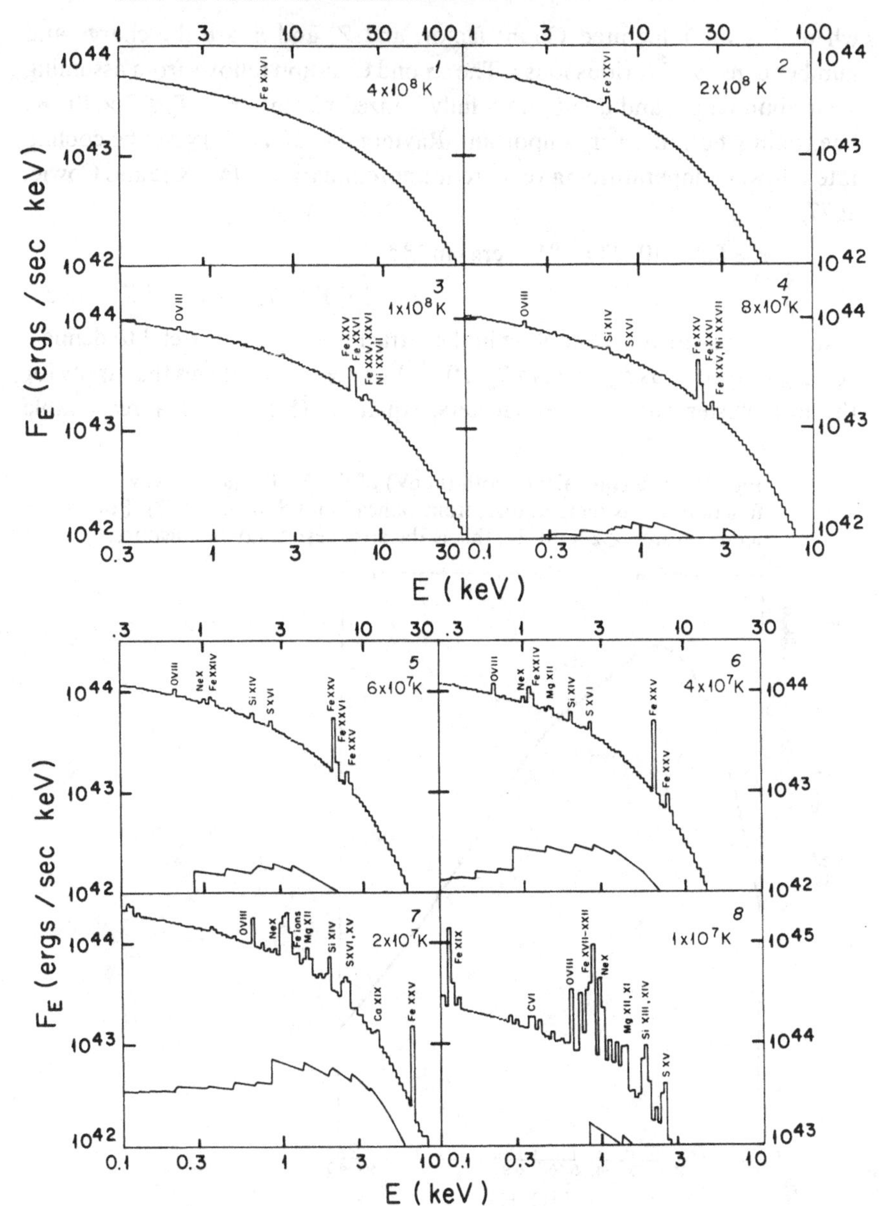

$T_g \gtrsim 3 \times 10^7$ K, the main emission mechanism is thermal bremsstrahlung, for which the total emissivity is

$$\varepsilon^{ff} = 1.435 \times 10^{-27} \bar{g} T_g^{1/2} n_e \sum_i Z_i^2 n_i \text{ erg cm}^{-3} \text{ s}^{-1}$$
$$\approx 3.0 \times 10^{-27} T_g^{1/2} n_p^2 \text{ erg cm}^{-3} \text{ s}^{-1}, \qquad (5.21)$$

where $\bar{g}$ is the integrated Gaunt factor, and Z_i and n_i are the charge and number density of various ions i. The second equation follows from assuming solar abundances and $\bar{g} = 1.1$ in a fully ionized plasma. For $T_g \lesssim 3 \times 10^7$ K, line cooling becomes very important. Raymond *et al.* (1976) give the cooling rate at lower temperatures; a very crude approximation is (McKee and Cowie, 1977)

$$\varepsilon \approx 6.2 \times 10^{-19} T_g^{-0.6} n_p^2 \text{ erg cm}^{-3} \text{ s}^{-1}$$
$$10^5 \text{ K} < T_g < 4 \times 10^7 \text{ K}. \qquad (5.22)$$

In assessing the role of cooling in the intracluster gas, it is useful to define a cooling time scale as $t_{cool} \equiv (d \ln T_g/dt)^{-1}$. For the temperatures that apply for the intracluster gas in most clusters, equation (5.21) gives a reasonable

Fig. 35. The equivalent width (in eV) of the Fe K line at 7 keV as a function of gas temperature, from Bahcall and Sarazin (1978). For gas temperatures $\gtrsim 2 \times 10^7$ K, this is the strongest X-ray line feature.

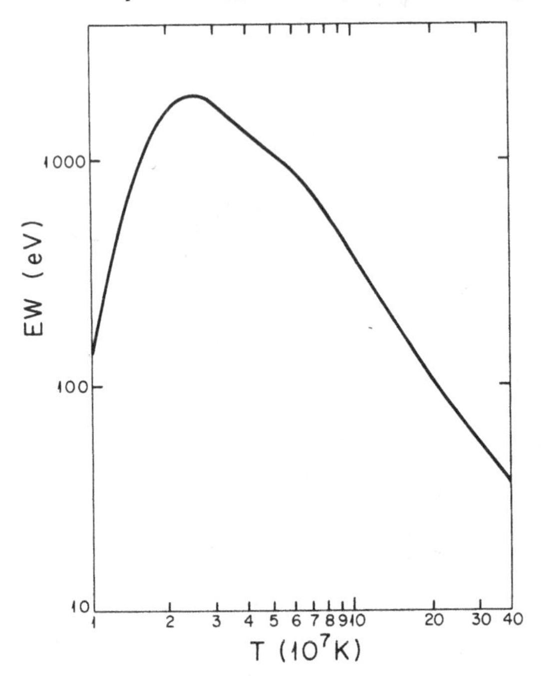

approximation to the X-ray emission. If the gas cools isobarically, the cooling time is

$$t_{cool} = 8.5 \times 10^{10}\,\text{yr}\left(\frac{n_p}{10^{-3}\,\text{cm}^{-3}}\right)^{-1}\left(\frac{T_g}{10^8\,\text{K}}\right)^{1/2}, \tag{5.23}$$

which is longer in most clusters than the Hubble time (age of the universe). Thus cooling is not very important in these cases. However, at the centers of some clusters the cooling time is shorter than the Hubble time, and these clusters are believed to have cooling flows (Section 5.7).

5.3.2 *Infall and Compressional Heating*

The heating of the intracluster gas will now be considered. The major point of this discussion is that, although the gas is quite hot, no major ongoing heating of the gas is generally necessary. This is true because the cooling time in the gas is long (equation (5.23)), and the thermal energy in the gas is comparable to or less than its gravitational potential energy. Almost any method of introducing the gas into the cluster, either from outside the cluster or from galaxies within the cluster, will heat it to temperatures on the order of those observed.

Heating of the gas due to infall into the cluster and compression will be considered here. First, imagine that the cluster was formed before the intracluster gas fell into the cluster, and that the intracluster gas makes a negligible contribution to the mass of the cluster (Section 4.4). If the gas was initially cold, and located at a large distance from the cluster, then its initial energy can be ignored. If the cluster potential remains fixed while the gas falls into the cluster and the gas neither loses energy by radiation nor exchanges its energy with other components of the cluster, then the total energy of the gas will remain zero. After falling into the cluster, the gas will collide with other elements of gas, and its kinetic energy will be converted to thermal energy. Thus infall and compression can produce temperatures on the order of

$$\frac{3}{2}\frac{kT_g}{\mu m_p} \approx -\phi, \tag{5.24}$$

where ϕ is the gravitational potential in the cluster. At the center of an isothermal cluster, $\phi \approx -9\sigma_r^2$, where σ_r is the line-of-sight velocity dispersion of the cluster. If this is substituted in equation (5.24), the derived temperature is

$$T_g \approx 5 \times 10^8\,\text{K}\left(\frac{\sigma_r}{10^3\,\text{km/s}}\right)^2, \tag{5.25}$$

which is a factor of 5–10 times larger than the observed temperatures (equation (4.10)). Of course, the gas that falls into a cluster was presumably bound to the cluster before it fell in, so that equation (5.25) overestimates the temperature. A similar calculation for gas bound to the cluster is given in Shibazaki *et al.* (1976).

The temperature may also be lower because of cooling during the infall, or because the gas fell in at the same time that the cluster was forming and thus experienced a smaller potential on average. If the gas fell in at the same time that the cluster collapsed and was heated by the rapid variation of the potential during violent relaxation (Section 2.9.2), then it might have the same energy per unit mass as the matter in galaxies,

$$\frac{3}{2}\frac{kT_g}{\mu m_p} \approx \frac{3}{2}\sigma_r^2, \tag{5.26}$$

which gives equation (5.14) for the temperature. This is in reasonable agreement with the intracluster gas temperatures determined from X-ray spectra (equation (4.10)).

These crude estimates are meant only to illustrate the point that the observed gas temperatures are consistent with heating due to infall into the cluster. More detailed models for infall are discussed in Section 5.10.1.

5.3.3 *Heating by Ejection from Galaxies*

The presence of a nearly solar abundance of iron in the intracluster gas (Sections 4.3.2 and 5.2.3) suggests that a reasonable fraction of the gas may have come from stars in galaxies within the clusters. The gas ejected from galaxies is heated in two ways. First, the gas may have some energy when it is ejected. Let ε_{ej} be the total energy per unit mass of gas ejected from a galaxy in the rest frame of that galaxy, but not including the cluster gravitational potential, and define $3kT_{ej}/2 \equiv \mu m_p \varepsilon_{ej}$. Second, the gas will initially be moving relative to the cluster center of mass at the galaxy's velocity. The ejected gas will collide with intracluster gas and thermalize its kinetic energy. On average this will give a temperature

$$kT_g \approx \mu m_p \sigma_r^2 + kT_{ej}. \tag{5.27}$$

If the ejection energy can be ignored, the temperature is given by equation (5.14), in reasonable agreement with the observations (equation (4.10)).

In a steady-state wind outflow from a galaxy, one expects $kT_{ej} \gtrsim \mu m_p \sigma_*^2$, where $\sigma_* \approx 200$ km/s is the velocity dispersion of stars within the galaxy. If the ejection temperature is near the lower limit given by this expression, then this form of heating will not be very important because $\sigma_*^2 \ll \sigma_r^2$. However, the ejection temperature could be considerably higher. For example, supernovae within galaxies could both produce the heavy elements seen in cluster X-ray spectra and heat the gas in galaxies until it was ejected. Supernovae eject highly enriched gas at velocities of $v_{SN} \approx 10^4$ km/s. The highly enriched, rapidly moving supernova ejecta would collide with the interstellar medium in a galaxy and heat the gas. If M_{SN} is the mass of ejecta from a supernova and M_{ej} is the resulting total gas mass ejected from the galaxy, then $T_{ej} \approx 2 \times 10^9 \text{ K}(v_{SN}/10^4 \text{ km/s})^2(M_{SN}/M_{ej})$, which will be significant if the supernova ejecta are diluted by less than a factor of about 100.

5.3.4 *Heating by Galaxy Motions*

Although ongoing heating of the intracluster gas may not be necessary to account for the observed gas temperatures, the estimates given above and the history of the gas are sufficiently uncertain that one cannot rule out ongoing heating as an important process. One way in which intracluster gas could be heated would be through friction between the gas and the galaxies that are constantly moving throughout the cluster (Ruderman and Spiegel, 1971; Hunt, 1971; Yahil and Ostriker, 1973; Schipper, 1974; Livio *et al.*, 1978; Rephaeli and Salpeter, 1980). The calculation of the magnitude of this drag force and of the consequent heating of the intracluster gas is complicated by the following problems. First, the motion of an average cluster galaxy through the intracluster medium is likely to be just transonic $M \approx 1$, where $M \equiv v/c_s$ is the Mach number, v is the galaxy velocity, and $c_s = 1480(T_g/10^8 \text{ K})^{1/2}$ km/s is the sound speed in the gas. If equation (4.10) for the observed gas temperature is assumed, and the average galaxy velocity is taken to be $\sqrt{3}\sigma_r$, then the average Mach number is $\langle M \rangle \approx 1.5$. Thus the galaxy motion cannot be treated as being either highly supersonic (strong shocks, etc.) or very subsonic (incompressible, etc.). In some cases shocks will be formed by the motion, and in some cases no shocks form. Second, the mean free path λ_i of ions in the intracluster medium due to Coulomb collisions (equation (5.34)) is similar to the radius of a galaxy $R_{gal} \approx 20$ kpc (Nulsen, 1982). Thus it is unclear whether the intracluster gas should be treated as a collisionless gas or as a fluid, and the role of transport processes such as viscosity (Section 5.4.4) is uncertain. For example, the Reynolds number of the flow about an object of radius R is $Re \approx 3(R/\lambda_i)M$ (equation (5.46) below), and thus is somewhat larger than unity if $R = R_{gal}$. It is therefore uncertain whether the flow will be laminar or turbulent. The magnetic field can affect transport processes (Section 5.4.3), but the coherence length of the field l_B estimated from Faraday rotation observations is also comparable to the size of a galaxy (Section 3.6). Finally, the nature of the drag force depends on whether the galaxy contains interstellar gas or not. If the galaxy contains no gas, it affects the intracluster medium only through its gravitational field. If the galaxy contains high density gas, it can give the galaxy an effective surface. For example, a gasless galaxy in supersonic motion probably will not produce a bow shock, while a gas-filled galaxy may (Ruderman and Spiegel, 1971; Hunt, 1971; Gisler, 1976).

It is convenient to write the rate of energy loss by the galaxy and the heating rate of the intracluster medium as

$$\frac{dE}{dt} = \pi R_D^2 \rho_g v^3, \tag{5.28}$$

where ρ_g is the intracluster gas density and R_D is the effective radius of the galaxy for producing the drag force. First, assume that the intracluster gas is collisionless. Then the drag is given by the dynamical friction force of equation (2.34) and

$$R_D^2 = R_A^2 \ln(\Lambda)[\text{erf}(x) - x \cdot \text{erf}'(x)], \tag{5.29}$$

where $R_A \equiv 2Gm/v^2$ is the accretion radius (Ruderman and Spiegel, 1971), m is the galaxy mass, Λ is given by equation (2.29), and $x \equiv \sqrt{5/6}M$. By the virial theorem applied to the galaxy, $R_A \approx R_{gal}(\sigma_*/\sigma_r)^2$, and the accretion radius is typically much smaller than the galaxy radius, since the galaxy velocity dispersion is smaller than that of a cluster. In the limit of hypersonic motion $M \gg 1$, the term in brackets reduces to unity. In this limit, Rephaeli and Salpeter (1980) have shown that equation (5.29) gives the drag force for any value of the mean free path or viscosity if no gas is present in the galaxy. The definition of Λ must be slightly modified (Ruderman and Spiegel, 1971).

For a $10^{11} M_\odot$ galaxy moving at 1000 km/s through intracluster gas with a proton number density of 10^{-3} cm^{-3}, the rate of heating of the gas is $\approx 10^{41}$ erg/s. If the cluster contained 1000 such galaxies, the total heating rate would be $\approx 10^{44}$ erg/s. While not trivial, this heating rate is too small to heat the intracluster gas in a Hubble time (Schipper, 1974; Rephaeli and Salpeter, 1980). If the total mass of intracluster gas is $\gtrsim 10^{14} M_\odot$ and the gas temperature is $\approx 6 \times 10^7$ K, then the time required to heat the gas at this rate would be $\gtrsim 7 \times 10^{11}$ yr.

The drag force can be considerably increased if the galaxy contains gas or magnetic fields that prevent the penetration of the galaxy by the intracluster gas (Ruderman and Spiegel, 1971; Yahil and Ostriker, 1973; Livio *et al.*, 1978; Shaviv and Salpeter, 1982; Gaetz *et al.*, 1987). However, the main effect of the drag force may be to remove the gas from the galaxy and heat it to the temperature determined by the galaxy kinetic energy per mass (Section 5.3.3). The stripping of gas from a galaxy is shown in Section 5.9 to be very efficient, and as discussed in Sections 2.10.2 and 4.6, galaxies in X-ray clusters are known to be very deficient in gas. It is possible that galaxies may retain a core of gas produced by stellar mass loss within the galaxy. Gaetz *et al.* (1987) give useful analytic fitting formulae for drag coefficients for galaxies with stellar mass loss, based on two-dimensional hydrodynamic simulations. Yahil and Ostriker (1973) and Livio *et al.* (1978) have suggested very large heating rates produced by having high rates of gas output in galaxies. Livio *et al.* assume a rather large cross section for galaxies and a rather low intracluster gas density. Both of these papers argue that the heating rate due to galaxy motions is so large that the intracluster gas is heated beyond the escape temperature from the cluster, and a cluster wind results (see Section 5.6).

There is a simple argument against the great importance of heating due to drag forces from the motions of galaxies. The mass associated with intracluster gas in a typical X-ray cluster is comparable to or greater than the mass associated with galaxies (Section 4.4.1). The average thermal velocities in the gas are comparable to or greater than the typical galaxy velocities (Sections 4.6, 5.3.2, and 5.5). Thus the total thermal energy in the gas is greater than or comparable to the total kinetic energy in the galaxies. It is difficult to believe that the galaxies heated the gas under these circumstances. If they did, then the massive galaxies would have lost most of their initial kinetic energy in the process; this should have produced quite extreme mass segregation, which is not observed (Section 2.7).

5.3.5 *Heating by Relativistic Electrons*

Clusters of galaxies often contain radio sources having steep spectral indices (Section 3.1); the radio emission from these sources is believed to arise from synchrotron emission by relativistic electrons. These electrons (particularly the lower energy ones) can interact with the intracluster gas and may heat this gas (Sofia, 1973; Lea and Holman, 1978; Rephaeli, 1979; Scott *et al.*, 1980). A relativistic electron passing through a plasma loss energy through Coulomb interactions with electrons within a Debye length of the particle, and through the interactions with plasma waves on larger scales. The rate of loss by an electron with total energy $\gamma m_e c^2$ is

$$-\frac{d\gamma}{dt} = \frac{\omega_p^2 r_e}{c}\left[\ln\left(\frac{m_e c^2 \gamma^{1/2}}{h\omega_p}\right) + 0.2\right], \tag{5.30}$$

where ω_p is the plasma frequency ($\omega_p^2 \equiv 4\pi n_e e^2/m_e$), $r_e \equiv e^2/(m_e c^2)$ is the classical electron radius, and n_e is the electron density in the plasma. The term in square brackets is ≈ 40 for most values of γ and n_e of interest, and does not vary significantly with either parameter. Numerically, the heating rate is $\approx 10^{-18} n_e$ erg/s, ignoring the variation of the term in square brackets. The total heating rate is determined by multiplying this rate per electron by the total number of relativistic electrons in the intracluster gas. If the relativistic electrons have a power-law spectrum (equation (5.2)), then the heating rate can be determined from the synchrotron radio emission rate given by equation (5.7). The heating rate is

$$\frac{dE}{dt} = \frac{1.2 \times 10^9 n_e}{a(p)\alpha_r} \left(\frac{dL_r}{d\nu_r} \nu_r^{\alpha_r} \right) \times \left(\frac{2.4 \times 10^{-7}}{B} \right)^{\alpha_r + 1} \gamma_l^{-2\alpha_r} \text{ erg/s}. \tag{5.31}$$

Here L_r is the radio luminosity at a frequency ν_r, α_r is the radio spectral index, B is the intracluster magnetic field, $a(p)$ is the function given by equations (5.4) and (5.8), and γ_l is the lower limit to the electron spectrum (equation (5.2)). Cgs units are to be used for L_r and B. The first quantity in parentheses is independent of the frequency ν_r at which the radio source is observed. Equation (5.31) includes only electrons and should be increased to include the heating by ions, which do not produce observable radio emission. While the radio flux and spectrum can be measured directly, the magnetic field strength must be estimated from the radio observations (Section 3.6), and the lower limit to the electron spectrum is generally unknown.

This heating rate is significant only for radio sources with steep spectra which extend to very, very low frequencies. Lea and Holman (1978) used low frequency radio observations of clusters to determine the radio flux and spectral index, and found that the electron spectrum must extend down to $\gamma_l \approx 10(B/\mu\text{G})^{-1}$ if the heating rate is to be comparable to the X-ray luminosity of the cluster. These low energy electrons would produce radio emission at a frequency of about 400 Hz, which is about 10^5 times too low to be observed. Extrapolating the radio spectrum from the lowest observed frequencies ($\approx$26 MHz) down to these low frequencies increases the total number of relativistic electrons and the corresponding heating rate by about 10^5. Thus the hypothesis that relativistic electrons provide significant heating to the intracluster gas according to equation (5.31) requires an enormous and untestable extrapolation of cluster radio properties.

Several authors have argued that the heating rate of equation (5.31) should be increased by collective plasma interactions between the relativistic electrons and the intracluster gas (Lea and Holman, 1978; Rephaeli, 1979; Scott *et al.*, 1980). In these models, it is assumed that the relativistic electrons are streaming away from a powerful radio source at the center of the cluster. Rephaeli (1979) assumed that the streaming speed is limited by the Alfvén velocity, as has generally been argued (Jaffe, 1977; Section 3.4). Then he finds that the relativistic electrons excite Alfvén waves and lose energy at a rate of $-d\gamma/dt \approx v_A \gamma / L_e$, where L_e is the scale length of the relativistic electron distribution. For $B \approx 1\,\mu$G, $L_e \approx 1$ Mpc, and $n_e \approx 10^{-3}\,\text{cm}^{-3}$, this gives a heating rate about $10^{-2}\gamma_l$ of that in equation (5.31), which is never very significant.

Alternatively, Scott *et al.* (1980) have assumed that the relativistic electrons stream at nearly the speed of light (Holman *et al.*, 1979). As discussed in Section 3.4, this hypothesis is controversial. They discuss a number of plasma instabilities that greatly increase the heating rate under these circumstances. The increase could be as much as a factor of 10^5, which would allow the electrons that produce the observed radio emission in clusters to heat the intracluster gas.

Models in which relativistic electrons heat the intracluster gas suffer from two general problems. First, the total energy requirements of $10^{63\text{-}64}$ erg are extreme for a single radio source, although it is possible that many cluster sources contribute to the heating over the lifetime of the cluster. Second, the radio sources generally occupy only a small fraction of the cluster; cluster-wide radio haloes are rare (Section 3.4). It is difficult to see how several discrete radio sources would heat the intracluster gas without producing very strong, observable variations in the X-ray surface brightness, which are not seen.

Vestrand (1982) has suggested that heating of the intracluster gas by relativistic electrons is important only in clusters having radio haloes. He noted that Coma, which has a prominent halo, also has an unusually high gas temperature for its velocity dispersion (Section 4.5.1).

5.4 Transport Processes

Processes that redistribute energy, momentum, or heavy elements within the intracluster gas will now be reviewed.

5.4.1 *Mean Free Paths and Equilibration Time Scales*

The mean free paths of electrons and ions in a plasma without a magnetic field are determined by Coulomb collisions. As in the stellar dynamical case, it is important to include distant as well as nearby collisions. The mean free path λ_e for an electron to suffer an energy exchanging collision with another electron is given by (Spitzer, 1956)

$$\lambda_e = \frac{3^{3/2}(kT_e)^2}{4\pi^{1/2} n_e e^4 \ln \Lambda}, \tag{5.32}$$

where T_e is the electron temperature, n_e is the electron number density, and Λ

is the ratio of largest to smallest impact parameters for the collisions. For $T_e \gtrsim 4 \times 10^5$ K, this Coulomb logarithm is

$$\ln \Lambda = 37.8 + \ln\left[\left(\frac{T_e}{10^8\,\mathrm{K}}\right)\left(\frac{n_e}{10^{-3}\,\mathrm{cm}^{-3}}\right)^{-1/2}\right], \tag{5.33}$$

which is nearly independent of density or temperature. Equation (5.32) assumes that the electrons have a Maxwellian velocity distribution at the electron temperature. The equivalent mean free path of ions λ_i is given by the same formula, replacing the electron temperature and density with the ion temperature T_i and density, dividing by the ion charge to the fourth power, and slightly increasing $\ln \Lambda$. In the discussion that follows the ions will generally be assumed to be protons, and the diffusion of heavy elements will be discussed in Section 5.4.5 below. Numerically,

$$\lambda_e = \lambda_i \approx 23\,\mathrm{kpc}\left(\frac{T_g}{10^8\,\mathrm{K}}\right)^2\left(\frac{n_e}{10^{-3}\,\mathrm{cm}^{-3}}\right)^{-1} \tag{5.34}$$

assuming that $T_e = T_i = T_g$.

In general, these mean free paths are shorter than the length scales of interest in clusters (≈ 1 Mpc), and the intracluster medium can be treated as a collisional fluid, satisfying the hydrodynamic equations. Note that the mean free paths are comparable to the size of a galaxy, however, and in the interaction between intracluster gas and individual galaxies the gas may be nearly collisionless, as mentioned previously (Section 5.3.4).

If a homogeneous plasma is created in a state in which the particle distribution is non-Maxwellian, elastic collisions will cause it to relax to a Maxwellian distribution on a time scale determined by the mean free paths (Spitzer, 1956, 1978). Electrons will achieve this equilibration (isotropic Maxwellian velocity distribution characterized by the electron temperature) on a time scale set roughly by $t_{eq}(e, e) \equiv \lambda_e/\langle v_e\rangle_{rms}$, where the denominator is the rms electron velocity,

$$t_{eq}(e, e) = \frac{3m_e^{1/2}(kT_e)^{3/2}}{4\pi^{1/2}n_e e^4 \ln \Lambda}$$

$$\approx 3.3 \times 10^5\,\mathrm{yr}\left(\frac{T_e}{10^8\,\mathrm{K}}\right)^{3/2}\left(\frac{n_e}{10^{-3}\,\mathrm{cm}^{-3}}\right)^{-1}. \tag{5.35}$$

The time scale for protons to equilibrate among themselves is $t_{eq}(p, p) \approx (m_p/m_e)^{1/2} t_{eq}(e, e)$, or roughly 43 times longer than the value in equation (5.35). Following this time, the protons and ions would each have Maxwellian distributions, but generally at different temperatures. The time scale for the electrons and ions to reach equipartition $T_e = T_i$ is $t_{eq}(p, e) \approx (m_p/m_e) t_{eq}(e, e)$, or roughly 1870 times the value in equation (5.35). For heavier ions, the time scales for equilibration are generally at least this short if the ions are nearly fully stripped, because the increased charge more than makes up for the increased mass. For $T_g \approx 10^8$ K and $n_e \approx 10^{-3}\,\mathrm{cm}^{-3}$, the longest equilibration time scale is only $t_{eq}(p, e) \approx 6 \times 10^8$ yr. Since this is shorter than the age of the cluster or the cooling time, the intracluster plasma can generally be characterized by a single kinetic temperature T_g. Under some circumstances, plasma instabilities may bring about a more rapid equilibration than collisions (McKee and Cowie, 1977).

5.4.2 *Thermal Conduction*

In a plasma with a gradient in the electron temperature, heat is conducted down the temperature gradient. If the scale length of the temperature gradient $l_T \equiv T_e/|\nabla T_e|$ is much longer than the mean free path of electrons λ_e, then the heat flux is given by

$$\mathrm{Q} = -\kappa \nabla T_e, \tag{5.36}$$

where the thermal conductivity for a hydrogen plasma is (Spitzer, 1956)

$$\begin{aligned}\kappa &= 1.31 n_e \lambda_e k \left(\frac{kT_e}{m_e}\right)^{1/2} \\ &\approx 4.6 \times 10^{13} \left(\frac{T_e}{10^8\,\mathrm{K}}\right)^{5/2} \left(\frac{\ln \Lambda}{40}\right)^{-1} \mathrm{erg\,s^{-1}\,cm^{-1}\,K^{-1}}.\end{aligned} \tag{5.37}$$

Because of the inverse dependence on the particle mass, thermal conduction is primarily due to electrons. This equation includes a correction for the self-consistent electric field set up by the diffusing electrons. If the very weak dependence of $\ln \Lambda$ on density is ignored, then κ is independent of density but depends very strongly on temperature.

If the scale length of the thermal gradient l_T is comparable to or less than the mean free path of electrons, then equation (5.36) overestimates the heat flux, since it would imply that the electrons are diffusing at a speed greater than their average thermal speed. Under these circumstances the conduction is said to 'saturate', and the heat flux approaches a limiting value Q_{sat}. Cowie and McKee (1977) calculate this saturated heat flux by assuming that the electrons have a Maxwellian distribution, and an infinitely steep temperature gradient, and that the correction for a self-consistent electric field is the same as in the unsaturated case. They find

$$Q_{sat} = 0.4 n_e k T_e \left(\frac{2kT_e}{\pi m_e}\right)^{1/2}. \tag{5.38}$$

A general expression for the heat flux, which interpolates between the two limits of equations (5.36) and (5.38), is then

$$\mathbf{Q} \approx -\frac{\kappa T_e}{l_T + 4.2\lambda_e} \frac{\nabla T_e}{|\nabla T_e|}. \tag{5.39}$$

The mean free path of electrons in the intracluster gas (equation (5.34)) is typically small compared to the cluster dimensions, and heat conduction within the intracluster gas itself is probably unsaturated. However, the mean free path is comparable to the size of a galaxy, and saturated heat conduction may be important in evaporation from or accretion to galaxies (Sections 5.7 and 5.9).

Within the intracluster medium, thermal conduction will act to transport heat from hot to cold regions and, in the absence of any competing effect, to make the temperature spatially constant (isothermal). Assuming equal ion and electron temperatures, the temperature in a Lagrangian element of the intracluster gas will vary as

$$\frac{3}{2} \frac{\rho_g k}{\mu m_p} \frac{dT_g}{dt} - \frac{kT_g}{\mu m_p} \frac{d\rho_g}{dt} = -\nabla \cdot \mathbf{Q}, \tag{5.40}$$

where ρ_g is the gas density. It is useful to define a conduction time scale as $t_{cond} \equiv -(d \ln T_e/dt)^{-1}$, which is on the order of $|t_{cond}| \approx (n_e l_T^2 k)/\kappa$. As a specific example, consider a cluster in which the gas is hydrostatic, adiabatic (isentropic), and extends to very large distances but is not in contact with any intercluster gas; such models are discussed in some detail in Section 5.5.2 below. If the cluster potential is given by the analytic King form (equation (5.59) below), the gas is assumed to cool isobarically (at constant pressure), and the variation of ln Λ is ignored, then the conduction time at a radius r is given by

$$\frac{1}{t_{cond}} = \frac{2\mu m_p \kappa_o}{5\rho_{go} r_c^2 k} g(r/r_c), \tag{5.41}$$

where κ_o and ρ_{go} are the conductivity and gas density at the cluster center, and r_c is the cluster core radius. The function $g(x)$ is

$$g(x) \equiv (x^2+1)^{-3/2} - \frac{5}{2x^2 f(x)} [f(x) - (x^2+1)^{-1/2}]^2, \tag{5.42}$$

where $f(x) \equiv \phi(r)/\phi_o$ is the ratio of the cluster potential to its central value and is given by equation (5.59) below. The function $g(x)$ is plotted in Figure 36. Because conduction only transports heat, the average temperature of the gas is not changed; in the inner parts the gas is cooled and in the outer parts the gas is heated. However, because the X-ray emission is proportional to the square of the density, temperatures determined from X-ray spectra are mainly affected by the innermost gas and are lowered by conduction. As is clear from

Figure 36, heat conduction is most effective in the cluster core, and $|t_{cond}|$ increases very rapidly with radius. Since $g(0) = 1$, the central value to the conduction time scale $t_{cond}(0)$ is given by the first term in equation (5.41), or

$$t_{cond}(0) = 3.3 \times 10^8 \text{ yr} \left(\frac{n_o}{10^{-3} \text{ cm}^{-3}} \right) \left(\frac{T_e}{10^8 \text{ K}} \right)^{-5/2} \times \left(\frac{r_c}{0.25 \text{ Mpc}} \right)^2 \left(\frac{\ln \Lambda}{40} \right), \tag{5.43}$$

where n_o is the central proton density, and solar abundances have been assumed. Thus heat conduction may be relatively effective in the core of a cluster. At radii $r \gtrsim 2r_c$, the conduction time is typically a factor of ≈ 100 longer, and conduction is only marginally effective in the outer parts of the cluster. The conduction time may be increased further by the presence of a magnetic field in the cluster.

5.4.3 *Effects of the Magnetic Field*

Charged particles gyrate around magnetic field lines on orbits with a radius (the gyroradius) of $r_g = (mv_\perp c/ZeB)$, where m is the particle mass, $v_\perp$ is the component of its velocity perpendicular to the magnetic field, Z is the particle charge, and B is the magnetic field strength. If $v_\perp = \sqrt{2kT_g/m}$, which is

Fig. 36. The conduction time as a function of position in an adiabatic model for the intracluster gas. The conduction time is relative to its central value, and the radius is in units of the cluster core radius r_c. The solid (dashed) curve indicates the portion of the cluster where the gas is cooled (heated) by conduction.

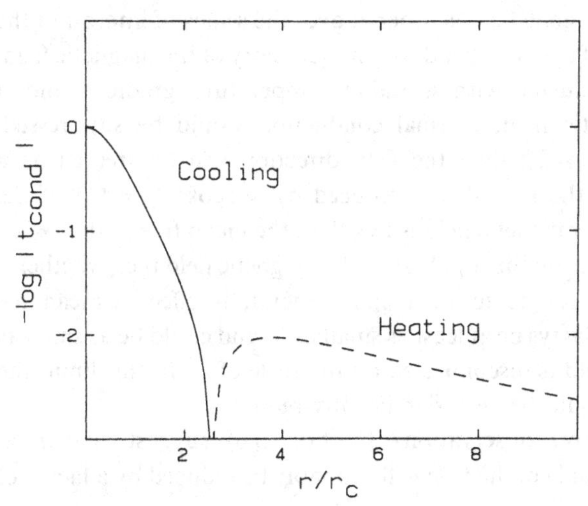

the rms value in a thermal plasma, then

$$r_g = \frac{3.1 \times 10^8 \text{ cm}}{Z} \left(\frac{T_g}{10^8 \text{ K}} \right)^{1/2} \left(\frac{m}{m_e} \right)^{1/2} \left(\frac{B}{1 \, \mu\text{G}} \right)^{-1} \tag{5.44}$$

which is much smaller than any length scale of interest in clusters, and is also much smaller than the mean free path of particles due to collisions $r_g \ll \lambda_e$. Then, the effective mean free path for diffusion perpendicular to the magnetic field is only on the order of r_g^2/λ_e (Spitzer, 1956). Because they have larger gyroradii, the ions are most effective in transport processes perpendicular to the magnetic field. In practice, the gyroradii are so small that diffusion perpendicular to the magnetic field can be ignored in the intracluster gas.

Consider the effect of the magnetic field on thermal conduction, when the temperature gradient lies at an angle θ to the local magnetic field direction. Only the component of the gradient parallel to the field is effective in driving a heat flux, and only the component of the resulting heat flux in the direction of the temperature gradient transports any net energy. If the conduction is unsaturated, the heat flux parallel to the thermal gradient is thus reduced by a factor of $\cos^2 \theta$. If the conduction is saturated, the heat flux is independent of the temperature gradient; the appropriate factor is just $\cos \theta$ (Cowie and McKee, 1977).

Observations suggest that the magnetic field in clusters may be tangled, so that the direction θ varies throughout the intracluster gas (Section 3.6). Let l_B be the coherence length of the magnetic field, so that the field will typically have changed direction by $\approx 90°$ over this distance. Let l_T be the temperature scale height (Section 5.4.2 above) and let λ_e be the mean free path of electrons. Consider first the case where $\lambda_e \ll l_T$, so that the conduction is unsaturated. Then, if $l_B \gtrsim l_T$, the magnetic field is ordered over the scales of interest in the cluster, and the value of $\cos^2 \theta$ depends on the geometry of the magnetic field. For example, for a cluster with a radial temperature gradient and a circumferential magnetic field, thermal conduction would be suppressed. Alternatively, if $l_T \gg l_B \gg \lambda_e$, then the field direction can be treated as a random variable, and the heat flux is reduced by $\approx \langle \cos^2 \theta \rangle = 1/3$. If the coherence length of the magnetic field is less than the mean free path $l_B \ll \lambda_e$, the conductivity depends on the topology of the magnetic field (i.e., whether it is connected over distances greater than l_B). In general, the effective mean free path for diffusion will always be at least as small as l_B, and could be as small as l_B^2/λ_e if the magnetic field is disconnected on the scale of l_B. In this limit, the thermal conduction would be very significantly reduced.

The Faraday rotation observations (Section 3.6) suggest that $l_B \approx \lambda_e \approx 20$ kpc. Thus thermal conduction will probably be reduced by a factor of at least 1/3, and the conductivity time scale (equations (5.41) and (5.43)) should be increased by at least this factor.

5.4.4 *Viscosity*

If there are shears in the velocity in a fluid, then viscosity will produce forces that act against these shears. A unit volume of the fluid will be subjected to a force given by

$$\mathbf{F}_{vis} = \eta \left(\nabla^2 \mathbf{v} + \frac{1}{3} \nabla \nabla \cdot \mathbf{v} \right), \tag{5.45}$$

where η is the dynamic viscosity, and the bulk viscosity has been assumed to be zero. For an ionized plasma without a magnetic field, η is given by

$$\begin{aligned} \eta &\approx \frac{1}{3} m_i n_i \langle v_i \rangle_{rms} \lambda_i \\ &\approx 5500 \text{ gm cm}^{-1} \text{ s}^{-1} \left(\frac{T_e}{10^8 \text{ K}} \right)^{5/2} \left(\frac{\ln \Lambda}{40} \right)^{-1}, \end{aligned} \tag{5.46}$$

where m_i, n_i, $\langle v_i \rangle_{rms}$, and λ_i are the mass, number density, rms velocity, and mean free path of ions, respectively. Like the thermal conductivity, the dynamic viscosity is independent of density and depends strongly on the temperature. However, because of the dependence on the particle mass, the viscosity is primarily due to ions, not electrons. The Reynolds number for flow at a speed v past an object of size l is defined as $Re \equiv \rho_g v l / \eta$, which can be written as

$$Re \approx 3M \left(\frac{l}{\lambda_i} \right), \tag{5.47}$$

where $M \equiv v / c_s$ is the Mach number and c_s is the sound speed. As noted in Section 5.3.4, this indicates that the flow around moving galaxies in a cluster is probably laminar, but not certainly so. The viscosity affects the rate of heating of the intracluster gas by galaxy motions (Section 5.3.4), and the rate of stripping by interstellar gas in cluster galaxies (Section 5.9).

As with thermal conduction, one can define a velocity scale length l_v so that the absolute value of the term in parentheses in equation (5.45) is $\langle v_i \rangle_{rms} / l_v^2$. Then, if the ion mean free path λ_i is shorter than l_v, equation (5.45) applies. However, if $l_v < \lambda_i$, then equation (5.45) requires that the viscous stresses exceed the ion pressure and that momentum be transported faster than the thermal speed of the ions. This is not possible, and the viscous stresses must saturate at a value comparable to the ion pressure. To my knowledge, this effect has not been included in calculations of astrophysical fluid flows. It should be particularly important in flows around galaxies, where previous calculations, particularly those involving Kelvin–Helmholtz instabilities, have applied equation (5.45) to flows with very large shears.

5.4.5 *Diffusion and Settling of Heavy Ions*

At the same temperature, heavy ions will move more slowly than light ions, and all ions will move more slowly than electrons. As a result, the heavy ions will tend to settle towards the center of the cluster. This settling is generally halted when it results in an electric field large enough to balance the extra gravitational force on the heavier ions. Models for the distribution of heavy elements in clusters, assuming that settling has occurred rapidly enough to reach this equilibrium, are discussed in Section 5.5.6 below. Here, the rate of settling will be considered.

Heavy ions that diffuse toward the cluster center rapidly reach a drift velocity at which gravitational acceleration is balanced by the slowing effect of collisions (Fabian and Pringle, 1977; Rephaeli, 1978). Using the expression for the mean free path of a slow moving heavy ion in a plasma (Spitzer, 1956), we obtain the drift velocity $\mathbf{v}_D$,

$$Am_p\mathbf{g} = \mathbf{v}_D \frac{16\sqrt{\pi}Z^2e^4m_p^{1/2}}{3(2kT_g)^{3/2}} \sum_i Z_i^2 A_i^{1/2} n_i \ln \Lambda_i, \tag{5.48}$$

where A and Z are the heavy ion atomic number and charge, $\mathbf{g}$ is the gravitational acceleration, T_g is the gas temperature, and the sum is over the various ions in the plasma. Assuming cosmic abundances, one finds

$$v_D = 2.9\left(\frac{A}{56}\right)\left(\frac{Z}{26}\right)^{-2}\left(\frac{|g|}{3\times 10^{-8}\,\mathrm{cm\,s^{-1}}}\right)\left(\frac{T_g}{10^8\,\mathrm{K}}\right)^{3/2} \times\left(\frac{n_p}{10^{-3}\,\mathrm{cm}^{-3}}\right)^{-1}\left(\frac{\ln\Lambda}{40}\right)^{-1}\ \mathrm{km/s}. \tag{5.49}$$

The values of A and Z in equation (5.49) are those for fully ionized iron. The X-ray lines of iron are the strongest lines observed from clusters, and have been detected in a large number of clusters (Section 4.3.2). These lines play an important role in considerations of the origin and energetics of the intracluster gas (Section 5.10), and thus the distribution of iron is particularly important. As noted by Rephaeli (1978), the value of the drift velocity in equation (5.49) is considerably smaller than the value given by Fabian and Pringle (1977) because they failed to include helium and heavier elements in the slowing time.

According to equation (5.49), the drift velocity depends on the cluster gravitational potential and the density and temperature of the intracluster gas. If the cluster potential is given by the King analytic approximation to the isothermal distribution (equations (2.13), (5.59)), then the largest value of g

occurs at roughly the core radius r_c (actually $r = 1.027 r_c$), and

$$g \lesssim 2.0 \times 10^{-8} \left(\frac{\sigma_r}{10^3 \text{ km/s}} \right)^2 \left(\frac{r_c}{0.25 \text{ Mpc}} \right)^{-1} \text{cm s}^{-2}, \tag{5.50}$$

where σ_r is the radial velocity dispersion of the cluster. The time to drift in a small distance dr is then $dt = dr/v_D$. If one sets $dr \approx r_c$, then the drift time is

$$t_D \approx 1.2 \times 10^{11} \text{ yr} \left(\frac{A}{56} \right)^{-1} \left(\frac{Z}{26} \right)^2 \left(\frac{T_g}{10^8 \text{ K}} \right)^{-3/2}$$

$$\times \left(\frac{n_p}{10^{-3} \text{ cm}^{-3}} \right) \left(\frac{\ln \Lambda}{40} \right) \left(\frac{\sigma_r}{10^3 \text{ km/s}} \right)^{-2} \left(\frac{r_c}{0.25 \text{ Mpc}} \right)^2 . \tag{5.51}$$

If the intracluster gas is adiabatic (constant entropy per particle; Section 5.5), then $n_p \propto T_g^{3/2}$, and the variation of these two quantities does not affect the drift time. If the gas is isothermal, then the reduction of density with distance from the cluster center shortens the drift time, although the increase in the distance to be traversed and the decrease in the gravitational acceleration more than make up for this. Rephaeli (1978) has integrated the drift time for a variety of equations of state for the gas, and finds that the drift times from radii greater than four core radii into one core radius is always greater than 10^{11} yr times the cluster parameters in equation (5.51), with the gas density and temperature evaluated at the cluster center. Since this means the drift time is generally at least ten times the probable age of the cluster, the settling of iron into the cluster center is unlikely to occur.

As noted by Rephaeli (1978), the presence of a magnetic field in the intracluster gas would inhibit the sedimentation of iron even further. The situation is similar to that for thermal conduction (Section 5.4.3); iron ions could only drift along the direction of magnetic field lines. Unless the magnetic field were predominantly radial, settling of iron would be very strongly inhibited. The effect is even larger than that on thermal conduction, because the sedimentation of iron ions requires that the individual ions drift over large distances, while heat can be conducted from the inner to the outer portions of a cluster without having an individual electron traverse these distances. For example, loop structures in the magnetic field would tend to prevent any sedimentation beyond the bottom of the loop.

Although it appears unlikely that heavy elements like iron would have settled to the cluster center, this does not mean that the abundance of heavy elements is necessarily uniform. Heavy elements might be concentrated in the cluster core if they were injected into the intracluster gas in the core. If the heavy elements were produced in stars within galaxies, as is usually assumed (Sections 4.3.2 and 5.10), then this gas might have been predominantly deposited in the cluster core because either ram pressure or collisional stripping will occur more easily there. Of course, if all the intracluster gas were produced by this mechanism, then no abundance gradients would be produced if all galaxies in the cluster contained gas with the same average abundances. If, however, the intracluster medium were partly due to heavy element containing gas stripped from galaxies, and partly due to heavy element lacking intergalactic gas that fell into the cluster, and if the stripping were concentrated at the cluster core, then a gradient in the abundance of heavy elements could result (Nepveu, 1981b). Subsequent mixing of the intracluster gas (Section 5.4.6 below) might still destroy such abundance variations.

Such an abundance gradient could greatly reduce the amount of iron needed to produce the observed iron lines. If the iron were concentrated at the cluster core, it would all be located in regions of high electron density, so that the iron line emissivity per iron ion would be higher than the average emissivity per hydrogen ion (Section 5.2). The amount of iron needed to produce the observed iron lines could be reduced by as much as a factor of 20 (Abramopoulos *et al.*, 1981). Since the production of a nearly solar abundance of iron is one of the major problems with galaxy formation in clusters and the origin of the intracluster gas, the possibility that the real abundance might be much lower is very important. Unfortunately, all of the available spectra of the 7 keV iron line in clusters have been made with low spatial resolution detectors, which do not provide sufficient information on the location of the iron. The observation of spatially resolved spectra of clusters including the 7 keV iron line are critically important to our understanding of clusters, and will be possible if the AXAF satellite is launched (Chapter 6).

5.4.6 *Convection and Mixing*

The diffusive transport processes discussed so far tend to cause the gas in a cluster to be isothermal, tend to damp out fluid motions, and tend to cause heavy elements to settle to the cluster center. On the other hand, mixing processes due to turbulent motions of the intracluster gas tend to make the specific entropy (the entropy per atom) equal within the cluster, tend to make the composition of the gas homogeneous, and drive fluid motions. It is conventional to describe a gas in which the entropy per atom is constant as an 'adiabatic' gas, If mixing occurs on a time scale that is rapid compared to the age of the cluster or the time scale for diffusive transport, the resulting intracluster gas distribution will tend to be adiabatic and to have constant heavy element abundances.

One possible source of mixing motions in the gas is convection. If the intracluster gas were hydrostatic but had a steep temperature gradient

$$-\frac{d\ln T_g}{dr} > -\frac{2}{3}\frac{d\ln n_p}{dr}, \tag{5.52}$$

it would be unstable to convective mixing. If the temperature gradient in equation (5.52) were exceeded by a significant amount, mixing would occur within several sound crossing times in the cluster. Since this is a rather short time (equation (5.54) below), it is reasonable to assume that the temperature gradient is smaller than that in equation (5.52).

Motions of galaxies through the intracluster gas may mix the gas. This process has not been treated in any great detail in the literature. It might be reasonable to assume that each galaxy mixes the gas within a wake of radius R_W. Very roughly, the gas within the whole cluster would be mixed on a time scale given by

$$t_{mix} \approx \frac{1}{n_{gal}\sigma_r \pi R_W^2}\left(\frac{r_c}{R_W}\right)^2, \tag{5.53}$$

where n_{gal} is the number density of galaxies, σ_r is their velocity dispersion, and r_c is the cluster core radius. Depending on whether the galaxy contained any interstellar medium or not, the wake radius might be ≈ 10 kpc, or only as large as the accretion radius (Section 5.3.4). The resulting value of t_{mix} is generally much longer than 10^{11} yr for any reasonable values of the cluster parameters, and mixing due to galaxy motions is probably not very important. Nepveu (1981b) discussed the mixing of gas ejected from cluster galaxies and showed that the mixing was not effective unless the galaxy motions were highly subsonic, which is not the case. His calculations indicated that the gas remained inhomogeneous on both small and large scales.

Large scale hydrodynamic motions during the formation of the cluster may be effective in mixing the intracluster gas. Similarly, mixing may occur when subclusters merge within the cluster.

5.5 Distribution of the Intracluster Gas – Hydrostatic Models

In general, the elastic collision times for ions and electrons (equations (5.32), (5.33)) in the intracluster gas are much shorter than the time scales for heating or cooling (Section 5.3) or any dynamical process, and the gas can be treated as a fluid. The time required for a sound wave in the intracluster gas to cross a cluster is given by

$$t_s \approx 6.6 \times 10^8\,\text{yr}\left(\frac{T_g}{10^8\,\text{K}}\right)^{-1/2}\left(\frac{D}{\text{Mpc}}\right), \tag{5.54}$$

where T_g is the gas temperature and D is the cluster diameter. Since this time is short compared to the probable age of a cluster of 10^{10} yr, the gas will be hydrostatic and the pressure will be a smoothly varying function of position unless the cluster gravitational potential varies on a shorter time scale or the gas is heated or cooled more rapidly than this. The cooling time due to thermal bremsstrahlung (equation (5.23)) is much longer than the sound crossing time, and the same is true for the time scales for heating by any of the processes that have been suggested (Section 5.3). Thus the gas distribution is usually assumed to be hydrostatic:

$$\nabla P = -\rho_g \nabla \phi(r), \tag{5.55}$$

where $P = \rho_g k T_g / \mu m_p$ is the gas pressure, ρ_g is the gas density, and $\phi(r)$ is the gravitational potential of the cluster. Now, the sound crossing time condition given above ensures that the pressure in the gas be uniform on small scales, but the same need not be completely true of the entropy. In the rest of this section, we will assume that the intracluster gas is locally homogeneous.

If, in addition, the cluster is assumed to be spherically symmetric, equation (5.55) reduces to

$$\frac{1}{\rho_g}\frac{dP}{dr} = -\frac{d\phi}{dr} = -\frac{GM(r)}{r^2}, \tag{5.56}$$

where r is the radius from the cluster center and $M(r)$ is the total cluster mass within r. If the contribution of the intracluster gas to the gravitational potential is ignored, then the distribution of the intracluster gas is determined by the cluster potential $\phi(r)$ and the temperature distribution of the gas $T_g(r)$. Further, then equation (5.56) is a linear equation for the logarithm of the gas density; the central density (or any other value) can be specified in order to determine the full run of densities. If self gravity is included, then the density scale, temperature variation, and cluster potential cannot be given independently.

In most models, the cluster potential is assumed to be that of a self-gravitating isothermal sphere (Section 2.7). For convenience, the analytic King approximation to the isothermal sphere is often assumed (equation (2.13)), so that the cluster total density, mass, and potential are given by

$$\rho(r) = \rho_o(1 + x^2)^{-3/2}, \tag{5.57}$$

$$M(r) = 4\pi\rho_o r_c^3\{\ln[x + (1 + x^2)^{1/2}] - x(1 + x^2)^{-1/2}\}, \tag{5.58}$$

$$\phi(r) = -4\pi G\rho_o r_c^2 \frac{\ln[x + (1 + x^2)^{1/2}]}{x}, \tag{5.59}$$

$$x \equiv \frac{r}{r_c}.$$

Here, ρ_o is the central density and r_c is the core radius of the cluster (Section 2.7). The central density and core radius are related to the line-of-sight velocity dispersion σ_r (Section 2.6) of an isothermal cluster by

$$\sigma_r^2 = \frac{4\pi G \rho_o r_c^2}{9}, \tag{5.60}$$

which follows from equation (2.9). Thus the central value of the cluster gravitational potential is

$$\phi_o = -9\sigma_r^2. \tag{5.61}$$

It is perhaps worth noting that although the analytic King model is a good fit to the inner portions of an isothermal sphere, the analytic King potential has a finite depth (given by equation (5.61)), while the isothermal sphere potential is infinitely deep. The total mass associated with the analytic King distribution diverges as the radius increases, and it is usual to cut the distribution off in some way (equations (2.11) and (2.12)). The effect of these cutoffs on the cluster potential and on the resulting hydrostatic gas models is discussed in Sarazin and Bahcall (1977) and Bahcall and Sarazin (1978).

In calculating the cluster potential, we have ignored the effect of individual galaxies. The potential wells associated with individual galaxies are considerably shallower than those associated with clusters. This is shown by the fact that the velocity dispersions of stars in galaxies ($\sigma_* \approx 300$ km/s; Faber and Gallagher, 1979) are much smaller than the velocity dispersions of galaxies in clusters ($\sigma_r \approx 1000$ km/s); the gravitational potentials increase with the square of the velocity dispersions. Thus an individual galaxy will not significantly perturb the distribution of intracluster gas (Gull and Northover, 1975).

5.5.1 *Isothermal Distributions*

The simplest distribution of gas temperatures would be an isothermal distribution, with T_g being constant. The intracluster gas would become isothermal if thermal conduction were sufficiently rapid (see equation (5.41) for the relevant time scale). Alternatively, the gas may have been introduced into the cluster with an approximately constant temperature, and its thermal distribution be unchanged since that time. Lea *et al.* (1973) fit the gas distributions in the Coma, Perseus, and M87/Virgo clusters, assuming that the gas distributions were self-gravitating and isothermal. If $\phi(r)$ is due only to ρ_g, then equation (5.56) and Poisson's equation for the gravitational potential are equivalent to equation (2.9), the equation of an isothermal sphere. Note, however, that this is not trivially true, since equation (2.9) was derived from the stellar dynamical equation for a collisionless gas, while equation (5.56) is the hydrostatic equation for a collisionally dominated fluid. Lea *et al.* therefore used equation (5.57) to fit the gas density ρ_g. This is not consistent, since the gas masses derived from these fits were generally less than 20% of the virial mass of the cluster, and the core radii derived were larger than those of the galaxies (Section 4.4).

More consistent isothermal models can be derived if the gravitational potential of the cluster is not assumed to come only from the gas (Cavaliere and Fusco-Femiano, 1976, 1978; Sarazin and Bahcall, 1977). Equation (5.56) can be written

$$\frac{d \ln \rho_g}{dr} = -\frac{\mu m_p}{kT_g}\frac{d\phi(r)}{dr}. \tag{5.62}$$

If the potential is given by equations (5.59) and (5.61), then the gas distribution is given by

$$\rho_g(r) = \rho_{go}\left[1 + \left(\frac{r}{r_c}\right)^2\right]^{-3\beta/2}. \tag{5.63}$$

Here

$$\beta \equiv \frac{\mu m_p \sigma_r^2}{kT_g} = 0.76\left(\frac{\sigma_r}{10^3\,\mathrm{km/s}}\right)^2\left(\frac{T_g}{10^8\,\mathrm{K}}\right)^{-1}, \tag{5.64}$$

where the numerical value follows if solar abundances are assumed so that $\mu = 0.63$. Another way to derive equation (5.63) is to note that if the galaxy velocity dispersion is isotropic, the galaxy density ρ_{gal} in a cluster will be given by equation (5.62), replacing $kT_g/\mu m_p$ with σ_r^2. Then eliminating the potential between the equations for ρ_g and ρ_{gal} yields $\rho_g \propto \rho_{gal}^{\beta}$, and equation (5.63) follows if the galaxy distribution is given by equation (5.57) (Cavaliere and Fusco-Femiano, 1976).

This self-consistent isothermal model (equation (5.63)) assumes that the gas and galaxy distributions are both static and isothermal and that the galaxy and total mass distributions are identical. While none of these assumptions is fully justified, and the gas is probably not generally isothermal, this model has the advantage that the resulting gas distribution is analytic and that basically all the integrals needed to compare the model to the observations of clusters are also analytic. For example, the total gas mass and emission integral (equation (5.20)) are

$$\begin{aligned} M_g &= \pi^{3/2}\rho_{go}r_c^3\,\frac{\Gamma[3(\beta-1)/2]}{\Gamma(3\beta/2)} \qquad (\beta > 1) \\ &= 3.15 \times 10^{12} M_\odot \left(\frac{n_o}{10^{-3}\,\mathrm{cm}^{-3}}\right) \\ &\quad \times \left(\frac{r_c}{0.25\,\mathrm{Mpc}}\right)^3 \frac{\Gamma[3(\beta-1)/2]}{\Gamma(3\beta/2)}, \end{aligned} \tag{5.65}$$

$$EI = \pi^{3/2}\left(\frac{n_e}{n_p}\right)n_o^2 r_c^3 \frac{\Gamma(3\beta - 3/2)}{\Gamma(3\beta)} \qquad (\beta > 1/2)$$

$$= 3.09 \times 10^{66}\,\mathrm{cm}^{-3}\left(\frac{n_o}{10^{-3}\,\mathrm{cm}^{-3}}\right)^2$$

$$\times \left(\frac{r_c}{0.25\,\mathrm{Mpc}}\right)^3 \frac{\Gamma(3\beta - 3/2)}{\Gamma(3\beta)}, \tag{5.66}$$

where n_o is the central proton density, Γ is the gamma function, and solar abundances have been assumed in deriving the numerical values. The values of β in the parentheses give the limits such that the appropriate integrals converge at large radii. Similarly, the X-ray surface brightness at a projected radius b is proportional to the emission measure EM, defined as

$$EM \equiv \int n_p n_e dl, \tag{5.67}$$

where l is the distance along the line-of-sight through the cluster at a projected radius b. Then, for the self-consistent isothermal model, the emission measure is

$$EM = \sqrt{\pi}\left(\frac{n_e}{n_p}\right)n_o^2 r_c \frac{\Gamma(3\beta - 1/2)}{\Gamma(3\beta)}(1 + x^2)^{-3\beta + 1/2}$$

$$(\beta < 1/6) \tag{5.68}$$

where $x \equiv b/r_c$. The microwave diminution (Section 3.5) at low frequencies is given by

$$\frac{\Delta T_r}{T_r} = -\frac{2\sqrt{\pi}\sigma_T k T_g}{m_e c^2}\left(\frac{n_e}{n_p}\right)n_o r_o$$

$$\times \frac{\Gamma(3\beta/2 - 1/2)}{\Gamma(3\beta/2)}(1 + x^2)^{-(3\beta - 1)/2} \qquad (\beta > 1/3) \tag{5.69}$$

where σ_T is the Thomson cross section, and T_r is the cosmic background radiation temperature. Numerically,

$$\Delta T_r = 0.10\,\mathrm{mK}\left(\frac{n_o}{10^{-3}\,\mathrm{cm}^{-3}}\right)\left(\frac{T_g}{10^8\,\mathrm{K}}\right)\left(\frac{r_c}{0.25\,\mathrm{Mpc}}\right)$$

$$\times \left(\frac{T_r}{2.7\,\mathrm{K}}\right)\frac{\Gamma(3\beta/2 - 1/2)}{\Gamma(3\beta/2)}(1 + x^2)^{-3\beta/2 + 1/2}. \tag{5.70}$$

Equation (5.68) has been used extensively to model the surface brightness $I(b)$ of the X-ray emission from clusters (Gorenstein *et al.*, 1978; Branduardi-Raymont *et al.*, 1981; Abramopoulos and Ku, 1983; Jones and Forman, 1984; Section 4.4.1). The most accurate data have come from the *Einstein* X-ray satellite, which was sensitive only to low energy X-rays ($h\nu < 4$ keV). For high temperature gas ($T_g \gtrsim 3 \times 10^7$ K), the low energy X-ray emissivity is nearly independent of temperature, and thus $I(b) \propto EM$ even if the gas temperature varies. Large surveys of X-ray distributions fit by equation (5.68) have been made by Abramopoulos and Ku (1983), who set $\beta = 1$ (equal gas and galaxy distributions), and Jones and Forman (1984), who allowed β to vary. Figure 16 shows the data on the X-ray surface brightness of three clusters from Jones and Forman (1984) and their best fit models using equation (5.68). Equation (5.68) is a good fit to the majority of clusters, but fails in the central regions of some clusters, possibly because these clusters contain cooling accretion flows (Sections 4.3.3, 4.4.1, and 5.7). The average value of β determined by fits to the X-ray surface brightness of a large number of clusters was found to be (Jones and Forman, 1984)

$$\langle \beta_{fit} \rangle = 0.65. \qquad (5.71)$$

Thus the X-ray surface brightness and implied gas density vary on average as

$$I_x(b) \propto \left[1 + \left(\frac{b}{r_c}\right)^2\right]^{-3/2}, \qquad (5.72)$$

$$\rho_g(r) \propto \left[1 + \left(\frac{r}{r_c}\right)^2\right]^{-1}. \qquad (5.73)$$

This indicates that the gas density should fall off less rapidly with radius than the galaxy density (in agreement with many other observations, such as Abramopoulos and Ku, 1983), and that the energy per unit mass is higher in the gas than in the galaxies (Jones and Forman, 1984). For this average value of β, the total X-ray luminosity converges, but the total gas mass given by equation (5.65) does not.

Unfortunately, this does not agree with the determinations of the X-ray spectral temperatures and the galaxy velocity dispersions of clusters (Mushotzky, 1984). For example, the observed correlation between σ_r and T_g in equation (4.10) implies that the average value β determined by gas temperatures and galaxy velocity dispersions is $\langle \beta_{spect} \rangle \approx 1.3$. From a sample of clusters with well-determined spectra, Mushotzky (1984) finds $\langle \beta_{spect} \rangle \approx 1.2$, which he notes is about twice the value determined from observations of the X-ray surface brightness. While Jones and Forman (1984) argue that their values of β_{fit} are in excellent agreement with the determinations from spectral observations, in fact their data show that the two values do not agree to within the errors in the majority of cases. For the clusters they studied $\langle \beta_{spect} \rangle \approx 1.1$. Thus the general result seems to be that

$$\langle \beta_{spect} \rangle \approx 1.2 \approx 2 \langle \beta_{fit} \rangle. \qquad (5.74)$$

A number of suggestions have been made as to the origin of this discrepancy. First, the gas may very well not be isothermal. However,

Mushotzky (1984) has argued that the same problem occurs for other thermal distributions in the gas. Second, it may be that the line-of-sight velocity dispersion does not represent accurately the energy per unit mass of the galaxies. Equation (5.63) assumes that the galaxy velocity distribution is isotropic. The distribution could be anisotropic, either if the cluster is highly flattened (Section 2.9.3) or if the cluster is spherical but galaxy orbits are largely radial (rather than having a uniform distribution of eccentricities; see Section 2.8). A detailed study (Kent and Sargent, 1983) of the positions and velocities of galaxies in the Perseus cluster has produced a more accurate description of the cluster potential and significantly reduced the discrepancy, although it still is significant. Third, it may be that many of the galaxy velocity dispersions measured for clusters are contaminated by foreground or background groups (Geller and Beers, 1982). The velocity dispersions may also be affected by subclustering or nonvirialization of the cluster. All of these effects will cause the data to overestimate the actual velocity dispersion (the cluster potential), and thus to overestimate β_{spect}. Perhaps one indication that such systematic errors in the velocity dispersion might be occurring is that Coma, the best studied regular cluster, does not show a β discrepancy. Another possible solution to the β discrepancy would be if the gas dominated the total cluster mass at large radii (Henriksen and Mushotzky, 1985; Section 5.5.5); this would invalidate the assumption that the galaxies and gas were test particle distributions in the missing mass potential.

Finally, let me point out an utterly trivial possible explanation of the β discrepancy. The particular form of the gas distribution in the self-consistent isothermal model as given above (equations (5.63) through (5.70)) depends on assuming that the cluster potential is fit by the *King approximation* to the isothermal sphere (equations (5.59) and (5.61)). This approximation breaks down at large radii, where the King model density varies as $\rho \propto r^{-3}$ while a real isothermal sphere density varies as $\rho \propto r^{-2}$. Of course, it is the gas distribution at large radii which has the greatest leverage in affecting the fit to equation (5.63). If the mass density in clusters is *really* isothermal, then equation (5.63) will not fit the observed gas distribution for the correct value of β. As an alternative, let us assume that the cluster mass distribution can be fit by the simple analytic form $\rho(r) = \rho_o(1 + x^2)^{-1}$, which is similar to equation (5.57) but has the correct asymptotic form for an isothermal sphere. Then all the formulae for the self-consistent isothermal sphere remain unchanged if we substitute $\beta \rightarrow (2/3)\beta$ in equations (5.63) through (5.70). Retaining the present definition of β, this is equivalent to $\beta_{fit} = (2/3)\beta_{spect}$, which is essentially consistent with the observations. The observed $\langle \beta_{fit} \rangle \approx 2/3$ implies $\langle \beta_{spect} \rangle \approx 1$ which is consistent with the spectral observations. Another way of expressing all this is to say that the observed gas distribution (equation (5.73)) is essentially that of an isothermal sphere.

This possible discrepancy between the globally determined X-ray temperatures of clusters and their surface brightness will probably only be resolved when spatially resolved X-ray spectra are available, allowing a simultaneous determination of the spatial variation of the gas density and temperature (Chapter 6).

One of the derivations of the self-consistent isothermal model involves noting that the galaxy distribution also solves the hydrostatic equation if the galaxy velocity dispersion is isotropic (see discussion following equation (5.64)). If the galaxy distribution is spherical, but the galaxy velocity dispersions in the directions parallel to the cluster radius (σ_r) and transverse to that direction (σ_t) are different, this derivation fails. However, for a constant (isothermal) galaxy velocity dispersion and a constant anisotropy of the velocity dispersion $\eta \equiv 1 - \sigma_t^2/\sigma_r^2$, the gas density is given by $\rho_g \propto (\rho_{gal} r^{2\eta})^\beta$ (White, 1985).

Finally, Henry and Tucker (1979) have pointed out the existence of a simple relationship $(L_x r_c)^{2/5} \propto T_g$ between X-ray temperatures, luminosities, and core radii in clusters based on the isothermal model.

5.5.2 *Adiabatic and Polytropic Distributions*

The intracluster gas will be isothermal if thermal conduction is sufficiently rapid (Section 5.4.2). On the other hand, if thermal conduction is slow, but the intracluster gas is well-mixed, then the entropy per atom in the gas will be constant (Section 5.4.6). In an adiabatic gas, the pressure and density are simply related,

$$P \propto \rho^\gamma, \tag{5.75}$$

where γ is the usual ratio of specific heats and is $\gamma = 5/3$ for a monatomic ideal gas. While the value of 5/3 would be expected to apply if the intracluster gas were strictly adiabatic, equation (5.75) is often used to parametrize the thermal distribution of the intracluster gas, with γ taken to be a fitting parameter. For example, $\gamma = 1$ implies that the gas distribution is isothermal. Intracluster gas models with an arbitrary value of γ are often referred to as 'polytropic' models, and γ is called the polytropic index. Intracluster gas models with the polytropic index $\gamma > 5/3$ are convectively unstable (equation (5.52)), and thus hydrostatic polytropic models must have $1 \leqslant \gamma \leqslant 5/3$.

Adiabatic and polytropic intracluster gas distributions were introduced by Lea (1975), Gull and Northover (1975), and Cavaliere and Fusco-Femiano (1976). Given equation (5.75), the hydrostatic equation (5.55) can be rewritten by noting that

$$\frac{1}{\rho_g}\nabla P = \frac{\gamma}{\gamma - 1}\frac{k}{\mu m_p}\nabla T_g, \tag{5.76}$$

so that

$$\frac{T_g}{T_{go}} = 1 + (\alpha - 1)\left[1 - \frac{\phi(\mathbf{r})}{\phi_o}\right], \tag{5.77}$$

$$\frac{\rho_g}{\rho_{go}} = \left(\frac{T_g}{T_{go}}\right)^{1/(\gamma - 1)}.$$

Here, ϕ_o, T_{go}, and ρ_{go} are the central values of the cluster gravitational potential, the intracluster gas temperature, and the density, respectively. In nonspherical clusters, 'central' means the lowest point in the cluster gravitational potential.

From equation (5.77), it is clear that the intracluster gas temperature will always decrease with increasing distance from the cluster center in adiabatic or polytropic models.

The integration constant α is defined as

$$\alpha \equiv \frac{T_{go}}{T_{g\infty}}, \tag{5.78}$$

where $T_{g\infty}$ is the gas temperature at infinity. If $\alpha > 0$, the gas distribution extends to infinity, and the gas is not gravitationally bound to the cluster. If $\alpha < 0$, the gas extends only to a finite distance at which $\phi(\mathbf{r})/\phi_o = \alpha/(\alpha - 1)$, and the gas is gravitationally bound to the cluster. In the $\alpha > 0$ models, the intracluster gas connects with and is confined by intercluster gas, whose temperature is given by $T_{g\infty}$. In general, the central gas temperature is given by

$$T_{go} = \frac{\gamma - 1}{\gamma}\frac{\mu m_p \phi_o}{k(1 - \alpha)}. \tag{5.79}$$

The enthalpy per particle in these models is a constant, both spatially and temporally, and is given by $h = \alpha[\gamma/(\gamma - 1)]kT_{go}$. Thus models with $\alpha \approx 0$ cool very rapidly (Bahcall and Sarazin, 1978). Reasonable bound, intracluster gas models therefore usually assume $\alpha \lesssim 0.1$.

If equation (5.59) is used for the cluster potential, then the gas density and temperature are analytic functions of radius in the cluster, but are sufficiently complex that the integrals for emission measures, masses, and so on are not analytic. Moreover, since the gas is not isothermal, the X-ray surface brightness and the emission measure are not simply related. Spectra, X-ray surface brightness profiles, masses, microwave diminutions, and a number of other quantities for these models are given in Sarazin and Bahcall (1977) and Bahcall and Sarazin (1978). Fits of these models to the surface brightness and spectra of a number of clusters are given in Bahcall and Sarazin (1977) and Mushotzky *et al.* (1978).

5.5.3 *More Complicated Distributions*

The models discussed above assume that clusters are spherical, that the gas is hydrostatic, that the cluster potential is known in advance, and that the entropy distribution in the gas is given by a very simple polytropic distribution. Intracluster gas models have been calculated which attempt to generalize each of these assumptions. Here, some more complicated hydrostatic models are reviewed.

Stimpel and Binney (1979) (see also Binney and Stimpel, 1978) showed how spheroidal models for the intracluster gas distribution could be derived, using the observed galaxy distribution to determine the shape of the cluster potential. These models were fit to the galaxy counts in the Coma cluster, and models for the distribution of X-ray emission and microwave diminution (Section 3.5) were derived. There is an uncertainty in the shape of the cluster potential because the galaxy counts themselves cannot determine whether the cluster shape is more nearly prolate or oblate. However, Stimpel and Binney show that the resulting X-ray distributions in the two cases are considerably different, and thus X-ray observations can be used to determine the true shapes of elongated clusters (Chanan and Abramopoulos, 1984).

In general, the entropy distribution in a cluster will depend on the origin of the intracluster gas and the history of the cluster. A number of authors have solved the hydrodynamic equations for simple models for the origin of the intracluster gas and of the subsequent history of the cluster; these calculations will be reviewed in Section 5.10. It is perhaps not surprising that these calculations do not generally lead to entropy distributions of the very simple sort (isothermal, adiabatic, or polytropic) considered above.

A number of authors have attempted to model these more detailed entropy distributions by allowing the polytropic index or the isothermal parameter β to vary with radius. One example is to allow the gas to be isothermal in the cluster core, where conduction is effective (Section 5.4.2), but adiabatic in the outer parts (Cavaliere and Fusco-Femiano, 1978; Cavaliere, 1980). Cavaliere and Fusco-Femiano (1978) also included the effect of the gas density on the cluster potential.

5.5.4 *Empirical Gas Distributions Derived by Surface Brightness Deconvolution*

The gas distributions in clusters can be derived directly from observations of the X-ray surface brightness of the cluster, if the shape of the cluster is known and if the X-ray observations are sufficiently detailed and accurate. This method of analysis also leads to a very promising method for determining cluster masses (Section 5.5.5). The X-ray surface brightness at a photon frequency ν and at a projected distance b from the center of a spherical cluster is

$$I_\nu(b) = \int_{b^2}^{\infty} \frac{\varepsilon_\nu(r)dr^2}{\sqrt{r^2 - b^2}}, \qquad (5.80)$$

where ε_ν is the X-ray emissivity. This Abel integral can be inverted to give the emissivity as a function of radius,

$$\varepsilon_\nu = -\frac{1}{2\pi r}\frac{d}{dr}\int_{r^2}^{\infty} \frac{I_\nu(b)db^2}{\sqrt{b^2 - r^2}}. \qquad (5.81)$$

Because of the quantized nature of the observations of the X-ray surface brightness (photon counts per solid angle) and the sensitivity of integral deconvolutions to noise in the data, the X-ray surface brightness data are often smoothed, either by fitting a smooth function to the observations or by applying these equations to the surface brightness averaged in rings about the cluster center.

Now, the emissivity is given by equation (5.19) and depends on the elemental abundances, the density of the gas, and the gas temperature. Thus the distribution of these three properties in the cluster could be determined from observations of the X-ray surface brightness $I_\nu(b)$. Basically, the continuum emission is due mainly to free–free emission (equation (5.11)) and is relatively insensitive to the heavy element abundances, while the line emission measures these abundances. For a given set of abundances, the emissivity can be written as

$$\varepsilon_\nu = n_e^2 \Lambda_\nu(T_g), \qquad (5.82)$$

where $\Lambda_\nu(T_g)$ is n_p/n_e times the sum on the right side of equation (5.19). Thus $\varepsilon_\nu(r)$ can be found from observations of the X-ray surface brightness as a function of photon frequency, and the frequency dependence and magnitude of ε_ν give the local temperature and density.

Unfortunately, observations of $I_\nu(b)$ are really not available for clusters. These observations require an instrument with good spatial *and* spectral resolution. Most of the X-ray spectra of clusters have been taken with instruments having very poor spatial resolution (comparable to the size of the cluster; Section 4.3.1). High spatial resolution observations of clusters have primarily come from the *Einstein* X-ray observatory (Section 4.4). The imaging instruments on *Einstein* had only limited spectral resolution. Moreover, the optics in *Einstein* could only focus soft X-rays ($h\nu \lesssim 4$ keV). At typical gas temperatures in clusters ($kT_g \approx 6$ keV), most of the X-ray emissivity is due to thermal bremsstrahlung, and the emissivity is nearly independent of frequency for $h\nu \lesssim kT_g$. Thus the *Einstein* surface brightness distributions cannot be used directly to determine the local temperature. If the limited spectral resolution of the *Einstein* imagers is ignored, their observations provide $\langle I_x(b) \rangle$, the surface brightness averaged over the sensitivity of the detector as a function of photon frequency. From $\langle I_x(b) \rangle$, the sensitivity averaged emissivity

$$\langle \varepsilon_x(r) \rangle = n_e^2 \langle \Lambda_x(T_g) \rangle \tag{5.83}$$

can be found from equation (5.81). Unfortunately, even if the elemental abundances are assumed to be known, this equation provides only one quantity at each radius, and it is impossible to determine both n_e and T_g.

In many analyses of X-ray cluster observations, a second equation for the density and temperature has been provided by assuming that the intracluster gas is hydrostatic and that the cluster potential is known. The hydrostatic equation in a spherical cluster (5.56) can be rewritten as

$$\frac{d \ln(n_e T_g)}{d \ln r} = -\frac{GM(r)\mu m_p}{rkT_g}, \tag{5.84}$$

where $M(r)$ is the total cluster mass and the mean atomic weight $\mu \approx 0.63$ is assumed to be independent of radius. Combining this with equation (5.83) gives

$$\left[1 - \frac{1}{2}\frac{d \ln\langle \Lambda_x \rangle}{d \ln T_g}\right]\frac{d \ln T_g}{d \ln r} = -\frac{GM(r)\mu m_p}{rkT_g} - \frac{1}{2}\frac{d \ln\langle \varepsilon_x \rangle}{d \ln r}. \tag{5.85}$$

This is an ordinary differential equation for $T_g(r)$, which can be integrated given a boundary condition. This has been taken to be the central temperature (White and Silk, 1980) or the intergalactic pressure (Fabian *et al.*, 1981a). Given $T_g(r)$, equation (5.83) gives the density profile.

Various versions of this method have been used to determine gas distributions in a large number of clusters using data from *Einstein* (White and Silk, 1980; Fabricant *et al.*, 1980; Fabian *et al.*, 1981; Nulsen *et al.*, 1982; Fabricant and Gorenstein, 1983; Canizares *et al.*, 1983; Stewart *et al.*, 1984a, b). In some cases, spectral information from *Einstein* or low spatial resolution spectra have been used to further constrain the temperature profiles or to determine the form of the cluster potential necessary for a consistent fit (Section 5.5.5). These analyses have provided information on the mass distribution in clusters and central galaxies (M87, in particular), the gas distributions in clusters, and the prevalence of cooling accretion flows in clusters. These topics will be discussed in more detail later.

Gas temperature and density profiles could be derived more directly if high spatial and spectral resolution data at photon energies up to 10 keV were available. The proposed AXAF satellite will have these capabilities (Chapter 6).

5.5.5 *Total Masses and Mass Distributions in Clusters – The Hydrostatic Method*

Masses for individual galaxies or for clusters of galaxies can be derived from the distribution of their X-ray emitting gas if this gas is in hydrostatic equilibrium. This is a reasonable approximation as long as the cluster is stationary (the gravitational potential does not change on a sound crossing time), forces other than gas pressure and gravity (magnetic fields, for example) are not important, and any motions in the gas are subsonic. Estimates of intracluster magnetic fields based on radio observations (Chapter 4) show that they are much too small to have a significant dynamical effect. In Section 5.6, we will show that supersonic expansion of the gas (cluster winds) is unlikely. Some clusters show evidence for a slow settling of the intracluster gas due to cooling at the cluster center (a cooling flow; Sections 4.3, 4.4, and 5.7), but these motions are very subsonic except possibly very near the cluster center ($r \lesssim 1$ kpc; equation (5.105)).

Under these circumstances, the gas obeys the hydrostatic equation (5.56), and the cluster mass $M(r)$ can be determined if the density and temperature of the intracluster gas are known. This method of determining the mass has a number of advantages over the use of the virial theorem (Section 2.8) or any other method which uses the galaxies as test particles. First, the gas is a collisional fluid, and the particle velocities are isotropically distributed. On the other hand, galaxies in clusters (or stars in galaxies) are collisionless, and uncertainties in the velocity anisotropy can significantly affect mass determinations. Second, the hydrostatic method gives the mass as a function of radius, rather than the total mass alone as given by the virial method. Third, the statistical accuracy of this method is not limited by the number of galaxies in the cluster; the statistical accuracy can be improved by lengthening the observation time. Fourth, better statistics in the X-ray measurements means that it is easier to avoid problems with background contamination, and to resolve possible uncertainties due to subclustering (Geller and Beers, 1982). Finally, hydrostatic mass determinations are not very sensitive to the shape of the cluster (Strimpel and Binney, 1979; Fabricant *et al.*, 1984).

The first applications of this method to determine mass distributions were by Bahcall and Sarazin (1977) and Mathews (1978). The method has been developed extensively by Fabricant *et al.* (1980, 1984) and Fabricant and Gorenstein (1983). Applications of the method have been reviewed by Sarazin (1986b). Ideally, one would measure the spatially and spectrally resolved X-ray surface brightness $I_\nu(b)$ to directly deconvolve the gas density and temperature as a function of the radius (Section 5.5.4; equations (5.80) through (5.83)). The mass is then given by the hydrostatic equation, which can be written as

$$M(r) = -\frac{kT_g(r)r}{\mu m_p G}\left(\frac{d\ln n_e}{d\ln r} + \frac{d\ln T_g}{d\ln r}\right).$$

Note that the mass depends only weakly on the gas density n_e (only its logarithmic derivative enters), but depends strongly on the gas temperature.

Unfortunately, as discussed in Section 5.5.4, the limited spectral response of the *Einstein* X-ray observatory has prevented the direct determination of temperature profiles for the intracluster gas. Accurate profiles of the gas density are known. In order to apply the hydrostatic method to clusters, some simple assumption must be made about the temperature distribution $T_g(r)$. Unfortunately, because the mass is strongly affected by T_g, this means that the resulting mass profiles will be very uncertain. Several analyses (Vallee, 1981; Fabricant *et al.*, 1984) have assumed that gas is isothermal (Section 5.5.1) or that the gas temperature and density are related by a simple polytropic equation (Section 5.5.2). These analyses give a somewhat smaller total cluster mass than previous virial estimates, and as noted in Section 5.5.1, somewhat higher gas masses.

As discussed in Section 4.3, excellent global cluster X-ray spectra exist from the HEAO-1 A-2 detectors. These spectra generally cannot be fit by emission at a single temperature (Henriksen, 1985). The spectra can be used to determine the amount of gas (or, more precisely, the amount of $EI = n_p n_e V$, where V is the volume (equation (4.3))) as a function of temperature, but cannot tell us where the gas is located because of their poor spatial resolution. The *Einstein* imaging observations give $n_e(r)$ (which can be integrated to give $n_p n_e V$), but give no information about the temperature structure. However, the comparison of these two results ((EI vs. T_g) and (EI vs. r)) allow the determination of (T_g vs. r), if *we assume that* T_g *is a monotonic function of the radius* r (Henriksen, 1985; Henriksen and Mushotzky, 1986; Cowie *et al.*, 1987). Since the observed gas densities vary at large distances like an isothermal sphere $\rho_g \propto r^{-2}$ (equation (5.73)), while these mass determinations require that the gas temperature decrease with radius, the total mass density will always decrease with radius more rapidly than the gas density. Since the total density is the sum of the gas density, the galaxy density, and the missing mass density, these determinations suggest that the missing mass is concentrated towards the cluster center. For the Coma cluster, this method gives a *missing mass density which is more centrally concentrated than that of the galaxies*, which is in turn more centrally concentrated than that of the intracluster gas. Values of the mass-to-light ratio of the entire cluster of $M/L_V \approx 100h_{50}$ are found, with the intracluster gas contributing about 30% of the mass.

If these monotonic temperature mass determinations are correct, one is led to a different picture of the missing mass than that presented in Section 2.8. If the missing mass is more concentrated than the visible matter in the cluster, it suggests that the missing mass has undergone dissipation. Combined with the smaller ratio of missing mass to visible mass, this suggests that the missing mass is baryonic matter, and not some weakly interacting species. The major uncertainty in these analyses is the assumption of a monotonic temperature gradient. While this seems quite plausible, there is no compelling physical argument requiring that this be true. While most detailed models for the ejection or infall of intracluster gas give monotonic gradients (Section 5.10), some do not.

When spatially and spectrally resolved X-ray surface brightness measurements are available, it will be possible to directly determine the mass profiles of clusters of galaxies and determine the distribution of the missing mass. The same method can be applied to the X-ray emitting gas in elliptical galaxies (Section 5.8). This capability should become available with the launch of AXAF (Chapter 6). Until that time, the present results on the distribution of the missing mass must be regarded as tantalizing but tentative.

5.5.6 *Chemically Inhomogeneous Equilibrium Models*

The heavier an ion is, the slower its thermal motion in the intracluster gas will be. As a result, heavy ions will tend to drift toward the center of the cluster. In Section 5.4.5 the drift rate was calculated and shown to be rather slow, although the precise value depends on cluster parameters. Further, the intracluster magnetic field may be very effective in retarding any sedimentation of heavy elements. Nonetheless, it is possible that heavier elements may have drifted to the cluster center in some cases, or been introduced into the intracluster gas quite near the center. Eventually, the motion of heavy ions towards the cluster center will stop when the resulting charge separation produces an electric field that cancels the effect of the gravitational field.

Abramopoulos *et al.* (1981) have calculated chemically inhomogeneous models for the intracluster gas in thermodynamic equilibrium. The temperature of the gas is therefore constant, both spatially and among the various ions and electrons. These models satisfy the hydrostatic equation (5.55) for the total pressure, but this does not uniquely define the chemically inhomogeneous model, since any value of the total pressure can be attributed to an infinite number of different combinations of partial pressures of different ions. In thermodynamic equilibrium, the number density of any ion is given by the Boltzmann equation

$$n_i(\mathbf{r}) = n_{io} \exp\left[-\frac{A_i m_p \phi(\mathbf{r}) + Z_i e \phi_E(\mathbf{r})}{kT_g} \right], \tag{5.86}$$

where A_i, Z_i, and n_{io} are the mass, charge, and central density of the ion i. The cluster gravitational potential ϕ and electrical potential ϕ_E are defined in this equation to be zero at the cluster center. The gravitational potential is assumed to be fixed and known, as in most of the hydrostatic models discussed in the previous sections. The electrical potential is given as a solution of Poisson's equation

$$\nabla^2 \phi_E(\mathbf{r}) = -4\pi \sum_i n_i(\mathbf{r}) Z_i e, \tag{5.87}$$

where the sum includes electrons as one of the ionic species. Because the coupling constant for electrical forces exceeds that of gravitational forces by $\approx 10^{40}$, while the fraction of the cluster mass due to intracluster gas is much greater than the inverse of this, the electrical potential will generally be quite small $\phi_E \approx -\phi m_p/e$, and a sufficient approximation to the solution of equation (5.87) is to assume that the gas is very nearly electrically neutral:

$$\sum_i n_i Z_i = 0. \tag{5.88}$$

The electric potential is then given by requiring a simultaneous solution of equations (5.86) and (5.88), and requiring that the total abundance $\int n_i dV$ of each of the ions be fixed.

In the case where the plasma can be assumed to consist only of protons and electrons, with the abundances of all heavy elements being so low that they do not contribute to the potential, the solution for the ion and electron densities is the same as the isothermal hydrostatic solution given above (equation (5.63)), the electric potential is $\phi_E = -m_p\phi/2e$, and the densities of any trace heavy elements will be

$$n_i = n_{io}\left(\frac{n_p}{n_{po}}\right)^{2A_i - Z_i}. \tag{5.89}$$

Unfortunately, the abundance of helium certainly cannot be treated as insignificant, and other heavier elements may contribute to the electric potential near the cluster core. Equations (5.86) and (5.88) have been solved numerically by Abramopoulos *et al.* (1981), assuming an analytic King potential for the cluster, taking the gas temperature to be $\beta = 2/3$ (equation (5.64)), and assuming a range of heavy element abundances from pure hydrogen and helium to solar. They find that the heavy elements are strongly concentrated to the cluster core. To balance the resulting increase in the charge density, the electrons also must be centrally condensed.

The concentration of heavy elements to the cluster core, where the electron density is high, results in an increase in the X-ray line intensity for a given set of heavy element abundances. As a result, Abramopoulos *et al.* (1981) argue that the observed line strengths could be produced by gas with iron abundances of only 1/20 of solar.

5.6 Wind Models for the Intracluster Gas

If the rate of heating of the intracluster medium is sufficiently large, the gas will be heated to the escape temperature from the cluster, and will form a wind leaving the cluster. Since the sound crossing time across a cluster is relatively short (equation (5.54)), the cluster would be emptied of gas rapidly unless gas were constantly being added to the cluster. Yahil and Ostriker (1973) suggested that gas is being injected into the intracluster medium at rates of 10^3–$10^4 M_\odot$/yr, and is sufficiently heated to produce a steady-state outflow of gas. These mass ejection rates are considerably larger than those expected from stellar mass loss from visible stars in galaxies (Section 5.10).

Let $\dot{\rho}$ be the rate of mass input per unit volume into the intracluster gas, and let $h(r)$ be the heating rate per unit volume. If the cluster is assumed to be spherically symmetric, with a gravitational potential $\phi(r)$, and the gas is injected into the cluster with no net velocity, but with the velocity dispersion of the galaxies, then the hydrodynamic equations for a steady-state wind are

$$\begin{gathered}
\frac{1}{r^2}\frac{d}{dr}(r^2\rho_g v) = \dot{\rho}, \\
\dot{\rho}v + \rho_g v\frac{dv}{dr} + \frac{dP}{dr} + \rho_g\frac{d\phi(r)}{dr} = 0, \\
\frac{1}{r^2}\frac{d}{dr}\left\{r^2\rho_g v\left[\frac{v^2}{2} + \frac{5}{2}\frac{P}{\rho_g} + \phi(r)\right]\right\} = h(r) + \dot{\rho}\left[\frac{3}{2}\sigma_r(r)^2 + \phi(r)\right],
\end{gathered} \tag{5.90}$$

where ρ_g, P, and v are, respectively, the gas density, pressure, and velocity, and σ_r is the one-dimensional velocity dispersion of the galaxies in the cluster.

Yahil and Ostriker argued that the source of mass input into the cluster was mass loss by stars in cluster galaxies, and that

$$\dot{\rho}(r) = \alpha_* \rho(r), \tag{5.91}$$

where ρ is the total mass density of the cluster and α_*^{-1} is the characteristic time scale for gas ejection from galaxies. They argued that $\alpha_*^{-1} \approx 10^{12}$ yr.

Yahil and Ostriker considered two mechanisms that might heat the gas. First, if the gas was ejected by supernovae it could be heated within galaxies before being ejected into the cluster (Section 5.3.3; equation (5.27)); they referred to this as the HIG (heating in galaxies) model. Defining $\lambda \equiv 3kT_{ej}/(2\mu m_p \sigma_r^2)$, the heating rate due to ejection from galaxies becomes

$$h(r) = \lambda \sigma_r^2 \dot{\rho}. \tag{5.92}$$

Alternatively, Yahil and Ostriker suggested that the heating might be due to the motion of galaxies through the cluster (Section 5.3.4); they referred to this as the HBF (heating by friction) model. Then, the heating rate is

$$h(r) = \pi \langle R_D^2 \Delta v^3 \rangle \rho_g \left[\frac{\rho(r)}{m_{gal}} \right], \tag{5.93}$$

where R_D is the drag radius (equation (5.28)), $m_{gal} \equiv \rho/n_{gal}$ is the total cluster mass per galaxy, Δv is the velocity of the galaxy relative to the gas (which is moving at v), and the average is over the Gaussian distribution of galaxy velocities. It is useful to define $\lambda(r)$ such that $h(r) \equiv \lambda(r)\sigma_r^2 \dot{\rho}$, so that λ is constant in the HIG model, and $\lambda(r) \approx \pi R_D^2 \rho_g \sigma_r / \alpha_* m_{gal}$ in the HBF model.

Let the cluster density be given by $\rho = \rho_o g(x)$, where ρ_o is the central density, $x \equiv r/r_c$, r_c is the cluster core radius, and $g(x)$ is a dimensionless function. Similarly, let the cluster potential be written as $\phi(r) = \phi_o \bar{\phi}(x)$, where ϕ_o is the central potential. If the mass distribution is assumed to be isothermal, so that σ_r is constant and is related to the cluster central potential ϕ_o and the central density ρ_o by equations (5.60) and (5.61), then the equation of continuity becomes

$$x^2 \rho_g v = \alpha_* \rho_o r_c G(x), \tag{5.94}$$

$$G(x) \equiv \int_0^x g(x')(x')^2 dx'.$$

If the King analytic form of the cluster density is assumed, then g, G, and $\bar{\phi}$ are given by the functions on the right-hand sides of equations (5.57), (5.58) and (5.59), respectively. The energy equation can also be integrated to give

$$\frac{1}{2} v^2 + \frac{5}{2} \frac{P}{\rho_g} = \sigma_r^2 \left[\Lambda(x) + \frac{3}{2} + 9\bar{\phi} - 9\Phi \right], \tag{5.95}$$

where Λ and Φ are the integrated heating rate and potential energy

$$\Lambda(x) \equiv \frac{\int_0^x g(x')\lambda(x')(x')^2 dx'}{G(x)}, \tag{5.96}$$

$$\Phi(x) \equiv \frac{\int_0^x g(x')\bar{\phi}(x')(x')^2 dx'}{G(x)}.$$

The left-hand side of equation (5.95) is positive definite, which implies that there is a minimum heating rate at any radius

$$\Lambda(x) > 9\Phi - 9\bar{\phi} - \frac{3}{2}. \tag{5.97}$$

If the King analytic model for the potential is assumed, then the maximum value for the right-hand side of equation (5.97) is 1.35, and this is the minimum value of λ in the HIG model (equation (5.92)). Similarly, the central temperature is given by

$$kT_o = \frac{2}{5}\sigma_r^2 \mu m_p \left[\Lambda(0) + \frac{3}{2}\right], \tag{5.98}$$

so that the minimum central temperature in the HIG model is

$$T_o(HIG) > 8.7 \times 10^7 \text{ K} \left(\frac{\sigma_r}{1000 \text{ km/s}}\right)^2. \tag{5.99}$$

Similar arguments for the HBF model indicate that $\pi R_D^2 \rho_{go} \sigma_r / (\alpha_* m_{gal}) \gtrsim 3/2$, and that the central temperature is

$$T_o(HBF) \gtrsim 3.4 \times 10^8 \text{ K} \left(\frac{\sigma_r}{1000 \text{ km/s}}\right)^2. \tag{5.100}$$

These analytic results were given in Bahcall and Sarazin (1978). The temperatures are lower than those in Yahil and Ostriker (1973) because they assumed $\mu = 1$, which is not valid for an ionized plasma.

The observed cluster temperatures (Section 4.3.1) are generally below those required by equations (5.99) or (5.100), so it does not seem that the gas is heated sufficiently to produce a wind. The energy required to support such a wind also seems to be prohibitively large, since it would require that considerably more than the whole thermal content of the gas (more because of the kinetic energy of the flow) be produced in a flow time, which for a transonic wind is comparable to a sound crossing time. In the HIG model, this would require a very large supernova rate or other energy source; in the HBF model, it would require that the galaxies have delivered to the gas much more kinetic energy than they currently possess (Section 5.3.4). As a result, it seems unlikely that wind models fit the distribution of the intracluster gas at all radii, but it remains possible that gas is flowing out from the outer portions of clusters.

Livio *et al.* (1978) argue that viscous drag between intracluster and interstellar gas can produce a very large heating rate due to galaxy motions if the intracluster gas is sufficiently hot, so that the Reynolds number is small (equations (5.46) and (5.47)). As discussed in Section 5.3.4, this requires that galaxies maintain large amounts of intracluster gas; it seems more likely that the drag forces will strip the gas from the galaxies. Unless the galaxies have a large rate of mass outflow, Rephaeli and Salpeter (1980) have shown that the drag does not increase for small Reynolds number (Section 5.4.4). Again, there is the problem that the heating required to support a strong wind is greater than the total kinetic energy content of the galaxies.

Lea and Holman (1978) discussed winds driven by heating by relativistic electrons (Section 5.3.5). They derived analytic heating limits like those of equation (5.59). To produce a steady-state wind, the heating rate must be much greater than that necessary to heat the gas in a Hubble time, and as a result the required values of γ_l are very low. They find that even in the Perseus cluster, which is one of the strongest radio clusters known, $\gamma_l \ll (B/\mu\text{G})^{-(\alpha_r+1)/2\alpha_r}$ is required for a steady-state wind. Moreover, this mechanism suffers the problem that extended halo radio sources, of the sort necessary to heat the intracluster gas, are relatively rare.

In summary, it seems unlikely that the intracluster gas in most clusters is involved in steady-state outflow, although gas may be flowing out of the outer portions of clusters.

5.7 Cooling Flows and Accretion by cDs

5.7.1 *Cooling Flows*

If the rate of cooling in the intracluster gas is sufficiently rapid, gas may cool and flow into the center of the cluster. Lea *et al.* (1973), Silk (1976), Cowie and Binney (1977), and Fabian and Nulsen (1977) noted that the cooling times t_{cool} (equation (5.23)) in the more luminous X-ray clusters are comparable to the Hubble time t_H (the age of the universe and approximate age of clusters). This might be a coincidence, but it would also be a natural consequence of the cooling of the intracluster gas. Imagine that a certain amount of gas was introduced into the cluster when it formed. Initially, the cooling time in this gas might be much longer than the age of the cluster, and cooling would be unimportant. However, as long as cooling was more rapid than heating, and the gas was not removed from the cluster, the cooling time would not increase, and eventually the age of the cluster would exceed the cooling time. Once this occurs, the gas will cool in the cluster core, and the pressure of the surrounding gas will cause the cool gas to flow into the cluster center. The surrounding hot intracluster gas will always have $t_{cool} \approx t_H$. Silk (1976) and Cowie and Binney (1977) showed that this cooling condition could explain a number of the observed correlations between the optical and X-ray properties of clusters, particularly the $L_x - \sigma_r$ relation (equation (4.8)). The subject of these cooling flows has been reviewed recently by Fabian *et al.* (1984b).

Silk (1976) suggested that the intracluster gas was introduced into the cluster when it formed, and that cooling has reduced the amount of intracluster gas so as to maintain $t_c \approx t_H$. Cowie and Binney (1977), on the other hand, argued that the intracluster gas is constantly being replenished by ejection from cluster galaxies. Then, the density of gas in the cluster center would increase until the age exceeded the cooling time. After this point, a stable steady-state would be achieved, in which the rate of ejection of gas by galaxies into the cluster was balanced by the rate of removal of gas through cooling in the cluster core (Cowie and Binney, 1977; Cowie and Perrenod, 1978). As long as the mass ejection rate is not too low, cooling would regulate the inflow, and the densest hot gas at the cluster center would always have $t_{cool} \approx t_H$.

The equations for a steady-state cooling flow in a spherically symmetric cluster are identical to those for a wind (equation (5.90)), except that the heating rate $h(r)$ is replaced by the cooling rate ε (equations (5.21) and (5.22)). Because the cooling rate is proportional to the square of the density, it is useful to define $\varepsilon \equiv \rho_g^2 \Lambda(T_g)$. Then,

$$\frac{1}{r^2}\frac{d}{dr}(r^2\rho_g v) = \dot{\rho},$$

$$\dot{\rho}v + \rho_g v\frac{dv}{dr} + \frac{dP}{dr} + \rho_g\frac{d\phi(r)}{dr} = 0, \tag{5.101}$$

$$\frac{1}{r^2}\frac{d}{dr}\left\{r^2\rho_g v\left[\frac{v^2}{2} + \frac{5}{2}\frac{P}{\rho_g} + \phi(r)\right]\right\} = -\rho_g^2\Lambda(T_g) + \dot{\rho}\left[\frac{3}{2}\sigma_r(r)^2 + \phi(r)\right],$$

and the other symbols are defined following equation (5.90). If the gas is injected continuously into the cluster ($\dot{\rho} \neq 0$), then the boundary conditions are set by requiring no inflow from outside the cluster. If the gas is not constantly being added to the cluster ($\dot{\rho} = 0$), then no cluster-wide steady-state flow will be possible. However, these equations will still apply approximately within the radius defined by setting the cooling time equal to the age of the cluster (the 'cooling surface'). In order to match to the hydrostatic intracluster gas distribution beyond this radius, the proper density and temperature is specified on this surface. Because the inflow is generally extremely subsonic except in the innermost parts, the outer parts of these cooling flows are very nearly hydrostatic (the middle equation in (5.101) reduces to (5.55)). The velocity is determined by the inflow rate $\dot{M} \equiv 4\pi\rho_g v r^2$, which must be constant in a steady-state inflow without sources or sinks for mass.

Thermal conduction (Section 5.4.2) has not been included in the cooling flow equations. Generally, the existence of cooling flows requires that conduction not be very important, since otherwise cooling in the cluster core will be balanced by heat transported inwards by conduction, rather than heat convected inwards by the cooling flow (Mathews and Bregman, 1978; Takahara and Takahara, 1979; Binney and Cowie, 1981; Nulsen *et al.*, 1982). In most cases, conduction must be much slower than the Spitzer rate (equation (5.37)) for a nonmagnetic plasma; the suppression of conduction might be due to a very tangled or circumferential magnetic field (Section 5.4.3).

These equations have been solved for cooling in clusters by Cowie and Binney (1977), assuming a King model for the cluster density (equation (2.13)), mass ejection by galaxies as given by Yahil and Ostriker (1973; equation (5.91)), pure bremsstrahlung cooling (equation (5.21)), and subsonic flow (dropping the quadratic velocity terms in equations (5.101)). Figure 37 shows the density and temperature variations in a typical model. The density rises continuously into the center; this contrasts with hydrostatic models, in which it levels off (Section 5.5). The temperature has a maximum at about two core radii and declines into the cluster center. The behavior of these solutions in the inner regions can be understood approximately. Assuming that the mass flux is fixed ($\dot{M} = \text{const}$), that the flow is subsonic, that the gravitational potential is not important (so that the pressure is nearly constant and the flow is driven by pressure and cooling and not by gravity), and that the cooling function is a power-law in the temperature $\Lambda \propto T_g^\alpha$, the temperature and density are found to vary as $\rho_g \propto 1/T_g \propto r^{-3/(3-\alpha)}$ (Nulsen *et al.*, 1982). For $T_g \gtrsim 4 \times 10^7$, cooling is due to thermal bremsstrahlung (equation (5.21)), $\alpha = 1/2$, and $\rho_g \propto r^{-6/5}$. At lower temperatures, $\alpha \approx -0.6$ (equation (5.22)) and $\rho_g \propto r^{-5/6}$. When the gas has cooled sufficiently for the gravitational potential to be important (or if the potential gradient is increased by a central galaxy), then the temperature tends to vary as $kT_g/\mu m_p \approx GM(r)/r$, where $M(r)$ is the total mass within r (Fabian and Nulsen, 1977).

Evidence for the existence of such cooling flows in X-ray clusters includes the detection of peaks in the soft X-ray surface brightness at the cluster center (Branduardi-Raymont *et al.*, 1981; Fabian *et al.*, 1981a; Canizares *et al.*, 1983; Fabricant and Gorenstein, 1983; Jones and Forman, 1984; Stewart *et al.*, 1984a, b), the measurement of inverted temperature gradients $dT_g/dr > 0$ (Gorenstein *et al.*, 1977; Ulmer and Jernigan, 1978; White and Silk, 1980), and the observation of central X-ray surface brightnesses and temperatures that imply cooling times much less than the Hubble time (Canizares *et al.*, 1983; Stewart *et al.*, 1984b). The strongest evidence comes from the detection of soft X-ray line emission from low ionization stages produced at temperatures of $T_g \approx 10^6$–10^7 K, coming from the cluster center (Canizares *et al.*, 1979, 1982; Canizares, 1981; Mushotzky *et al.*, 1981; Lea *et al.*, 1982; Nulsen *et al.*, 1982; Culver *et al.*, 1983; Mushotzky, 1984; Section 4.3.3).

Theoretical models for cooling flows can be used to estimate the rates of cooling in clusters from these X-ray observations. A simple estimate may also be derived by noting that for steady-state, isobaric cooling, the luminosity emitted in any temperature range dT_g is

$$dL(T_g) = \frac{5}{2} \frac{\dot{M}}{\mu m_p} k dT_g \tag{5.102}$$

if the gravitational potential does not change significantly during the cooling (Cowie, 1981). Here, μ is the mean atomic mass. This equation can be used to estimate the luminosity in any spectral feature produced in the cooling flow by integrating $dL(T_g)$ over the fraction of the total emission in that feature as a function of temperature. A more accurate semi-empirical method of

Fig. 37. The gas temperature and density in a cluster with a steady-state cooling flow, from the models in Cowie and Binney (1977). The gas density is normalized to its value at the cluster core radius, and the gas temperature is normalized to $\mu m_p \sigma_r^2/k$, where σ_r is the cluster velocity dispersion and μ is the mean atomic mass. The radius from the cluster center is in units of the cluster core radius.

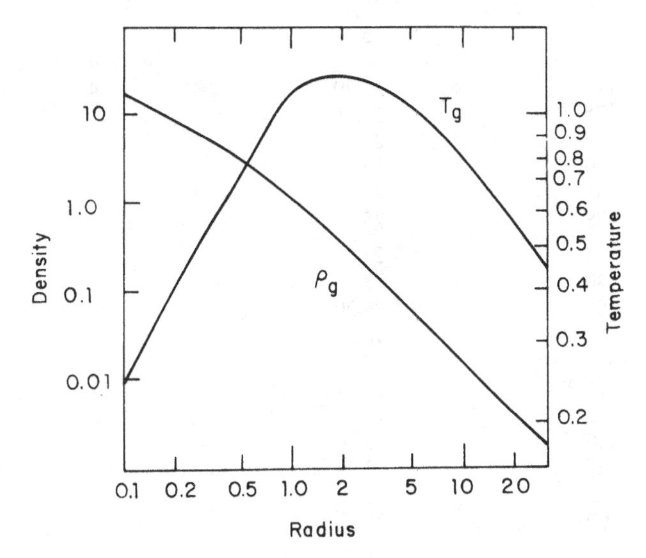

determining cooling rates for clusters has been derived by Canizares *et al.* (1983), Stewart *et al.* (1984b), and Fabian *et al.* (1984b). This method is essentially that of Section 5.5.4. First, the X-ray surface brightness is deconvolved to give the X-ray emissivity as a function of position (equation (5.81)). Second, the hydrostatic equation is used to derive the gas temperature as a function of position, assuming some form for the gravitational potential

Table 4. *Clusters that show evidence for cooling flows*
Cooling rates scale as h_{50}^{-2}. First, Abell clusters are listed, then non-Abell clusters. The references for the X-ray or optical emission line observations are listed below.

Cluster	$\dot{M}$ ($M_\odot$/yr)	References: X-ray	References: Optical
A85	100	25	13,4
A262	28	25	14
A400	2	25	
A426 Perseus	300	6,21,1,20	15,4
A496	200	22,20	13,7,4
A576	40	26,23	
A978	500		13
A1060	6	25	
A1126	500		13,14
A1795	400	25,20	13,4
A1983	7	25	
A1991	115	25	
A2029	250	25,20	4
A2052	120		13,14,5
A2063	26	25	
A2107	18	25	
A2142	28	16	4
A2199	110	25,20	4
A2319	75	25,26	4
A2415	15	25	
A2626	10	25	
A2657	36	25	
A2670	78	25	
SC0107-46	4	25	
AWM7	40	3	
2A0335 + 096	280	24,20	
MKW4	7	18,3	
SC0745-191	1000	9	9
M87/Virgo	3-20	10,11,2,17,20	12
Centaurus	22	19	8
MKW3s	100	3	
AWM4	25	3	
SC1842-63	3	25	
SC2059-247	500	27	

(equation (5.85)). (Note that the hydrostatic equation holds for the outer portions of these cooling flows because they are very subsonic; see Section 5.7.2.) Third, the cooling time is calculated in the gas, and the gas is assumed to form a steady-state cooling flow within the cooling surface at which $t_{cool} \approx t_H$. Finally, the steady-state energy equation (the last of equations (5.101)) is used to determine the cooling rate $\dot{M}$.

Table 4 gives a list of clusters showing evidence for cooling flows, with estimates for the cooling rate $\dot{M}$. The values for $\dot{M}$ range from 3–1000$M_\odot$/yr. The Perseus cluster has one of the largest observed cooling rates (see also Section 4.5.2). A very small cooling rate of 3–20$M_\odot$/yr is found in M87 in the Virgo cluster; such a small cooling flow would be difficult to observe in more distant clusters. It is particularly interesting that cooling flows have now been detected in poor clusters (Canizares *et al.*, 1983), as well as in a large number of rich clusters (Stewart *et al.*, 1984b).

It should perhaps be pointed out that all of the above evidence shows that there is *cool* gas in the cores of these clusters, but it does not directly show that there is *cooling* gas flowing into the cluster core. That is, the motion of the gas has not been detected directly, as X-ray spectral observations have insufficient resolution to detect Doppler shifts produced by the slow inflow. It has occasionally been argued that the gas might actually be in hydrostatic and thermal equilibrium, with the extra emission at the cluster center being due to some central heat source (see, for example, Tucker and Rosner, 1983). Since the cooling rate of a gas at constant pressure decreases with temperature at all temperatures of interest for X-ray emission ($10^5 \lesssim T_g \lesssim 10^9$ K), the higher the

Table 4 *continued*

References			
1	Canizares (1981)	2	Canizares *et al.* (1979, 1982)
3	Canizares *et al.* (1983)	4	Cowie *et al.* (1983)
5	Demoulin-Ulrich *et al.* (1984)	6	Fabian *et al.* (1981a)
7	Fabian *et al.* (1981b)	8	Fabian *et al.* (1982a)
9	Fabian *et al.* (1985)	10	Fabricant *et al.* (1980)
11	Fabricant and Gorenstein (1983)	12	Ford and Butcher (1979)
13	Heckman (1981)	14	Hu *et al.* (1985)
15	Kent and Sargent (1979)	16	Lea *et al.* (1981)
17	Lea *et al.* (1982)	18	Malumuth and Kriss (1986)
19	Matilsky *et al.* (1985)	20	Mushotzky (1984)
21	Mushotzky *et al.* (1981)	22	Nulsen *et al.* (1982)
23	Rothenflug *et al.* (1984)	24	Schwartz *et al.* (1980b)
25	Stewart *et al.* (1984b)	26	White and Silk (1980)
27	White *et al.* (1981a)		

heating rate (or the nearer the cluster center), the cooler the gas would be in thermal equilibrium. However, such hydrostatic, thermal equilibrium models (in which one heats the gas in order to cool it) can generally be shown to be thermally unstable on the cooling time scale (Stewart *et al.*, 1984a; Fabian *et al.*, 1984b). The gas must either heat up and expand out of the core, or cool down and form a cooling flow on this time scale. Moreover, such a thermal equilibrium model for the X-ray emission would not explain the observed association of cooling flows with optical line emitting filaments, as discussed below.

5.7.2 *Accretion by Central Galaxies*

Many clusters of galaxies have luminous galaxies located at their centers (Section 2.10.1), and these galaxies appear to be moving relatively slowly compared to the average cluster galaxy (Quintana and Lawrie, 1982). In fact, such a luminous galaxy is found at the center of every cooling flow that has been observed (Jones and Forman, 1984). The cooling intracluster gas may, then, be accreted by the central galaxy in the cluster when its temperature has fallen to the point where it can be bound to the galaxy, and the gas can cool further and flow into the galaxy center (Silk, 1976; Cowie and Binney, 1977; Fabian and Nulsen, 1977; Mathews and Bregman, 1978).

It is unlikely that the presence of the central galaxy causes the cooling flow (that is, increases the cooling rate significantly) because the gravitational potential associated with a galaxy is much smaller than that associated with a cluster (cluster velocity dispersions are much larger than those of galaxies). Thus the presence of a central galaxy will not cause a large perturbation in the density of hydrostatic intracluster gas (Section 5.5) and will not affect the cooling rate significantly. Initially, the inflow of cooling gas is driven by the pressure of the surrounding hot medium; only when the gas has cooled significantly does the gravitational potential of the galaxy influence the flow. If the mass of the central galaxy is small, the flow may be pressure-driven over most of its extent (Fabian and Nulsen, 1977; Binney and Cowie, 1981).

In order to accrete cooling gas, a galaxy must be moving slowly through the intracluster medium. The gas must be able to cool before the galaxy has moved away. Thus the velocity is limited to

$$v_{gal} \lesssim \frac{R_{gal}}{t_{cool}} \approx 100\ \mathrm{km/s}\left(\frac{R_{gal}}{100\ \mathrm{kpc}}\right)\left(\frac{t_{cool}}{10^9\ \mathrm{yr}}\right)^{-1}. \tag{5.103}$$

Thus a typical cluster galaxy with $v_{gal} \approx \sigma_r \approx 10^3$ km/s will have great difficulty accreting. Central dominant galaxies are expected to have velocities of at least ≈ 100 km/s on average, due to gravitational encounters with other galaxies (Section 2.9.1). Based on the asymmetry in the nonthermal radio emission and optical line filaments around the galaxy M87 in the Virgo cluster, De Young *et al.* (1980) argued that this accreting galaxy was moving about 200 km/s to the north. Dones and White (1985) showed that this was inconsistent with the observed temperature structure in the cooling flow; that is, it would strongly violate this cooling limit.

Models for the accretion of cooling gas onto central galaxies have been given by Fabian and Nulsen (1977), Mathews and Bregman (1978), and Binney and Cowie (1981). All of these calculations assume the $\dot{\rho}=0$ in equations (5.101); the sources of the gas are external to the central galaxy, so that $\dot{M}$ is constant within the galaxy. Mathews and Bregman integrate equations (5.101) inward through a sonic radius r_s at which the inflow becomes supersonic. The structure of the flow is determined by $\dot{M}$ and the gas temperature at the sonic radius T_s. The sonic radius is a solution of the following equation (Mathews and Bregman, 1978):

$$r_s = \frac{3\mu m_p}{10kT_s}\left[GM(r_s) + \frac{\dot{M}\Lambda(T_s)\mu m_p}{10\pi k T_s}\right], \tag{5.104}$$

where $M(r)$ is the total galactic and cluster mass within the radius r. This equation can have multiple solutions, although usually only one of these at most corresponds to an astrophysically interesting density. For reasonable values of $\dot{M}$ and T_s, one finds $r_s \approx 0.1$–2 kpc (Mathews and Bregman, 1978). A simple estimate of the sonic radius can be derived in a number of ways. First, if the inflow is driven by cooling rather than gravity, then the second term in equation (5.104) will be more important. This is essentially equivalent to arguing that the inflow time and cooling time must be roughly equal. Second, even if gravity is important to the inflow, $kT_g/\mu m_p \approx GM(r)/r$ (see discussion after equations (5.101)), and the first term in equation (5.104) only increases r_s by several times. Thus an estimate for r_s is

$$r_s \approx 0.1\,\text{kpc}\left[\frac{\dot{M}}{100 M_\odot/\text{yr}}\right]\left[\frac{T_s}{10^7\,\text{K}}\right]^{-2.6}, \tag{5.105}$$

where the cooling function in equation (5.22) has been used. Note that r_s is proportional to $\dot{M}$.

On the other hand, Fabian and Nulsen (1977) and Cowie *et al.* (1980) argued that the gas cannot flow in as far as r_s because the inflowing gas must have at least a small amount of angular momentum. Let $\mathbf{l} \equiv \mathbf{r} \times \mathbf{v}$ be the angular momentum per unit mass of the incoming gas, and assume that this is conserved. Then the inflow will stop and the gas will form a rotating disk at a stagnation radius r_{st} given by

$$r_{st} = \frac{l^2}{GM(r_{st})} = 2.3\,\text{kpc}\left[\frac{l}{10^3\,\text{kpc}\cdot\text{km/s}}\right]^2\left[\frac{M(r_{st})}{10^{11}M_\odot}\right]^{-1}. \tag{5.106}$$

Ideally, the angular momentum of inflowing gas in a spherical cluster would be zero, but realistically the gas must have some nonradial velocity. Cowie *et al.* argued that the gas was ejected by cluster galaxies, and that the angular momentum of the central cooling gas would be determined by the rms residual velocity of galaxies in the cluster core,

$$l \approx \frac{\sigma_r}{\sqrt{N_{gal}}} r_c \approx 2 \times 10^4\,\text{kpc}\cdot\text{km/s}, \tag{5.107}$$

where σ_r is the cluster velocity dispersion, $N_{gal} \approx 100$ is the number of galaxies in the cluster core, and r_c is the cluster core radius. Alternatively, if the gas had no nonradial motion relative to the cluster, l might be due to the central galaxy orbital velocity, which for $v_{gal} \approx 100$ km/s and a galaxy orbital radius of ≈ 30 kpc gives $l \approx 3 \times 10^3$ kpc·km/s. However, both of these estimates imply gas velocities relative to the central galaxy that exceed the cooling time scale limit (equation (5.103)). Taking this limit at a cooling radius of 200 kpc gives $l \lesssim 4 \times 10^3$ kpc·km/s. Thus, unless the inflow is much more radial than these estimates suggest, it is likely that the flow will stagnate before becoming supersonic ($r_{st} \gtrsim r_s$).

Viscosity (Section 5.4.4) might transport angular momentum out of the flow and reduce l. If ionic viscosity is effective, thermal conduction will probably also be important and may suppress the cooling flow (Cowie *et al.*, 1980). However, the magnetic field may couple the inner and outer parts of the flow and provide an effective viscosity; a circumferential field could both transport angular momentum out of the inflow, and suppress thermal conduction. Alternatively, if the magnetic field were too weak to be dynamically important and the ionic viscosity were small, the Reynolds number for the flow would be large and the flow could become turbulent. Turbulent viscosity might transport angular momentum out of the flow (Nulsen *et al.*, 1984), and, at the same time, the turbulence could tangle the magnetic field and inhibit thermal conduction.

Cowie *et al.* (1980) treated the inflow as radial (which is reasonable for $r \gg r_{st}$), and included the effect of angular momentum by adding a repulsive centrifugal potential $\phi_{cent} = l^2/2r^2$ to the gravitational potential ϕ in equations (5.101). Moreover, they assumed that $r_{st} \gg r_s$, and therefore they dropped all quadratic or higher order terms in the flow velocity v.

If the accretion rate in the flow decreases inward ($d\dot{M}/dr > 0$), the radial flow can continue into the center of the galaxy without passing through a sonic radius (Nulsen *et al.*, 1984). There is some evidence that $\dot{M}$ does decrease inwards in M87/Virgo and NGC 1275/Perseus (Fabian *et al.*, 1984b; Stewart *et al.*, 1984a). It may be that some of the cooling gas is being converted into stars (Section 5.7.4 below).

Many central galaxies in clusters with cooling flows have nuclear nonthermal radio sources, which may be powered by the further accretion of a small portion of the cooling gas onto the central 'engines' (black holes?) of these sources (Burns *et al.*, 1981b; Bijleveld and Valentijn, 1983; Valentijn and Bijleveld, 1983; Jones and Forman, 1984; Nulsen *et al.*, 1984). There is evidence for a correlation between the radio luminosity and accretion rate in these galaxies (Valentijn and Bijleveld, 1983; Jones and Forman, 1984).

5.7.3 *Thermal Instability and Optical Filamentation*

As the gas cools, any inhomogeneities in the gas density will tend to be amplified (Fabian and Nulsen, 1977). In a region of higher than average density, the temperature will tend to be lower, to preserve pressure equilibrium. Both higher density and lower temperature will speed up the cooling rate, and lowering the temperature will increase the density contrast. Mathews and Bregman (1978) analyzed the growth of density inhomogeneities and the thermal instability of the cooling gas. They considered only radial comoving, isobaric perturbations (no change in the velocity or pressure as compared to the unperturbed flow at the same position), and assumed $\dot{M}$ is constant. They found that the gas is thermally unstable if

$$\Upsilon(T_g) \equiv \frac{2\Lambda(T_g)}{kT_g} - \frac{d\Lambda(T_g)}{dkT_g} > 1. \tag{5.108}$$

This is true for any interesting gas temperature ($T_g \gtrsim 10^5$ K). Assuming that there is an initial perturbation of relative amplitude $(\Delta\rho/\rho)_i$ at some large radius r_i, the amplified perturbation at an inner radius $r < r_i$ is

$$\left(\frac{\Delta\rho}{\rho}\right) = \left(\frac{\Delta\rho}{\rho}\right)_i \exp\left[\frac{2}{5}\frac{\mu m_p}{4\pi}\int_r^{r_i} \Upsilon(T_g)\dot{M}\frac{dr}{(vr)^2}\right] \tag{5.109}$$

in the linear regime ($(\Delta\rho/\rho) \ll 1$). Mathews and Bregman found that amplification factors of 10^{3-4} were likely for flows into the sonic radius. White and Sarazin (1987a) generalized the perturbation analysis to arbitrary perturbations and flows with sinks for gas ($\dot{M} \neq$ const), and showed that the flow is always unstable if inequality (5.108) is satisfied, and that the fastest growing mode is radial, comoving, and isobaric, and is amplified according to equation (5.109). Thermal conduction could suppress the thermal instability of cooling flows, unless conduction is itself suppressed by magnetic fields.

Cowie *et al.* (1980) generalized this analysis by considering the motion of finite sized 'blobs' with a finite density perturbation, without assuming that the blobs comove with the flow. They argue that the motion of blobs relative to the unperturbed gas (buoyancy) will stabilize some perturbations, but that the gas will still be unstable for some blob sizes and densities. Because they assume the flow stagnates, the inflow must become very unstable for $r \approx r_{st}$.

Given that the gas in cooling flows is thermally unstable and that the growth time for the instability is comparable (or somewhat shorter than) the cooling time, these flows should become very inhomogeneous unless the inflowing intracluster gas were very smooth. Thus it may not be reasonable to model the flows with a single phase; multiphase models for cooling flows are discussed in Section 5.7.5.

Optical emission line filaments are often seen near the centers of clusters having cooling flows, often within the central galaxies that are accreting the cooling gas (Ford and Butcher, 1979; Kent and Sargent, 1979; Stauffer and Spinrad, 1979; Heckman, 1981; Fabian *et al.*, 1981b, 1982a; Cowie *et al.*, 1983; Hu *et al.*, 1983, 1985; van Breugel *et al.*, 1984). The size of these extended filament systems ranges roughly from 1–100 kpc. These filaments emit the Balmer lines of hydrogen, as well as forbidden lines from heavier elements. Their spectra are similar to those seen in astrophysical shocks. The emission from these filaments is generally believed to be the result of the same cooling flows; the gas is visible in optical line emission as it cools through the temperature range $T_g \approx 10^4$ K. The clumpy nature of these filaments is due, at least in part, to thermal instability in the cooling gas as discussed above. Figure 20 shows the optical filaments around NGC 1275 in the Perseus cluster, which is one of the best studied examples of this optical filamentation (Lynds, 1970). Clusters with cooling flows having optical filamentation are listed in Table 4.

Cowie *et al.* (1980, 1983) have derived the total luminosity in the Hα line expected in a cooling flow; they find

$$L(\mathrm{H}\alpha) \approx 5 \times 10^{40}\ \mathrm{erg/s}\left(\frac{\dot{M}}{100 M_\odot/\mathrm{yr}}\right). \tag{5.110}$$

This is somewhat larger than would be derived from equation (5.102), which assumes isobaric cooling. Thermally unstable blobs of gas probably cool nearly isobarically until $T_g \approx 10^6$ K; at lower temperatures the cooling time is shorter than the sound crossing time for the filaments, and the gas initially cools isochorically (at constant density). The cool clumps reach pressure equilibrium with the surrounding hot gas by the passage of repressurizing shocks (Mathews and Bregman, 1978; Cowie *et al.*, 1980). The Hα flux may also be increased by photoionization by X-rays and ultraviolet radiation from the surrounding hot gas or from the nuclear source in the accreting galaxy.

Cowie *et al.* (1983) have suggested, based on extensive observations of optical line emission, that these filament systems consist of two components: a cluster core component, which is 20–100 kpc in size, and a galaxy component, which is smaller. In some clusters both components are present; in some, only one or the other is observed. The larger cluster components tend to consist of highly elongated filaments, which may be stretched by tidal effects in the cluster or along the magnetic field lines. The galaxy components are more homogeneous, and are also elongated. Cowie *et al.* suggest that they are gas disks at the stagnation radius. In many clusters the filament systems show velocity shears consistent with centrifugally supported disks (Hu *et al.*, 1985), although velocity measurements in NGC1275 indicate that the gas is not rotating fast enough in this galaxy (Hu *et al.*, 1983). These emission line gas disks can be used to estimate the masses of the inner portions of the cD galaxies, just as the rotation velocities of spiral galaxy disks are used to determine their masses (Hu *et al.*, 1985).

Hu *et al.* (1985) find that the presence of optical emission line filaments in the central galaxies of clusters correlates very strongly with the central density of the intracluster gas. The emission lines are only found in clusters with central proton number densities $n_o \gtrsim 5 \times 10^{-3} h_{50}^{1/2}$ cm^{-3}. This is just what is expected if the emission lines are due to cooling flows, since a lower limit on the density implies an upper limit on the cooling time. For this density limit and an intracluster gas temperature of 7×10^7 K, equation (5.23) gives $t_{cool} \lesssim 10^{10} h_{50}^{-1/2}$ yr, which implies that the gas could cool in the lifetime of the cluster.

The filament system around NGC1275 in the Perseus cluster is the most luminous such system known, and is considerably brighter than would be expected given the accretion rate determined from the X-ray measurements. Hu *et al.* (1983) argue that it is also too luminous to be due to photoionization by the nuclear source in NGC1275. They show that the optical filaments at the velocity of NGC1275 are elongated at the same position and in the same direction as the filaments due to the spiral galaxy that is moving towards NGC1275 at 3000 km/s (see Section 4.5.2). They suggest that this galaxy is actually colliding with the base of the accretion flow onto NGC1275, and that the kinetic energy in this collision powers the emission line filaments. Two problems with this model, which are discussed by Hu *et al.*, are the lack of intermediate velocity gas and the difficulty of a gas-rich galaxy penetrating into the core of the Perseus cluster without having its gas stripped (Sections 2.10.2 and 5.9).

5.7.4 *Accretion-Driven Star Formation*

What is the duration of these cooling flows in clusters? Both optical and X-ray clusters have been observed at moderately high redshifts, and the gas distributions in clusters are relaxed and smooth; both suggest that the intracluster medium has been present in clusters for a significant fraction of the Hubble time. As described above, the cooling times at the centers of these flows are less than the Hubble time, which suggests that the flows have persisted for a significant fraction of the cluster lifetime. Thus the total mass of gas that has cooled and been accreted by the central galaxy in the cluster should be

$$M_{acc} \approx 10^{12} M_{\odot} \left(\frac{\dot{M}}{100 M_{\odot}/\mathrm{yr}} \right). \tag{5.111}$$

This assumes that the accreted mass is not ejected from the cluster center.

Although this is a small fraction of the total gas mass in a cluster ($\approx 10^{14} M_{\odot}$), it is comparable to the mass in luminous material in a very large (cD) galaxy. It is important to understand where this gas goes after it cools. Many possibilities can be ruled out. First, the gas could remain gaseous but cooler. However, the observations of the X-ray and optical line emission give upper limits on the total amount of ionized gas well below the total accreted mass. Observations of the 21 cm hydrogen line from central galaxies in clusters with cooling flows (Section 3.7) give upper limits on the total amount of neutral atomic hydrogen, typically $M_{HI} \leqslant 10^9 M_{\odot}$ (Burns *et al.*, 1981a; Valentijn and Giovanelli, 1982). Because of the instrumental sensitivity, current observations do not rule out the possibility that the accreted gas could be molecular hydrogen, although it seems likely that the large amounts of molecular gas required would lead to an efficient conversion into stars. As mentioned above, many of the accreting galaxies are radio sources (Burns *et al.*, 1981b), and this may indicate that some of the accreted gas flows into the galactic nucleus and is used to power the central engine of the radio source. However, these sources require only $\lesssim 10^{-2} M_{\odot}/\mathrm{yr}$ of gas to provide their observed radio power (Burns *et al.*, 1981b), so it is likely that the fraction of the gas accreted by the central engine is small. Moreover, there could not be as much as $10^{12} M_{\odot}$ in the nucleus of these galaxies, given their observed stellar velocity dispersions (Sarazin and O'Connell, 1983).

Generally, conversion of gas into stars is quite efficient within galaxies, and this conversion could provide both a stable reservoir for the accreted gas and a partial explanation for the existence of central dominant galaxies. Burns *et al.* (1981a) argued that star formation cannot be occurring at a high rate in these galaxies, because they would then contain more neutral hydrogen than is observed (see above). However, Fabian *et al.* (1982b) and Sarazin and O'Connell (1983) showed that the cooling times for neutral hydrogen were short enough that all the accreted gas could be cooling through this temperature range and forming stars. Within the disk of our own galaxy, stars are formed with a very wide range of masses extending up to $\approx 100 M_\odot$ (Salpeter, 1955). Stars more massive than $\approx 10 M_\odot$ produce Type II supernovae when they die, and the rate of these supernovae would probably heat the accreted gas sufficiently to prevent the formation of cooling flows (Wirth *et al.*, 1983). Moreover, if the spectrum of stellar masses formed from the cooling gas (the 'initial mass function') were similar to that in our own galaxy, the central galaxies in the accretion flows would be considerably bluer and brighter than they are observed to be (Fabian *et al.*, 1982b, 1984a; Sarazin and O'Connell, 1983). Burns *et al.* (1981a) gave similar arguments and concluded that star formation cannot be the ultimate reservoir for the cooling gas.

However, there is really no reason why the initial mass function for star formation in these cooling flows should be the same as that in the disk of our galaxy. If the forming stars had low masses $\lesssim 1 M_\odot$, these stars would not be very different from the stars found in typical elliptical galaxies (Cowie and Binney, 1977). Since star formation is very poorly understood and there is no successful quantitative theory for this process, one cannot calculate the initial mass function directly. However, Fabian *et al.* (1982a) and Sarazin and O'Connell (1983) have given a simple plausibility argument as to why the initial mass function for star formation in cooling flows might be limited to low mass stars; a similar argument for elliptical galaxies in general was given by Jura (1977). It is assumed that stars form eventually from the thermally unstable clouds of gas that are seen as optical filaments. Star formation is assumed to start when these clouds become gravitationally unstable and can no longer support themselves against their own gravity and the pressure of the surrounding medium. Clouds become gravitationally unstable when their mass exceeds the Jeans' mass, which for a spherical, static, nonmagnetic isothermal cloud of temperature T immersed in a low density medium of pressure P is given by (Spitzer, 1978)

$$M_J = 1.2\left[\left(\frac{kT}{\mu m_p}\right)^4 \frac{1}{G^3 P}\right]^{1/2}$$

$$= 0.64 M_\odot \left(\frac{T}{10\,\mathrm{K}}\right)^2 \left(\frac{\mu}{\mathrm{amu}}\right)^{-2} \left(\frac{P}{10^7 k\,\mathrm{cm}^{-3}\,\mathrm{K}}\right)^{-1/2}. \qquad (5.112)$$

Once a cloud starts to collapse, the pressure within the cloud will increase and the Jeans' mass may be reduced; this can cause the cloud to fragment and result in lower mass stars being formed. It is difficult to produce stars more massive than M_J, however, because before a suitably massive cloud could be assembled, it would become unstable and collapse. Thus it is possible that the Jeans' mass may provide an upper limit on the mass of the largest stars that form. In the disk of our galaxy, the interstellar medium typically has a pressure of $P \approx 2 \times 10^3 k \text{ cm}^{-3} \text{ K}$, and equation (5.112) gives $M_J \approx 50 M_\odot$. In the cooling flows in clusters, the pressures derived from models for the X-ray emission or determined directly from the optical line emitting filaments are 10^{3-4} times larger ($P \approx 10^{6-7} k \text{ cm}^{-3} \text{ K}$), and thus the Jeans' mass is $M_J \approx 1 M_\odot$. Thus it is possible that only low mass stars are formed from the cooling gas in clusters. A similar argument was given earlier for low mass star formation in normal elliptical galaxies by Jura (1977). In Fabian *et al.* (1982) and Sarazin and O'Connell (1983), this conclusion is shown to be unaffected by the temperature dependence in equation (5.112).

Fabian *et al.* (1982b) also point out that star formation in cooling flows may be different than in the disk of our galaxy because the star forming regions in these flows are unlikely to contain dust grains. In star forming regions in our galaxy, most of the refractory heavy elements are in the form of solid dust grains, and these grains absorb starlight, emit infrared radiation, and act as a heat source for the gas. Dust grains are destroyed in high temperature gas. Since the gas in cooling flows is initially very hot, any grains would have evaporated (Cowie and Binney, 1977; Fabian *et al.*, 1982b), and it is very difficult for grains to form in low density gas, even if it is cool. Thus it is unlikely that grains will be present in the cooling flows, even in the coolest, star forming clouds. The lack of grains probably lowers the gas temperature, which also tends to reduce M_J. Further, any attempt to estimate the star forming rate in these galaxies from the infrared emission of dust (Wirth *et al.*, 1983) is likely to greatly underestimate the real rate.

Sarazin and O'Connell (1983) have calculated the expected colors and optical spectra of central galaxies assuming that their stellar populations are a mixture of a normal giant elliptical population with a continuously forming population due to accretion-driven star formation. A variety of values for the upper mass limit and the shape of the initial mass function were used. They found that the accreting galaxies should have spectra and colors measurably different from those of nonaccreting giant ellipticals (color differences of typically $\Delta(U - V) \approx -0.3$ mag). They also found that with accretion rates of typically $\dot{M} \gtrsim 100 M_\odot/\text{yr}$, the entire stellar population of the central galaxies in many clusters could be due to accretion-driven star formation.

Valentijn (1983) has attempted to measure the colors of the stellar populations in seven cD galaxies and their spatial variations by surface photometry in two colors (B and V). He finds very large color gradients within these galaxies of typically $\Delta(B-V) \approx 0.4$ mag, with the galaxy centers being extremely red. Valentijn argues that these gradients are the result of accretion-driven star formation, and that as the pressure increases inwards in the cooling flow, the Jeans' mass is lowered (equation (5.112)) and the stellar population becomes redder. One problem is that the innermost regions of these galaxies are so red that the required stellar population would have a very small mass-to-light ratio and could not provide enough light (for the observed accretion rates) to account for the observed galaxy luminosity. The color gradients observed by Valentijn are very large (larger than Sarazin and O'Connell predicted), and it is very important that they be confirmed by further observations. Valentijn's photometry appears to disagree, in some cases, with that of other observers (Hoessel, 1980; Malumuth and Kirshner, 1985).

Color gradients might also result from dust extinction, abundance gradients in an old stellar population, or mergers of galaxies having different colors. A more direct way to detect a stellar population due to accretion is to observe absorption features due to that population in the spectrum of the cD galaxy. The galaxy NGC1275 in the Perseus cluster (Figure 20; Section 4.5.2) is particularly interesting in this regard. Its stellar surface brightness distribution is similar to that of a typical giant elliptical galaxy (Oemler, 1976). However, the stellar population is very blue, and has an A-star absorption spectrum (Kent and Sargent, 1979), whereas typical giant elliptical galaxies are dominated by K stars. Sarazin and O'Connell (1983) show that the colors and spectrum of this galaxy can be understood if the luminous portion of the galaxy is entirely due to accretion-driven star formation at a rate of $\approx 300 M_{\odot}$/yr as given by the X-ray observations, and the upper mass cutoff of the stars formed is $2.8 M_{\odot}$. (By contrast, Wirth *et al.* (1983) present a model for NGC1275 in which the initial mass function for star formation is similar to that in our galaxy, and very massive O stars are formed. However, in this model the star formation rate is more than an order of magnitude less than the observed rate of accretion onto NGC1275.)

One concern with NGC1275 is the presence of the foreground spiral galaxy. Hu *et al.* (1983) have suggested that this galaxy is colliding with the cooling flow and that this collision powers the optical line emission. It is also possible that this collision might affect the rate and initial mass function of star formation in the galaxy. It is thus very important to observe the spectra of other accreting central dominant galaxies, and see if a younger stellar population can be detected in them. Recently, O'Connell *et al.* (1987) obtained spectra for the inner regions of a number of accreting galaxies. The cD in A1795, which has a very large accretion rate $\dot{M} \approx 400 M_{\odot}/\mathrm{yr}$ (Table 4), has an F-star stellar spectrum, consistent with the entire galaxy being the result of accretion-driven star formation with an upper mass cutoff of about $1.5 M_{\odot}$.

Central dominant cluster galaxies appear, in many cases, to have a very large number of globular star clusters (spherical clusters of $\approx 10^{5-6}$ stars) associated with them (Harris *et al.*, 1983b). Fabian *et al.* (1984b) have suggested that these globular clusters might be produced by accretion-driven star formation. They note that the mass of a globular cluster is similar to the Jeans' mass of gas in the cooling flows at a temperature of 10^4 K, the temperature at which thermally unstable clouds are repressurized. However, Fall and Rees (1985) showed that the cooling time was much shorter than the free-fall time for Jeans' unstable clumps at this temperature, preventing the gravitational instability of the gas. The cooling time depends on the abundance of heavy elements, which may have been considerably lower when galaxies first formed (Section 5.10). Thus Fall and Rees argue that globular clusters formed out of cooling flows during the formation of galaxies.

5.7.5 Cooling Flow Models with Star Formation

If cooling gas is being converted into stars in cooling flows, then terms representing the loss of gas should be added to the equations for the flow (equations (5.101)). Regardless of the ultimate fate of the cooling gas, the fact that the gas is thermally unstable means that it is not reasonable to treat the cooling flow as homogeneous. In homogeneous models for cooling flows, the gas remains hot enough to emit X-rays until it is within the sonic radius (Section 5.7.2). However, in an inhomogeneous flow, thermal instability will cause denser lumps of gas to cool below X-ray emitting temperatures while the more diffuse gas is still quite hot. Thus, even if star formation were not occurring, thermal instabilities would reduce the amount of *hot* gas as it moves towards the center of the galaxy.

If the mass flow rate $\dot{M}$ of hot gas decreases with distance from the center of the flow, this should result in an observable reduction in the amount of X-ray emission near the center of the flow. Such a reduction does appear to be required by the X-ray surface brightness data of cooling flow clusters. Using the semi-empirical method to determine $\dot{M}$ from X-ray surface brightness profiles (Section 5.7.1), Fabian *et al.* (1984b) and Stewart *et al.* (1984a) found that the cooling rate $\dot{M}$ increased with radius in M87/Virgo and NGC1275/Perseus. Unfortunately, their method of determining $\dot{M}$ is inconsistent if $\dot{M}$ is not constant (White and Sarazin, 1987a, b). First, their particular form of the energy equation (last of equations (5.101)) requires that

$\dot{M}$ be constant. Second, they assumed that only the hot diffuse gas contributed to the X-ray emission. But the gas being removed from the flow by thermal instabilities cools radiatively, producing X-ray emission. Thus the X-ray emissivity should include a term proportional to the rate of gas loss through thermal instabilities. Semi-empirical methods to determine $\dot{M}(r)$ including more consistent treatments of mass loss have been given by Fabian *et al.* (1985, 1986a) and White and Sarazin (1987a, b). These studies indicate that $\dot{M}(r)$ increases with increasing r in the best studied cooling flow clusters. Fabian *et al.* argue that the variation of $\dot{M}(r)$ is well represented by $\dot{M}(r) \propto r$, while White and Sarazin contend that the variation of $\dot{M}(r)$ is too sensitive to the assumed form of the gravitational potential to allow any strong statements to be made.

If the gas in cooling flows is inhomogeneous, it is much more difficult to model the dynamics of the flows. In principle, in a correct treatment of the flow the gas would be represented by a continuous range of densities ρ_g. Rather than giving a single set of thermodynamic variables, say ρ_g, T_g, and v, as a function of position r, one should specify a distribution of densities. For example, $f(\rho_g, r)$ might be the fraction of the volume or mass in the flow at r which is in the form of gas at a density ρ_g. Correspondingly, $T_g(\rho_g, r)$ and $v_g(\rho_g, r)$ would be the mean temperature and velocity of gas having a density ρ_g located at r. Obviously, this would vastly increase the complexity of the hydrodynamical modeling of the flow. For one thing, there is no clear physical argument which specifies the boundary conditions on the inhomogeneities (for example, the value of $f(\rho_g)$ at the cooling radius r_{cool}). Perturbations in the flow probably cool isobarically until they are too cool to produce a significant amount of X-ray emission ($T_g \lesssim 10^6$ K), so that it is probably reasonable to assume that all density phases have the same pressure at each position. Then the temperature is just $T_g(\rho_g, r) = P(r)/\rho_g$. However, the possibility that the different densities phases would have different velocities still is an enormous complication, since the hydrodynamical interactions between lumps of differing density and velocity would be extremely complex.

Two opposite approximations have been made to deal with this problem. First, Fabian *et al.* (1985, 1986a) assumed that all the density phases comove, so that both v and P are functions only of position r and not of density ρ_g. As noted above (Section 5.7.3), the fastest growing linear perturbations $\Delta\rho_g/\rho_g) \ll 1$ in a homogeneous flow do comove, which supports this idea. However, one might expect that once perturbations grow nonlinearly $(\Delta\rho_g/\rho_g) \gg 1$ they might drop out of the flow. This led White and Sarazin (1987a, b) to an opposite approximation. They argued that the isobaric cooling time decreases quite rapidly with decreasing temperature. Thus once a lump has cooled significantly below the average temperature, it will cool below X-ray emitting temperatures rapidly. As the density of the lump increases, its surface area will decrease and it can decouple from the flow and fall ballistically. Since the flow time is determined by the cooling time of the diffuse gas, the cooling and decoupling of higher density lumps can occur before the flow has moved inwards by a significant amount. In this limit, the cooling of dense lumps of gas can be treated as a local sink for the diffuse gas, and the flow equations revert to equations (5.101) with loss terms. Numerical models for cooling flow including loss terms which are proportional to either the cooling time or the growth time of thermal instabilities have been given by White and Sarazin (1987a, b).

These inhomogeneous models for cooling flows can be used to predict the X-ray surface brightness profiles and spectral variations of cooling flows (White and Sarazin, 1987b). If clumps of gas cool rapidly from X-ray emitting temperatures to $T_g \approx 10^4$ K, they can also predict the surface brightness of optical line emission. If these cooling condensates form stars quickly or if the cool lumps are decoupled from the flow and falling ballistically (and form stars eventually), these models can give the predicted distribution of these stars. In Section 5.7.4, attempts to detect the presence of a younger stellar population due to accretion in the optical spectra of central galaxies in cooling flows were discussed. It is also very important to study the spatial distribution of this population. As discussed in Section 2.10.1, central dominant galaxies in clusters appear to be composed of an extended giant elliptical interior, surrounded in the case of rich cluster cDs by a very extended halo. These cDs may also have dark, missing mass haloes (Section 5.8.1). Which of these components could be the result of accretion-driven star formation? If accretion produces the giant elliptical interiors, why do these resemble the stellar distributions in other nonaccreting giant ellipticals? Giant ellipticals have light distributions that are reasonably fit by de Vaucouleurs or Hubble profiles (Section 2.10.1), and recent numerical studies have suggested that these form naturally in violent relaxation (Section 2.9.2). Accretion-driven star formation is a slow process; will it give a similar distribution?

As discussed above, semi-empirical determinations of the hot gas inflow rates by Fabian *et al.* (1985, 1986a) are consistent with $\dot{M} \propto r$, although White and Sarazin (1987a, b) have argued that $\dot{M}(r)$ is extremely uncertain. If the cooling lumps form stars rapidly and if the orbits of the newly formed stars are not affected by the galaxy potential, this might imply that the density of the new stars varied as $\rho_* \propto r^{-2}$. This is much flatter than the density distribution of the luminous stars in elliptical galaxies. This is just the density distribution of an isothermal sphere, and is similar to the density distributions inferred for the missing mass haloes of spiral galaxies (Section 2.8). This led Fabian *et al.* (1986a, b) to suggest that the missing mass is very low mass stars formed in cooling flows; this requires that the initial mass function for star formation in cooling flows produce mainly very low mass stars ($M_* \lesssim 0.1 M_\odot$).

On the other hand, White and Sarazin (1987a, b) found that the predicted stellar distributions in their cooling flow models with star formation were very similar to those of the light from giant elliptical galaxies. In calculating the stellar distributions, they self-consistently included the effects of the stellar orbits in the galaxy and cluster gravitational potential.

Accretion-driven star formation may result in different stellar orbits in cD galaxies than in nonaccreting giant ellipticals. If the flows have little angular momentum (see above) and are radial, the resulting stellar orbits may be radial. If the processes of clumping and star formation impart significant random velocities to the star forming regions, the orbits might be isotropic. If the flow stagnates and forms a disk, stellar disks may be found within cD galaxies.

5.7.6 *Evolution of Cooling Flows and Active Galaxies*

One important question is what effect the evolution of clusters has on these cooling flows. An important type of evolution is the merging of subclusters; the observed double clusters (Section 4.4.2) may be systems undergoing this process. Such a merger will probably heat up the gas and disrupt any existing cooling flows (McGlynn and Fabian, 1984). Along these lines, Stewart *et al.* (1984b) speculate on the possibility that the Coma cluster, with its pair of D galaxies, is the result of such a merger of two subclusters. Perhaps each of the subclusters had a cooling flow, and accretion-driven star formation produced the two D galaxies. The merger would disrupt the flow, which would explain why Coma apparently lacks a cooling flow despite its high X-ray luminosity. The heating of the gas during the merger might explain the unusually high temperature of the intracluster medium in Coma (Section 4.5.1). In this way, accretion could have formed the central galaxies in clusters that do not currently have accretion flows.

Although this review is concerned with clusters, elliptical galaxies in irregular clusters and groups also appear to have considerable quantities of X-ray emitting gas, which may also form cooling flows (Section 5.8.3). These cooling flows are probably powered by stellar mass loss within the galaxies, rather than accretion of intracluster gas. For giant ellipticals stellar mass loss rates of $\approx 1 M_\odot$/yr are expected. In the past, rates of stellar mass loss in elliptical galaxies were probably higher than they are today (Section 5.10.1). Since cooling flows in cluster centers were probably weaker in the past, this suggests that individual early type galaxies may make an important contribution to the emission from clusters at high redshift (Fabian *et al.*, 1986a).

As was noted above, accreting central galaxies are often strong radio sources, indicating that some small portion of the cooling gas is accreted by the nucleus of the galaxy. Quasars appear to be galaxies with extremely luminous nonthermal nuclei. It is possible that they are also powered by accretion through cooling flows (Sarazin and O'Connell, 1983; Fabian *et al.*, 1986a). It is possible that the evolution of cooling flows in clusters or individual galaxies may help to determine the evolution of active galactic nuclei (Section 5.10). Active galaxies often show strong optical line emission; perhaps the lower excitation line emission seen in some active galaxies may be due to thermal instabilities in cooling flows, the mechanism that produces the line emission in nearby cooling flow galaxies.

In summary, it appears possible that most of the accreted gas in cooling flows is converted into low mass stars, and that this accretion-driven star formation may provide much of the luminous or nonluminous mass of central galaxies in clusters with cooling flows. This process may compete with mergers and tidal effects as a mechanism for the formation of cD galaxies (Section 2.10.1). It may also affect the evolution of active galaxies.

5.8 X-Ray Emission from Individual Galaxies

A number of topics concerning the X-ray emission from hot gas associated with individual galaxies in clusters will now be discussed. Material on the accretion of gas by central galaxies in clusters with cooling flows was reviewed in the previous section.

5.8.1 Massive Haloes around M87 and Other Central Galaxies

The X-ray emission from the M87/Virgo cluster is considerably different from the emission from richer and more regular clusters (Section 4.5.3). The gas is much cooler and less extended, and the X-ray luminosity is rather small. Initially, this was explained by noting that the X-ray luminosity and temperature of clusters appeared to correlate with the cluster velocity dispersion (equations (4.8) and (4.10)), although this argument was somewhat circular, since M87/Virgo was one of the strongest pieces of evidence in favor of these correlations.

Bahcall and Sarazin (1977) and Mathews (1978b) suggested an alternative; the gas in M87/Virgo might be hydrostatic and bound to the galaxy M87, rather than to the cluster as a whole. To bind the gas, the galaxy would have to have a much deeper gravitational potential well than it would appear to have from its optical emission. Based on the best observations at that time, Bahcall and Sarazin showed that M87 must have a massive halo, with a total mass of $\approx 3 \times 10^{13} M_{\odot}$ extending out to ≈ 100 kpc from the galaxy center. Bahcall and Sarazin assumed a number of reasonable equations of state for the gas (adiabatic, polytropic, etc.; see Section 5.5). Mathews (1978b) reached a similar conclusion, although he assumed that the gas was exactly isothermal and that the mass distribution was the King analytic form for an isothermal distribution (5.57). Because such isothermal gas models diverge unless the gas is cool compared to the potential (equations (5.65) and 5.66)), Mathews required much more mass, $\gtrsim 10^{14} M_{\odot}$. This high mass limit is very sensitive to the assumption that the gas is exactly isothermal (Bahcall and Sarazin, 1978; Fabricant *et al.*, 1980).

There is considerable evidence that M87/Virgo has a cooling flow (Section 5.7). However, at large distances from the galaxy center the flow is highly subsonic, and the hydrostatic equation applies. Thus the cooling flow does not alter the requirement that the mass be large.

Binney and Cowie (1981) suggested that the mass of M87 might be much lower ($\approx 3 \times 10^{11} M_{\odot}$). They argued that the gas around M87 is bound by the *pressure* of surrounding, hotter gas, rather than by the *gravity* of M87. This hotter gas was in turn assumed to be extended and bound by the cluster gravitation potential of Virgo. The hot, diffuse, confining, intracluster gas is required to have a density and temperature of $n_p \approx 10^{-3}\,\mathrm{cm}^{-3}$ and $T_g \approx 10^8$ K. The cool gas is produced by a cooling flow, which is drawn from the hot gas.

A key point in the Binney and Cowie model is that there must be enough hot, diffuse gas to provide a pressure that confines the cool gas around M87. Because this gas would be less dense than the cooler gas, it would have a lower surface brightness, but it should be detectable in hard X-rays. Davison (1978) and Lawrence (1978) found a hard X-ray component from M87/Virgo, which they suggested was extended. However, more recent observations indicate that this hard component is not extended, has a power-law spectrum, and is centered on M87 (Lea *et al.*, 1981, 1982). This hard source is probably nonthermal emission from the nucleus of M87. The present limits on the amount of extended hard X-ray emission are, at best, marginally consistent with the Binney and Cowie model (Fabricant and Gorenstein, 1983).

The *Einstein* satellite observations of the X-ray emission from M87 have been analyzed in detail by Fabricant *et al.* (1980), Fabricant and Gorenstein (1983), and Stewart *et al.* (1984a). All of these papers used basically the method of analysis outlined in Sections 5.5.4 and 5.5.5; that is, the observed surface brightness was deconvolved to give the variation of emissivity with radius from the galaxy center. At the time of the first paper, the spectral response of the IPC on *Einstein* was poorly calibrated, and the density was derived from the emissivity assuming that the gas was isothermal. From the hydrostatic equation, the total mass $M(r)$ interior to a radius r is

$$M(r) = -\frac{kT_g r}{G\mu m_p}\left(\frac{d\log\rho_g}{d\log r} + \frac{d\log T_g}{d\log r}\right). \tag{5.113}$$

Beyond 3 arcmin from the center of M87, the surface brightness is well represented as a power-law $\propto r^{-1.62}$ (Fabricant and Gorenstein, 1983). Thus, as long as the temperature gradient is not significant, the density derivative in this equation is a constant, and the mass increases with the radius. Fabricant *et al.* (1980) found a total mass of $2\text{–}4 \times 10^{13} M_\odot$ within a radius of 230 kpc.

Binney and Cowie (1981) analyzed the same *Einstein* data and found consistency with their low mass model for M87. They argued that the disparity between their conclusion and that of Fabricant *et al.* (1980) was due to two differences. First, Fabricant *et al.* assumed that the surface brightness approached zero far from M87, whereas Binney and Cowie invoke extended intracluster gas. More importantly, in the Binney and Cowie cooling flow model the gas temperature decreases rapidly towards the center, $d\log T_g/d\log r \approx 0.8$. This large temperature gradient nearly cancels the density gradient (i.e., the pressure is nearly constant), and the resulting mass is significantly reduced.

The *Einstein* data were reanalyzed by Fabricant and Gorenstein (1983) and Stewart *et al.* (1984a), who found that the mass of M87 is $3\text{–}6 \times 10^{13} M_\odot$ within a radius of 260 kpc. By this time, the spectral response of the IPC had been calibrated and the temperature gradient could be derived directly from the data, albeit with large errors. (Note that this is possible for M87/Virgo and not for most cluster sources because the temperature in M87/Virgo is low enough to be measured with the soft X-ray sensitivity of *Einstein*). In addition, the temperature gradient was constrained by the wide field proportional counter observations from previous satellites, and by data from the SSS and FPCS X-ray spectrometers on *Einstein*. The observed temperature gradient appears to be inconsistent with that required by Binney and Cowie. These data appear to rule out the Binney and Cowie model.

Thus M87 appears to have a very massive halo, which extends well beyond the region where the galaxy is luminous. Like the massive haloes around spiral galaxies (Section 2.8), the mass appears to increase roughly in proportion to the radius in the outer portions of M87. However, spiral galaxies have rotation velocities nearly independent of radius, on the order of 300 km/s (Faber and Gallagher, 1979). The inferred circular orbital velocity in the halo of M87 is much higher, about 750 km/s. This is also much higher than the orbital velocities of stars in the visible portion of M87, and thus the orbital velocities in M87 are not independent of radius over the span from the luminous portions of the galaxy to the outer halo.

One important question is whether other elliptical galaxies have massive dark haloes. This question will be addressed in Section 5.8.3.

5.8.2 Other Models for M87 and Other Central Galaxies

In Sections 5.7 and 5.8.1, a standard model for M87 (and other X-ray emitting central galaxies) has been given, in which there is a cooling flow leading to accretion by M87. The outer parts of the flow are very subsonic, and the gas is nearly hydrostatic and bound either by the massive halo of M87 or by the pressure of surrounding gas. One feature of this standard model is that thermal conduction must be suppressed in order to ensure that the losses due to cooling in the center of the galaxy are balanced by the enthalpy flux of inflowing gas rather than by heat conducted inwards. Two alternative models are now discussed in which conduction is not suppressed.

Takahara and Takahara (1979, 1981) suggested that gas was actually being thermally evaporated and flowing out from M87, rather than flowing inward. They argued that the gas in the evaporative flow was supplied by mass loss from stars in M87, at a rate of about $1 M_{\odot}/\text{yr}$. This gas is heated by thermal conduction from surrounding, extended hot intracluster gas, in which the galaxy is assumed to be immersed.

One problem with this model is that the range of temperatures in the gas is rather small, and the gas is generally hotter than 2×10^7 K. This model cannot produce the very strong soft X-ray line emission seen in M87 (Canizares *et al.*, 1979, 1982; Lea *et al.*, 1982; Stewart *et al.*, 1984a). It does not provide a very good fit to the X-ray surface brightness in the inner regions of the galaxy, and does not explain the origin of the line emitting filaments in M87 (Section 5.7.3).

Tucker and Rosner (1983) suggested that the gas in M87 (and other central dominant galaxies) is actually hydrostatic (not moving). In their model, the cooling of the gas is balanced by heating from thermal conduction in the outer regions and by heating by the relativistic electrons associated with the radio source in the inner parts. As discussed in Section 5.7.1, the extra heating by relativistic electrons actually results in thermal equilibrium being attained at a lower temperature, since cooling increases rapidly as the temperature is reduced. Such a model is necessarily thermally unstable (Stewart *et al.*, 1984a), and might form a cooling flow due to the growth of thermal fluctuations. This model requires a fairly large rate of heating from relativistic electrons (Scott *et al.*, 1980; Section 5.3.5), and there is considerable uncertainty about such large heating rates. The model does have the advantage that the similarity in the morphology of the X-ray and diffuse radio emission in M87 is explained.

In Section 5.7.2 the suggestion was made that the radio emission of central cluster galaxies was powered by having a small portion of the accreting gas reach the nucleus. In the Tucker and Rosner model, the gas is static. Thus there is no inflow to power the radio source. Tucker and Rosner suggest that these sources might be episodic. Initially, gas cooling in the core of the cluster might produce an accretion flow on the central galaxy, and start the radio source. The radio source would produce large numbers of relativistic electrons, which would heat the gas and turn off the accretion flow. This would stop the radio source, and the heating would decline as the relativistic electrons lost their energy. This would allow the accretion flow to restart, and the process might oscillate in this fashion indefinitely.

5.8.3 X-Ray Emission from Noncentral Cluster Galaxies

Recent analyses of the X-ray observation of early-type galaxies made with the *Einstein* X-ray observatory show that many non-cD ellipticals which are not at the centers of rich clusters are moderately strong X-ray sources (Nulsen *et al.*, 1984; Forman *et al.*, 1985; Trinchieri and Fabbiano, 1985; Canizares *et al.*, 1987; Trinchieri *et al.*, 1986). The X-ray luminosities range from $L_x \approx 10^{39}$–10^{42} erg/s. Many of these X-ray emitting elliptical and S0 galaxies are in the Virgo cluster (Forman *et al.*, 1979; Forman and Jones, 1982; Section 4.5.3). A number of peaks in the X-ray surface brightness of A1367 correspond to the positions of galaxies (Bechtold *et al.*, 1983; Section 4.5.4). There is a strong correlation between the optical and X-ray luminosities of these galaxies, with $L_x \propto L_B^{1.5-2.0}$, where L_B is the optical (blue) luminosity (Forman *et al.*, 1985; Trinchieri and Fabbiano, 1985). The X-ray emission is spatially extended, with a typical maximum radius for the brighter galaxies being $R_x \approx 50$ kpc, and the X-ray surface brightness is reasonably fit by $I_x \propto r^{-2}$ in the outer parts, where r is the radius from the galaxy center (Forman *et al.*, 1985).

There are a number of strong arguments, given in the papers just cited, which indicate that the X-rays arise as thermal emission from hot, diffuse gas. In the lower luminosity ellipticals, emission by binary X-ray sources may also play a role. The crude X-ray spectral information available suggests that the gas temperatures are on the order of $T_g \approx 10^7$ K (Forman *et al.*, 1985). The observed radial variation of the X-ray surface brightness requires that the gas density vary roughly as $\rho_g \propto r^{-3/2}$. The required gas densities in the inner parts are $\gtrsim 0.1$ atom/cm^3. The total mass of hot gas in these galaxies is roughly $M_g \approx 10^9 - 10^{11} M_\odot$ (Forman *et al.*, 1985). Assuming a stellar mass-to-light ratio of $(M/L_B)_* \approx 8 M_\odot / L_\odot$, the ratio of gas mass to stellar mass is $M_g/M_* \approx 0.02$.

Previous to the detection of this hot gas, elliptical were believed to be gas-poor systems. It was assumed that the gas ejected by stars in ellipticals was effectively heated by supernovae, forming a strong galactic wind which quickly removed the gas from the galaxy (Mathews and Baker, 1971). The X-ray observations show that gas is *not* being removed rapidly from these galaxies; apparently, the galactic winds do not generally occur, at least at the present time. The expected X-ray luminosities from winds are $\lesssim 10^{-4}$ of those observed in elliptical galaxies.

A simple model which fits the properties of these X-ray emitting galaxies is that the bulk of the gas forms a cooling flow, in which gas lost by stars in the galaxy is heated by supernovae, the motions of the gas-ejecting stars, and adiabatic compression, and then cools and slowly flows into the galaxy center (Nulsen *et al.*, 1984; Fabian *et al.*, 1986b; Canizares *et al.*, 1987; Sarazin, 1986b, c). There are several simple arguments which support this cooling flow model. First, the amount of gas seen is consistent with the current rates of stellar mass loss in ellipticals. The rate of stellar mass loss is proportional to the stellar density ρ_*, so that $\dot{\rho}_g = \alpha_* \rho_*$. For the stellar population in ellipticals, α_* is calculated to be $\alpha_* \approx 4.7 \times 10^{-20}\,\mathrm{s}^{-1} = 1.5 \times 10^{-12}\,\mathrm{yr}^{-1}$ (Faber and Gallagher, 1976). Thus, over $\approx 10^{10}$ years, the total mass of gas ejected at the present rate is $M_g \approx 0.015 M_*$, which is consistent with the observed gas masses. Second, the cooling times in the gas are quite short. Near the centers of the observed X-ray images, the gas densities and temperatures derived from the observations indicate that $t_{cool} \lesssim 10^7$ yr, and the cooling times typically reach 10^{10} yr only at the very edge of the observed X-ray images. Thus, any gas introduced into the galaxy will tend to cool. Even if the gas could be heated sufficiently to balance the average cooling rate, gas at temperature $\approx 10^7$ K is very thermally unstable (Section 5.7.3), and it would form clumps which would cool rapidly. Third, cooling flow models may lead to a natural explanation for the correlation observed between the X-ray and optical luminosities of elliptical galaxies (Nulsen *et al.*, 1984; Canizares *et al.*, 1987; Sarazin, 1986b, c). Finally, the observed radial variation of the X-ray surface brightness of ellipticals can be understood under the cooling flow hypothesis (Sarazin, 1986b, c).

One important aspect of these models is that the X-ray luminosities and their correlation with the optical luminosities can only be understood if the heating of the gas is primarily due to the motions of the gas-losing stars and to adiabatic compression during the inflow, and not due to supernova heating

(Canizares *et al.*, 1987; Sarazin, 1986b, c). This implies that the supernova rates in elliptical galaxies are much smaller than had previously been thought. The low supernova rates would also explain why these galaxies lack galactic winds.

In Section 5.5.5, a method was described which allows the mass in galaxies or clusters to be determined if they contain hot, hydrostatic gas. One very important application of this method would be to measure the masses of elliptical galaxies out to large distances. At present it is unclear whether elliptical galaxies possess extended dark matter. X-ray observations of M87 at the center of the Virgo cluster show that it does have such a halo (Bahcall and Sarazin, 1977; Mathews, 1978; Fabricant *et al.*, 1980; Fabricant and Gorenstein, 1983; Section 5.8.1), but it is unclear whether this mass is associated with the cluster center or with M87. Dark haloes have been deduced for spiral galaxies from the rotational velocities of the neutral hydrogen in the disks of the galaxies far outside of their optically luminous regions. Elliptical galaxies do not possess much natural hydrogen, and thus this technique cannot be applied to them. The masses of elliptical galaxies can be determined from the orbital velocities of their stars (actually the line-of-sight component of the stellar velocity dispersion; see Dressler, 1979, 1981). Unfortunately, these observations are very difficult and cannot be done in the outermost portions of the galaxies. Moreover, the masses are uncertain because the shapes of the orbits of the stars are not known (they may be radial or isotropic, for example; see Section 2.8). The orbital velocities of globular star clusters can also be used to determine the masses of elliptical galaxies (Hesser *et al.*, 1984). Masses can be derived from studies of binary galaxies, but here the orbital characteristics are even more uncertain (Faber and Gallagher, 1979).

The use of X-ray emitting gas to measure the masses of elliptical galaxies out to large distances has a number of important advantages (Section 5.5.5). For example, this gas can be observed out to very large distances from the center of the galaxy. Also, the orbits of gas particles are known to be isotropic because the gas is a collisional fluid. Thus X-ray measurements can give important information on the possible existence of massive haloes around elliptical galaxies.

Forman *et al.* (1985) attempted to derive mass profiles by applying the hydrostatic method to the X-ray observations of these normal elliptical galaxies. They concluded that these galaxies do indeed have heavy haloes. However, the errors in the temperature determinations for these galaxies are quite large, and the derived masses are very strongly affected by errors in the temperatures (Section 5.5.5). Trinchieri *et al.* (1986) find that the temperature errors are so large that the masses cannot be determined with sufficient accuracy to decide whether normal ellipticals have missing mass haloes. In any case, this is an ideal problem for a future X-ray observatory, such as AXAF (Chapter 6).

The environment of an early type galaxy may also affect the distribution of its X-ray emitting gas. Intracluster gas might either aid in the retention of gas in a galaxy by providing a confining pressure, or aid in the removal of the gas through ram pressure ablation and other stripping processes (Section 5.9). Of particular interest in this regard are M84 and M86, which are two of the most X-ray luminous ellipticals in the Virgo cluster (Figure 26), and the X-ray emitting galaxies in A1367. In A1367, 11 galaxies were detected in X-rays by Bechtold *et al.* (Section 4.5.4; Figure 27b), of which 8 were found to be spatially extended. The luminosities of these galaxies range from $L_x \approx 1\text{–}7 \times 10^{41}$ erg/s. It is possible that pressure confinement plays a role in the X-ray emission from these galaxies (Forman *et al.*, 1979; Fabian *et al.*, 1980; Bechtold *et al.*, 1983). The galaxies detected in A1367 have unusually high X-ray luminosities for their optical luminosities, and do not show the correlation between X-ray and optical luminosities seen in other elliptical galaxies. This suggests that the gas is not gravitationally bound, but rather is confined by the intracluster gas in A1367 (Bechtold *et al.*, 1983). However, the density of the intracluster gas in A1367 is comparable to that in the Virgo cluster, where the galaxies have normal X-ray luminosities. Alternatively, Canizares *et al.* (1987) have suggested that the clumps of X-ray emission attributed by Bechtold *et al.* to galaxies are really just fluctuations in the intracluster gas, and are only projected near galaxies by coincidence.

In Virgo, the X-ray emission from M86 is more extended than the emission from M84, and forms a plume extending to the north of the galaxy (Figure 26). Based on their radial velocities, M84 appears to be moving fairly slowly, while M86 is moving rapidly. A galaxy that is moving slowly through the intracluster medium might be able to retain more of the gas produced by stellar mass loss, while a more rapidly moving galaxy would tend to lose its gas due to stripping (Section 5.9; Forman *et al.*, 1979; Fabian *et al.*, 1980; Takeda *et al.*, 1984). Because its velocity is considerably larger than the average in the Virgo cluster, Forman *et al.* and Fabian *et al.* suggest that the orbit of M86 will carry it far outside the cluster core, where the intracluster gas density is low. In these outer regions of the cluster, stripping of gas may be ineffective, and the galaxy will accumulate gas (Takeda *et al.*, 1984). Forman *et al.* and Fabian *et al.* estimate an orbital period for M86 of roughly 5×10^9 yr. Thus the galaxy could amass roughly $5 \times 10^9 M_\odot$ of gas during each orbit. This gas would be stripped during each passage through the cluster core. In this interpretation, the plume to the north of M86 is the trail of gas being stripped from M86 as it enters the core of the Virgo cluster (Forman *et al.*, 1979; Fabian *et al.*, 1982).

5.9 Stripping of Gas From Galaxies in Clusters

The galaxies found in rich clusters generally have considerably less gas than galaxies in the field (Section 2.10.2). Their 21 cm radio line luminosities indicate that they have less neutral gas than more isolated galaxies (see Section 3.7 and the many references therein), and they also appear to have weaker optical line fluxes from ionized gas (Gisler, 1978). The galaxies in rich, regular, X-ray luminous clusters are predominantly ellipticals and S0s (Sections 2.5 and 2.10.1, Table 1), and the spirals that are seen in clusters often have weak ('anemic') spiral arms (van den Bergh, 1976) and tend to be found at large projected distances from the cluster center (Gregory, 1975; Melnick and Sargent, 1977). The fraction of spiral galaxies is anticorrelated with the X-ray luminosity of the cluster (Bahcall, 1977c; Melnick and Sargent, 1977; Tytler and Vidal, 1979; Figure 30; equation (4.9)) and with the local galaxy density (Dressler, 1980b). A primary difference between spiral galaxies and ellipticals or S0s is that spirals contain large amounts of cool gas.

These observations suggest that gas in galaxies in clusters may be removed by some process. In Section 2.10.2 the possibility that the differences in the galaxian populations of clusters and the field might result from the stripping of gas from spiral galaxies was discussed. The hypothesis is very controversial, and it appears unlikely that spiral stripping alone accounts for all the differences between different types of galaxies. Nonetheless, even within individual galaxy classes (Hubble types), galaxies in clusters appear to have less gas (Sullivan *et al.*, 1981; Giovanardi *et al.*, 1983; Section 3.7). This suggests that some process must remove gas from galaxies in clusters, even if this stripping does not solely determine the morphology of galaxies.

In addition to this indirect evidence for gas removal, the X-ray observations of M86 suggest that it is currently being stripped of gas (Forman *et al.*, 1979; Fabian *et al.*, 1980; Sections 4.5.3 and 5.8.3). There are also a number of cases where optical or 21 cm radio observations may show gas currently being stripped from galaxies (see, for example, Gallagher, 1978).

Mathews and Baker (1971) showed that, once an isolated elliptical or S0 galaxy had been stripped of its gas, supernova-driven winds would keep it gas free. If the interstellar gas density is low, then the energy input from supernovae cannot be radiated away, and heats the gas until it flows out of the galaxy (Sections 5.3.3 and 5.6). If the galaxy is immersed in intracluster gas, this mechanism is less effective, since the wind must overcome the confining pressure of the intracluster gas.

Spitzer and Baade (1951) suggested that collisions between spiral galaxies in the cores of compact clusters remove the interstellar medium from the disks of these galaxies. Their basic argument was that the stellar components of galaxies could pass through one another, because two-body gravitational interaction between galaxies or their component stars is very ineffective (Section 2.9.1). However, the mean free paths of gas atoms or ions (Section 5.4.1) are relatively short, and in the center of mass frame for the collision, the kinetic energy of the gas will be thermalized. If the gas cannot cool rapidly, it will be heated to a temperature at which it is no longer bound to either galaxy, since galaxies within a cluster move much more rapidly than stars move within galaxies. If the gas cools rapidly, it will still be left at rest in the center of mass frame, while the galaxies move away from it. In either case, the gas is no longer bound to either galaxy, since it has more kinetic energy (either thermal or bulk) than the maximum galaxy binding energy.

Spitzer and Baade estimated the rate of gas removal in the Coma cluster assuming all the galaxies were equivalent, and that the galaxies all had radial orbits, and found that a galaxy should be stripped in less than 10^8 yr. This is probably an underestimate of the time scale, since galaxy orbits are probably not radial, and a typical off-center collision between two galaxies of dissimilar mass and density may not remove all of the gas from both galaxies. Moreover, subsequent increases in the extragalactic distance scale have had the effect of increasing the time scale for this process. Sarazin (1979) estimated the mean time scale for collisional stripping of galaxies in a cluster, averaging over the cluster velocity dispersion, the spatial distribution of galaxies, and the diameters and masses of galaxies, and found a rate about 100 times smaller.

Gunn and Gott (1972) suggested that galaxies lose their interstellar gas by ram pressure ablation because of the rapid motion of the galaxies through the intracluster gas. They were primarily concerned with stripping gaseous disks from spiral galaxies to make S0s. Based on a static force balance argument, they suggested that a gaseous disk would be removed when the ram pressure $P_r = \rho_g v^2$ exceeded the restoring gravitational force per unit area in the disk $2\pi G \sigma_D \sigma_{ISM}$ (where ρ_g is the intracluster gas density, v is the galaxy velocity, σ_D the surface density of the spiral disk, and σ_{ISM} the surface density of interstellar gas in the disk). Taking $v^2 = 3\sigma_r^2$ for a typical galaxy in a cluster with a line-of-sight velocity dispersion σ_r, and assuming the disk has a uniform surface density, with a radius r_D and mass M_D, this condition becomes

$$\left(\frac{n_g}{10^{-3}\,\mathrm{cm}^{-3}}\right)\left(\frac{\sigma_r}{10^3\,\mathrm{km/s}}\right)^2 \gtrsim 3\left(\frac{M_D}{10^{11} M_\odot}\right)^2\left(\frac{r_D}{10\,\mathrm{kpc}}\right)^{-4}\left(\frac{M_{ISM}}{0.1 M_D}\right), \quad (5.114)$$

where n_g is the number density of atoms in the intracluster medium and M_{ISM} is the mass of interstellar medium in the disk.

Tarter (1975) and Kritsuk (1983) have given a simple semianalytical treatment of the stripping of gas disks in spiral galaxies in clusters by integrating the net force due to ram pressure and gravity, assuming one-dimensional motion. For a given intracluster gas distribution, they found the smallest distance from the cluster center at which a galaxy could retain its gas. The calculations by Tarter indicated that most spirals should be stripped fairly easily in X-ray clusters, and that remaining spirals would only be found in the outer parts. On the other hand, Kritsuk suggested that molecular clouds might be very difficult to strip from spirals.

It is interesting to calculate the expected dependence of the spiral fraction and the typical distance of spirals from the cluster center on the cluster X-ray luminosity, if one assumes that all clusters originally had the same spiral-rich galactic populations at all positions. If one also assumes that the reduction in the fraction of spirals is due to ram pressure ablation, that a spiral is stripped whenever the ram pressure exceeds a critical amount (given by equation (5.114) or something similar), that the galaxies in clusters have an analytic King isothermal distribution (equation (5.57)), and that the gas is isothermal and hydrostatic (equation (5.63)), then the spiral fraction varies as $f_{Sp} \approx A - B \log L_x$, where L_x is the X-ray luminosity and A and B are constants. This is just the empirical relationship found by Bahcall (1977c) (equation (4.9); Figure 30). The radius of a typical spiral varies as $r_{Sp}/r_c \propto (L_x)^{1/6\beta}$, where r_c is the core radius and β is the ratio of gas and galaxy temperatures (equation (5.64)).

The majority of galaxies presently in X-ray clusters are ellipticals and S0s that have a more spherical stellar distribution. Sarazin (1979) gave a semi-analytic treatment of the stripping of gas from spherical galaxies, using basically the same formulation as Tarter (1975). He found that galaxies would be stripped if the ram pressure were greater than about $2GM_{gal}\sigma_{ISM}/R^2$, where M_{gal} and R are the galaxy mass and radius, respectively. Takeda *et al.* (1984) performed numerical hydrodynamic simulations and derived a critical ram pressure of $2\rho_{ISM}\sigma_*^2$. Assuming that the ram pressure is much greater than this limit, the time scale for ram pressure stripping of a galaxy, defined as $t_{rp} \equiv (d \ln M_{ISM}/dt)^{-1}$, is

$$\begin{aligned} t_{rp} &\approx \frac{R}{v}\left(\frac{2\rho_{ISM}}{\rho_g}\right)^{1/2} \\ &\approx 3 \times 10^7 \, \mathrm{yr} \left(\frac{\rho_{ISM}}{\rho_g}\right)^{1/2} \left(\frac{v}{10^3 \, \mathrm{km/s}}\right)^{-1} \left(\frac{R}{20 \, \mathrm{kpc}}\right), \end{aligned} \tag{5.115}$$

where ρ_{ISM} is the average interstellar gas density, ρ_g is the intracluster gas density, v and M_{gal} are the galaxy velocity and mass, respectively. The calculations of Tarter and Sarazin assume that there are no sources of gas in the galaxy.

Gisler (1976, 1979) pointed out that it would be more difficult to strip a galaxy if the stars in the galaxy were constantly resupplying it with gas. Then, the ram pressure must overcome the momentum flux due to this mass input, as well as the gravitational attraction of the galaxy. He derived an approximate analytical relationship for the ram pressure needed to strip gas out of a galaxy out to a projected radius b, ignoring pressure forces in the gas and assuming one-dimensional motion. If the rate of gas loss by stars in the galaxy is $\dot{\rho}_{ISM} = \alpha_* \rho_*$ where ρ_* is the mass density of stars in the galaxy, then he found that gas would be stripped at any projected radius b such that the ram pressure exceeded $\approx 8\alpha_* \Sigma_*(b)\sigma_*$. Here, $\Sigma_*(b)$ is the projected stellar mass density at b, and σ_* is the line-of-sight velocity dispersion of stars in the galaxy. For a typical giant elliptical galaxy in a typical X-ray cluster, Gisler found that all the gas is lost if the rate of input is less than $0.1 M_\odot$/yr, and that all the gas is retained if it is greater than about $3 M_\odot$/yr. For intermediate values, a core of interstellar gas is retained by the galaxy.

Gisler's treatment ignored pressure forces, which must be important in elliptical galaxies (Jones and Owen, 1979; Takeda *et al.*, 1984). Jones and Owen argued that extended gas in elliptical galaxies must be hot (the sound speed in the gas must be comparable to the velocity dispersion of stars), and that pressure forces in the gas cannot be ignored. A pressure gradient may be set up in the interstellar gas that opposes the ram pressure; they argued that this increases the region of a galaxy that is shielded from stripping by about a factor of ten over Gisler's result. Then elliptical galaxies will typically retain cores of interstellar gas with radii of about 10 kpc, and Jones and Owen argue that this gas can explain the existence of well-defined radio jets in the inner parts of head-tail radio sources (Section 3.3; Figure 9).

Takeda *et al.* (1984) recently rederived Gisler's condition for continuous stripping of gas from galaxies with stellar gas loss, including pressure forces. They find that the correct condition for stripping is that the ram pressure exceed

$$\rho_g v^2 \gtrsim \frac{2}{v} \int \alpha_* \rho_* |\phi_*| dl \approx \frac{16\alpha_* \Sigma_*(b) \sigma_*^2}{v}, \qquad (5.116)$$

where ϕ_* is the galaxy gravitational potential. Equation (5.116) differs from Gisler's result by a factor of $\approx 2\sigma_*/v$.

Most of these calculations of stripping have been based on semianalytic estimates, which assume the motion to be one-dimensional, generally ignore the finite temperature and compressibility of the gases, and ignore any viscous forces. To avoid these assumptions and test the efficiency of ram pressure ablation, a large number of numerical hydrodynamic simulations have been made by Gisler (1976), Lea and De Young (1976), Toyama and Ikeuchi (1980), and Nepveu (1981a). These studies have all taken the galaxy to be spherical, and the simulations have been two-dimensional, with the flows assumed to be axially symmetric. Lea and De Young started with a galaxy with a significant amount of interstellar gas, moving in the midst of intracluster gas. On the other hand, Gisler was mainly interested in the effects of stellar mass loss on ram pressure ablation, and he started with an empty galaxy. The numerical calculations indicate that stripping occurs even more easily than the simple analytic force balance arguments suggest. Unless the rate of mass input due to stars is high, a typical galaxy will be stripped almost completely during a single passage through the core of a cluster if the intracluster gas density exceeds roughly 10^{-4} atom/cm^3.

Recently, Shaviv and Salpeter (1982) and Gaetz *et al.* (1987) made two-dimensional hydrodynamic calculations of ram pressure ablation for galaxies with stellar mass loss, including the cooling of the gas. Gaetz *et al.* also included star formation. They found that gas was ablated from the outer portions of the galaxy, but was retained in the inner portions and formed a cooling flow. One very useful feature of Gaetz *et al.* is that the hydro simulations were used to derive analytic fitting formulae for many physical quantities associated with the gas flows.

All of these numerical calculations start with the galaxy in the middle of a uniform cluster. Recently, Takeda *et al.* (1984) calculated the stripping of gas from a galaxy moving on a radial orbit from the outer parts of the cluster into the core. Stellar mass loss was assumed to add to the interstellar gas in the galaxy. The stripping was determined from two-dimensional hydrodynamic simulations. After the first passage through the cluster core, the behavior was periodic, with the galaxy accumulating interstellar gas when far from the core and losing nearly all of it when it passed through the core on each orbit. A large fraction of the accumulated interstellar gas was pushed out in a single 'blob' on each passage through the cluster core. These calculations may provide a model for the galaxy M86, other gas containing galaxies, and the galaxyless extended X-ray sources in A1367 (Sections 4.5.3, 4.5.4 and 5.8.3). The latter might be blobs released from stripped galaxies (Takeda *et al.*, 1984).

Under many circumstances, transport processes such as viscosity and turbulent mixing can be more important than ram pressure in removing gas from a galaxy in a cluster (Livio *et al.*, 1980; Nepveu, 1981b; Nulsen, 1982).

Nulsen has calculated the rate of laminar viscous stripping and stripping due to turbulence, and finds a viscous stripping time scale

$$t_{vs} \approx \frac{4R}{3v}\left(\frac{\rho_{ISM}}{\rho_g}\right)\left(\frac{12}{Re}+1\right)^{-1}, \tag{5.117}$$

where Re is the Reynolds number (equation (5.47)). In many cases, this is faster than the rate of ram pressure ablation given by equation (5.115). However, one caution is that the viscous stresses may saturate, since the mean free paths of ions are similar to the sizes of galaxies (Section 5.4.4).

Livio *et al.* (1980) suggest that the Kelvin–Helmholtz instability at the boundary between interstellar and intracluster gas can produce 'spikes' of interstellar gas protruding into the intracluster gas, which are sheared off, increasing the stripping rate. They estimate a mass loss rate roughly $\dot{M}_{KH} \approx \rho_{ISM}\pi R^2 \lambda_{KH}/t_{KH}$, where $\lambda_{KH} \approx 10^{21}$ cm and $t_{KH} \approx 10^6$ yr are the wavelength and growth time of the fastest growing modes. This stripping rate is considerably smaller than that given by equation (5.117). Moreover, Nulsen (1982) has argued that Livio *et al.* overestimated this rate because they ignored the effect of compressibility on the instability, and because they did not include the effect of the mass loss on the instability. Another problem is that the wavelength of the fastest growing mode is shorter than the ion mean free path $\lambda_{KH} \ll \lambda_i$, and the viscosity expression used is therefore not valid (Section 5.4.4). Nulsen showed that the instability was suppressed when viscosity is significant, and that the mass loss is not controlled by the fastest growing modes but by the induced velocities on the largest scales.

Another mechanism for removing gas from galaxies, which can operate even when the galaxies are moving slowly through the intracluster gas, is evaporation (Gunn and Gott, 1972; Cowie and McKee, 1977; Cowie and Songaila, 1977). Heat is conducted into the cooler galactic gas from the hotter intracluster gas, and if the rate of heat conduction exceeds the cooling rate, the galactic gas will heat up and evaporate. If cooling is assumed to be small, the evaporation rate with unsaturated conduction for a spherical galaxy immersed in intracluster gas of temperature T_g is (Cowie and Songaila, 1977)

$$\begin{aligned}\dot{M}_{ev} &\approx \frac{16\pi\mu m_p \kappa R}{25k} \\ &\approx 700 M_{\odot}/\mathrm{yr}\left(\frac{T_g}{10^8\,\mathrm{K}}\right)^{5/2}\left(\frac{R}{20\,\mathrm{kpc}}\right)\left(\frac{\ln\Lambda}{40}\right)^{-1},\end{aligned} \tag{5.118}$$

where κ is the thermal conductivity (Section 5.4.2; equation (5.37)) and Λ is the Coulomb logarithm (equation (5.33)). The stripping rate is a factor of $(2/\pi)$ smaller for a disk galaxy with the same radius. The evaporation rate will be significantly reduced if the conductivity saturates (Section 5.4.2), as is probably the case at least for disk galaxies. Unfortunately, thermal conductivity also depends critically on the magnetic field geometry (Section 5.4.3). If the conductivity is not suppressed by the magnetic field, this mechanism can play an important role in stripping gas from galaxies.

As pointed out by Nulsen (1982), there is a simple connection between mass loss by evaporation and mass loss by laminar viscosity (the *Re* term in equation (5.117)). At low velocities, this term dominates, and the viscous stripping rate is nearly independent of velocity, because the Reynolds number is proportional to velocity. Since both thermal conduction and ionic viscosity are transport processes and the ion and electron mean free paths are essentially equal, the rates of stripping from these two processes are simply related:

$$\dot{M}_{ev} \approx 3.5 \dot{M}_{vs}. \tag{5.119}$$

While this expression was derived for unsaturated conduction and viscosity and no magnetic suppression of either, it probably will remain approximately true even when these effects are included, since both thermal conduction and viscosity are affected similarly.

5.10 The Origin and Evolution of the Intracluster Medium

Where does all of this X-ray emitting intracluster gas come from? Did it fall into clusters from intergalactic space, or was it ejected from galaxies? When did it first fill the great volumes of space between galaxies in clusters? What can we learn about the origin of galaxies and clusters from the intracluster gas? Theories of the origin of the intracluster medium attempt to explain the current properties of X-ray clusters, to make predictions about their past history that can be tested with observations of clusters at high redshift, and to relate the history of the intracluster gas to theories of the origin of structure in the universe.

There are two basic constraints on such theories from the observations of present day X-ray clusters. First, these clusters typically contain $\approx 10^{14} M_{\odot}$ of intracluster gas, an amount at least comparable to the mass of luminous material in galaxies and at least one-tenth of the total mass of the cluster. Second, this gas produces X-ray lines from iron which require that the abundance of that element be about one-half of its solar value if the gas is homogeneous.

At present, there are very few observations of high redshift X-ray clusters, and thus the constraints that can be imposed on theories of the evolution of X-ray clusters by these observations are fairly weak. The only safe statement one can make at the present time is that X-ray clusters do not show any evidence of very dramatic evolution out to redshifts of $z \approx 0.5$ (Section 4.8). Another constraint on the evolution of X-ray clusters comes from the diffuse X-ray background (Fabian and Nulsen, 1979); the total emission from high redshift clusters cannot exceed the background brightness.

The question of the origin of the intracluster gas is strongly tied to the question of the origin of clusters and galaxies. This is a major area of astronomical research, and I shall not even attempt to summarize the theories in this area. Instead, I shall deal very narrowly with theories of the intracluster medium and shall largely ignore their dependence on models for galaxy formation, except when the X-ray cluster models provide information that affects these theories. Discussions of the effects that X-ray cluster observations have on theories of galaxy formation include Silk (1978), Binney and Silk (1978), White and Rees (1978), Fabian and Nulsen (1979), Binney (1980), and Field (1980).

There is an increasingly large amount of observational information on the evolution of galaxies; here only a few points will be noted. First, galaxies contain stars with a range of ages and heavy element abundances. Elliptical and S0 galaxies contain mainly older stars, with ages of around 10^{10} yr (Fall, 1981). Although galaxies do contain some stars with very low heavy element abundances, there is no significant population of stars with *no* heavy elements, and the number of low abundance stars is less than might be expected if all the heavy elements were formed in stars like the present stellar population (Carr *et al.*, 1984). No mechanism is known to produce heavy elements in any quantity outside of stars. This has led to the suggestion that there was an early (before 10^{10} yr ago) generation of stars, which produced the minimum level of heavy elements seen in stars today (Carr *et al.*, 1984). Second, there is some evidence that the population of galaxies has evolved in some clusters since redshifts of $z \approx 0.5$; this is the 'Butcher–Oemler effect' (Section 2.10.2). These clusters have an excess of blue galaxies, which suggests that they were undergoing fairly rapid star formation at that time. Finally, violent activity in the nuclei of galaxies appears to have been much more common in the past (Schmidt, 1978); this activity produces Seyfert galaxies, radio galaxies, and quasars. There were many more very luminous quasars at a time corresponding to a redshift of 2 than are seen around us today. On the other hand, since few quasars are known with redshifts greater than 3.5 (Osmer, 1978), something may have occurred then to start the violent nuclear activity in galaxies. Also, if quasars are all located in the nuclei of galaxies, then galaxies must have formed by a redshift of 3. Although this nuclear activity in galaxies is very poorly understood, most current theories require that gas be supplied to the nucleus (Rees, 1978); thus its variation may be related to the supply of intracluster gas and the stripping of gas from galaxies.

5.10.1 *Infall Models*

Gunn and Gott (1972) suggested that the intracluster medium was primordial intergalactic gas that had fallen into clusters. This intergalactic gas was never associated with stars or galaxies, and thus could be expected to have no heavy elements. They also noted that some of the intracluster medium could come from ram pressure stripping of interstellar gas out of galaxies. Assuming that clusters were immersed in uniform intergalactic medium, they estimated the amount that would fall into clusters. By comparing this to the amount of intracluster gas deduced from the early X-ray observations of clusters, they could give an upper limit on the density of the intergalactic gas. (This is an upper limit because some of the intracluster gas could have come out of galaxies.) These limits are usually given in terms of ρ_c, the density of matter necessary to close the universe

$$\rho_c \equiv \frac{3H_o^2}{8\pi G} = 4.7 \times 10^{-30} h_{50}^2 \,\text{gm/cm}^3, \tag{5.120}$$

where H_o is the Hubble constant. If ρ_{IG} is the density of intergalactic gas, then Gunn and Gott found $\rho_{IG}/\rho_c \lesssim 0.01$. They used a rather low value for the gas mass in clusters, and more recent calculations (for example, Cowie and Perrenod, 1978) give $\rho_{IG}/\rho_c \lesssim 0.2$. Gunn and Gott also noted that infall would heat the intracluster gas to about the observed temperatures (Section 5.3.2).

To determine the final configuration and evolution of the intracluster gas in the infall model, hydrodynamic simulations of the infall have been done by a number of authors (Gull and Northover, 1975; Lea, 1976; Takahara *et al.*, 1976; Cowie and Perrenod, 1978). These calculations all assumed that the cluster potential was fixed; the gas was assumed to fall into the cluster after the cluster itself had collapsed. All of these calculations were one-dimensional simulations of spherical clusters, although a number of different techniques were used to solve the hydrodynamic equations. With the exception of Lea's calculations, these simulations have given similar results.

As the gas first collapses into the core, its density increases and a shock propagates outward from the cluster center and heats the gas. This shock passes through the cluster in $\approx 10^9$ yr, essentially the sound crossing time for the cluster (equation (5.54)). After the passage of the shock, the hot intracluster gas is nearly hydrostatic, and its further evolution is quasistatic. As the shock moves into the outer parts of the cluster, it weakens; less gas is added to the cluster, and the cluster luminosity is nearly constant. On the other hand, Lea found that the shock heating caused the gas pressure to increase until the inflow was reversed and the intracluster gas expanded. This cooled the gas adiabatically, lowered its pressure, and it collapsed again. This process repeated itself, producing a large number of pulsations with a period of about 5×10^9 yr. During these pulsations, the X-ray luminosity oscillated wildly between roughly 10^{41} and 10^{48} erg/s. The other calculations of the infall of intracluster gas have failed to find these oscillations (Gull and Northover, 1975; Takahara *et al.*, 1976; Cowie and Perrenod, 1978; Perrenod, 1978b), and they are probably an artifact of Lea's numerical method. Such oscillations are in violent disagreement with the observed X-ray luminosity function of clusters (Schwartz, 1978).

Gull and Northover (1975) found that the shock strength was nearly constant as it propagated outward; they argued that this occurred because the shock speed was always about the free-fall speed in the cluster. They found that the resulting intracluster gas distribution was nearly adiabatic (Section 5.5.2). On the other hand, more detailed calculations by Cowie and Perrenod (1978) and Perrenod (1978b) found that the cluster gas distributions were not well represented by any polytropic distribution, unless thermal conduction was so effective that the gas was isothermal.

In the absence of significant cooling or thermal conduction, Cowie and Perrenod (1978) showed that the infall models with a fixed cluster potential are characterized by a single parameter, which gives the depth of the cluster potential well, $K \equiv (\sigma_r / H_o r_c)^2$ where σ_r and r_c are the cluster velocity dispersion and core radius, respectively. Models with significant cooling are also characterized by $B \equiv t_{cool} H_o$, where the cooling time is evaluated at the cluster center. If thermal conduction is important, the cluster evolution is also determined by the value of $C \equiv \kappa T_g / r_c \rho_g c_s^3$, where κ is the thermal conductivity, and T_g, ρ_g, and c_s are the gas temperature, density, and sound speed. When conduction saturates, the models are independent of C. In general, the gas temperatures in these models scale with σ_r^2.

Cowie and Perrenod found that models without significant cooling or conduction showed a very small decrease in the X-ray luminosity with time, less than a 40% reduction from $z = 1$ to the present. This decrease in luminosity resulted from the slow reexpansion of the intracluster gas as the shock weakened. In models with significant cooling, the cluster evolved to a nearly steady-state cooling flow (Section 5.7). In models with conduction, the X-ray luminosity increased slowly with time, by about 40% from a redshift $z = 1$ to the present. This occurred because conduction lowered the temperature in the cluster core (Section 5.4.2). The core then contracted so that the increasing density could maintain the pressure in the core. Since the X-ray luminosity increases more rapidly with density than does the pressure (equation (5.21)), the luminosity went up.

These models all assumed that the cluster potential was static; the cluster was assumed to collapse before any gas fell into it. Of course, there is no reason why intergalactic gas should wait until the cluster has formed before gaseous infall can occur. Perrenod (1978a, b) calculated the evolution of the intracluster gas in infall models in which the cluster potential varied in time. The cluster potential was taken from White's (1976c) N-body calculations of the collapse of a Coma-like cluster (Section 2.9; Figure 5). In White's models, the cluster first collapses with violent relaxation, then contracts slowly due to two-body interactions between galaxies. This contraction causes the cluster potential well to become deeper, and as a result the intracluster gas temperature and density increase with time. In contrast to the static potential models, Perrenod finds that the X-ray luminosity of his model clusters increases by about an order of magnitude from $z = 1$ to the present. The sizes of the gas distributions also shrink considerably. If thermal conduction is important, the further contraction in the gas distributions it produces makes them smaller than the observed sizes of X-ray clusters.

One interesting aspect of Perrenod's models is that many of the infall models have a temperature inversion ($dT_g/dr > 0$) in the cluster core, even if there is no significant cooling. This occurs because gas in the core has fallen through a shallower gravitational potential than gas further out. If such a temperature inversion were observed, it might be confused with a cooling flow (Section 5.7).

In White's N-body models, the cluster shows very strong subclustering at the beginning of its collapse, and forms two roughly equal subclusters, which merge as the cluster undergoes violent relaxation. Several double X-ray clusters are known (Section 4.4.2; Figure 18) that appear to be in just this stage of evolution (Forman *et al.*, 1981). Obviously, such subclustering cannot be treated in one-dimensional, spherical, hydrodynamic simulations. Gingold and Perrenod (1979) have made simplified three-dimensional hydro simulations of the evolution of clusters. When applied to the cluster potential from White's N-body models, these verified the previous one-dimensional calculations of Perrenod (1978b). They found that there was no significant enhancement of the X-ray emission from merging subclusters, beyond that predicted by single cluster models. Similar calculations were made by Ikeuchi and Hirayama (1979).

One major concern about all the Perrenod varying-potential models is the use of White's (1976c) N-body calculations for the cluster potential. In this particular set of models by White, the total virial mass of the cluster was assumed to reside in the individual galaxies. This gave the galaxies large masses, which increased their two-body interactions (Section 2.9.1), and caused the cluster core to contract rapidly. However, associating the missing mass in clusters with individual galaxies appears to produce more two-body relaxation in clusters than is observed (Sections 2.8 and 2.9.4); in fact, this was one of White's conclusions from his models. Thus it seems likely that Perrenod's calculations may significantly overestimate the increase with time of the X-ray luminosity and gas temperature and the decrease in the gas core size.

Clusters of galaxies are the largest organized structures in the universe, and X-ray emission from them should be recognizable to large redshifts (Chapter 6). They might therefore be useful as probes of the cosmological structure of the universe. Several cosmological tests have been proposed using X-ray clusters (Schwartz, 1976; Silk and White, 1978); although some of these tests are relatively insensitive to X-ray cluster evolution, most are strongly affected. These models suggest that it will be difficult to apply any tests that require that X-ray clusters have remained unchanged since $z = 1$ (Perrenod, 1978b; Falle and Meszaros, 1980). On the other hand, in Perrenod's models the luminosity and size of X-ray clusters depend strongly on the density of material in the universe, since this determines the speed with which clusters contract. In principle, this dependence of cluster evolution on density might provide useful cosmological information; in practice, the evolution models are too uncertain to be used reliably for this purpose.

The models described so far have dealt with the evolution of single clusters. Perrenod (1980) has attempted to predict the evolution of the luminosity function of X-ray clusters (Section 4.2). He assumed that galaxies formed before clusters, and that clusters were formed by the gravitational attraction of galaxies. He argued that the merging of clusters tends to produce larger clusters with deeper potential wells, and as a result the average X-ray luminosity increases. White (1982) showed that this argument is incorrect; the increase in the depth of cluster potential wells is more than offset by the decrease in their characteristic densities. Perrenod found a very rapid evolution of the luminosity function to higher luminosities; he predicted that there should be few luminous X-ray clusters at redshifts $z \gtrsim 1/2$. This evolution depends strongly on the average density of matter in the universe, and Perrenod proposed using it as a cosmological test. However, his basic model for clustering is apparently incorrect (White, 1982).

5.10.2 *Ejection from Galaxies*

The observation that the intracluster medium contains a significant abundance of heavy elements shows that it cannot be *entirely* due to the infall of primordial gas (Sections 4.3.2 and 5.2). The only way known for producing reasonable quantities of heavy elements is through nuclear reactions in stars.

Since there is no significant luminous stellar population outside of galaxies at present, there are two possibilities. First, there may have been an early generation of pregalactic stars (Carr *et al.*, 1984), or second, it may be that some portion of the intracluster gas was ejected from galaxies. This section considers the second possibility.

Are the present rates of mass loss from stars in galaxies sufficient to produce the required amount of gas if all the stellar mass loss is added to the intracluster gas? Let us assume that the intracluster gas is chemically homogeneous (Section 5.4.5), so that the inferred heavy element abundance is about half of the solar value. Then, since this is comparable to the present abundances in the stars in elliptical galaxies in clusters, ejection from galaxies would have to supply a significant portion of the observed intracluster gas. The total mass of intracluster gas is at least as large as the total mass of stars in galaxies in a luminous X-ray cluster ($\gtrsim 10\%$ of the virial mass). Now the current rate of mass loss expected from the stellar populations seen in elliptical or S0 galaxies is

$$\alpha_* \approx 1.5 \times 10^{-12}/\text{yr}. \tag{5.121}$$

Thus, only a few per cent of the intracluster gas could be supplied in a Hubble time $\approx 10^{10}$ yr at the current rate.

In doing estimates of this sort, it is very important to remember that mass loss from stars is due to their evolution as they exhaust their nuclear fuel, and thus the mass loss from a stellar population is primarily due to the most luminous stars. In estimating the rate of mass loss, the rate per luminous star should be multiplied by the mass of luminous matter, and not the total (virial) mass. Another way of stating this is that the rate of mass loss is proportional to the luminosity of a stellar population and not to its mass; thus the rate of mass loss per unit mass varies inversely with the mass-to-light ratio (Section 2.8).

The rate of mass loss from stars in galaxies must have been higher in the past, if stellar mass loss has contributed significantly to the intracluster medium. One simple way this can have occurred is for elliptical and S0 galaxies to have contained more massive stars in the past, since massive stars have higher rates of mass loss. At present, these galaxies have only relatively low mass stars $M_* \lesssim 0.8 M_\odot$ (Fall, 1981). Since stellar lifetimes decrease with mass, and the presently observed stars have lifetimes comparable to the Hubble time, any higher mass stars produced at the time of galaxy formation would no longer exist. High mass stars tend to die as supernovae which are very effective in producing and dispersing heavy elements, so these stars might provide the heavy elements in the intracluster gas and in the stars seen in the galaxies today without leaving any very low abundance stars (Carr *et al.*,

1984). Moreover, the supernovae could aid in the removal of gas from the galaxies into the intracluster medium.

It is often suggested that all the stars in an elliptical or S0 galaxy formed at one time during the formation of the galaxy itself. Stars with a wide range of masses are usually assumed to have been made in protogalaxies, and the present stellar population is only those lower mass stars whose lifetimes exceed the age of the galaxy. Usually, the distribution of the masses of stars that form (the initial mass function or IMF) is taken to be a power-law, and the star formation rate is assumed to decline exponentially with the age of the galaxy:

$$\frac{\partial^2 N_*}{\partial M_* \partial t} \propto M_*^{-a} \exp(-t/t_*) \qquad M_L \leqslant M_* \leqslant M_U \tag{5.122}$$

where N_* is the number of stars formed of mass M_*, the lower and upper limits to the IMF are M_L and M_U, and t_* is the time scale for star formation. A power law with 2.35 is called the 'Salpeter IMF' (Salpeter, 1955) and fits the current star formation in the disk of our galaxy. The time scale for star formation is often taken to be comparable to the dynamical time in a galaxy, $t_* \approx 3 \times 10^8$ yr. The resulting model for the gas loss from a galaxy depends on the IMF assumed and on the values of t_* and t_{st}, the time scale for the removal of gas from the galaxy.

Larson and Dinerstein (1975) calculated the properties of the intracluster gas based on this model. They assumed a Salpeter IMF. The only gas loss process they considered was supernova heating from the same stellar population; this may underestimate the ejected mass if collisions or ram pressure ablation also contribute. Based on this model, they found that the majority of the gas was not removed in galaxies more massive than about $10^{10} M_\odot$. The total gas mass ejected from galaxies in a cluster was about 30% of the mass in stars and galaxies, and the ejected mass had roughly solar abundances. Because these calculations preceded the detection of the iron X-ray lines in clusters, they successfully predicted that nearly solar abundances would be found.

Fairly similar results were found by Ikeuchi (1977), Biermann (1978), De Young (1978), and Sarazin (1979). Biermann assumed most of the galaxies were spirals that were stripped by collisions and ram pressure; this resulted in a more complete removal of interstellar gas, but over a longer time ($t_{st} \geqslant t_*$). Biermann found somewhat lower heavy element abundances of 0.1 to 0.5 of solar, with the gas mass ranging from 1 to 0.1 of that of the stars. De Young noted that the supernova energy was sufficient to unbind the gas produced by stellar mass loss in galaxies, and thus assumed that nearly all the gas was ejected quite rapidly ($t_{st} < t_*$). He also considered a wider range of IMFs, and generally found heavy element abundances that were larger, 1–3 times solar. This was primarily sensitive to the exponent a in the IMF. The ejected gas masses were 0.15 to 0.5 of the stellar mass. One exception to the agreement among these authors was Vigroux (1977), who claimed that galaxies could not make enough iron during the course of their normal evolution. The account he gave of his calculations was rather sketchy, so it is difficult to compare them to the others. With this exception, the general conclusion was that galaxies could eject an amount of gas about half the stellar mass with roughly solar abundances during their normal stellar evolution. If this gas were diluted with roughly an equal amount of unprocessed primordial gas, either within the forming galaxies or in the cluster, the observed mass and heavy element abundances in the intracluster gas would be reproduced.

In most of these models, the time scale of star formation is assumed to be short, $t_* \lesssim 10^9$ yr. During this time, most of the stars in the galaxy are formed, and the more massive and luminous stars live and die explosively in supernovae. As a result, it is expected that the newly formed galaxies would be very bright during this era (De Young, 1978; Bookbinder *et al.*, 1980).

The evolution of the intracluster gas in models with ejection from galaxies depends on the length of time it takes the newly enriched gas to be stripped, t_{st}. De Young (1978) noted that the energy input from the supernovae produced by a stellar population which would give the needed iron abundance would be sufficient to unbind the interstellar gas. This assumes that the supernova energy is efficiently converted to kinetic energy in the gas.

On the other hand, if the supernovae energy is radiated away, the gas may remain bound to the galaxy. Norman and Silk (1979) and Sarazin (1979) showed that this was likely to be the case, because the large quantity of intracluster gas in clusters implies that there was a large density of gas in protogalaxies. At high densities the gas cools rapidly, and individual supernova remnants radiate away their energy before they overlap. Under these circumstances, the galaxies may retain their gas. Norman, Silk, and Sarazin suggested that galaxies retain much of their initial gas content as extended hot coronae. If this gas cannot be removed by supernovae, then collisions or ram pressure remain as stripping mechanisms (Section 5.9). They further assumed that galaxy formation is very efficient, in the sense that nearly all the gas in a cluster was initially contained in galaxies. Then, there would be very little intracluster gas at first, and ram pressure ablation would not be effective. The galaxies would first lose gas slowly through collisions, and when the intracluster density was high enough, ram pressure stripping would start. Because ram pressure ablation both increases the gas density and increases with increasing gas density, this leads to a runaway stripping of cluster galaxies. The evolution of the gas in a cluster would then occur in two extended stages, with a rapid transition between them. First, all the gas would be bound to galaxies. Then, it would be rapidly stripped and remain distributed in the intracluster medium after that time. This was proposed as an explanation of the Butcher–Oemler effect (Section 2.10.2); the Butcher–Oemler clusters were still in this first stage. Unfortunately, this model predicts that these clusters have very little intracluster gas, when in fact they were subsequently observed to be luminous X-ray sources (Section 4.8). Larson *et al.* (1980) argued that the disks of spiral and S0 galaxies are produced by infall from coronae of gas bound to galaxies, and that a spiral galaxy becomes an S0 when the corona is stripped and the gas supply to the disk stops. In this way, there would be a longer interval between the stripping of the corona and the cessation of star formation in the disk. Perhaps Butcher–Oemler clusters are within this interval. One problem with these models for gaseous coronae is that they require a rather delicate and unstable balance between supernova heating and cooling.

Biermann (1978) proposed a similar model in which gas produced by disk galaxies is stored in their disks, and is eventually stripped by collisions and ram pressure. Himmes and Biermann (1980) gave a somewhat more detailed model, in which elliptical galaxies in a cluster lose their interstellar gas rapidly by supernova heating and provide an initial amount of intracluster gas, which begins the process of ram pressure stripping of spiral galaxies. They argued that this model can reproduce the present intracluster gas masses, iron abundances, and dependence of the galactic population on X-ray luminosity (Section 4.6). In this model, the spiral fraction in cluster decreases continuously with time, and the variation from a redshift of $z \approx 0.4$ to the present is consistent with the Butcher–Oemler effect.

One general feature of these models in which gas is ejected from galaxies over a long period of time is that the luminosities of X-ray clusters are expected to increase with time. Unfortunately, this is the same prediction made by Perrenod's infall models with deepening cluster potentials, as discussed in the previous section.

A number of one-dimensional, spherically symmetric, hydrodynamic simulations have been made of the evolution of intracluster gas, including ejection from galaxies. Cowie and Perrenod (1978) calculated models with a fixed cluster potential and assumed that the rate of gas ejection from galaxies varied inversely with time $\alpha_* \propto 1/t$. There was no primordial intracluster gas in these models. The gas was ejected at zero temperature (Section 5.3.3) and assumed to mix immediately with the intracluster gas. When the gas ejection rate is large, the models evolve to steady-state cooling flows (Cowie and Binney, 1977; Section 5.7.1). In models with lower ejection rates, the X-ray luminosity either is roughly constant (no thermal conduction) or increases by about a factor of two (thermal conduction) from a redshift $z = 1$ to the present. Perrenod (1978b) calculated ejection models in a varying cluster potential; the results were very similar to those described above for infall models. He found a better fit to the present gas distributions with these models than with infall models, and the ejection models were less sensitive to the assumed initial conditions and model parameters. These models showed a rapid increase in X-ray luminosity with time, as did the infall models.

Ikeuchi and Hirayama (1980) ran hydro models with no primordial gas, in which gas is ejected from all the galaxies simultaneously and very rapidly ($t_{st} \lesssim 10^7$ yr). This seems rather unlikely, since this time scale is less than the sound crossing time for a single galaxy. Because of this assumption of rapid ejection, they chose the following initial conditions: at the start of their calculation, the ejected gas was placed in the cluster in a nonhydrostatic distribution determined by their ejection model, and then 'let go'. The gas then adjusted to the cluster potential on a sound crossing time (Section 5.5). These models have very large X-ray luminosities $\approx 10^{48}$ erg/s during this initial relaxation time $t \approx 3 \times 10^8$ yr. It seems very unlikely that such high luminosities would be realized, since the actual gas ejection must take considerably longer than was assumed.

It has generally been assumed that the ejected gas mixes rapidly (both chemically and thermally) with the intracluster gas (De Young, 1978). However, Nepveu (1981b) argues that this will not occur (although the only mechanism he considers is turbulent mixing), and that the ejected and intracluster gases must be treated as two separate fluids.

Hirayama (1978) and Nepveu (1981b) have given hydrodynamic models for the evolution of the intracluster medium including both gas ejected from galaxies and primordial gas. In both cases, the primordial gas is initially relaxed, and galaxy gas is injected at a constant rate. In both of these calculations, the ejected gas is concentrated to the cluster center ($R \lesssim 2$ Mpc), and there is a large gradient in the heavy element abundance across the cluster. As noted previously (Sections 5.4.5 and 5.5.6), such a concentration of heavy elements to the cluster center will increase the strengths of the X-ray lines from these elements. Thus the abundances derived from the X-ray spectra of clusters could overestimate the real abundances. However, because most of the X-ray emission in these models comes from radii of less than 2 Mpc, this effect is not very serious. Spatially resolved X-ray spectra across a cluster might detect such a gradient, and this would allow one to deduce the proportions of ejected and intracluster gas.

6

PROSPECTS FOR THE FUTURE AND AXAF

What advances can be expected in the near future in the study of X-ray clusters? In this brief look at future prospects, I shall concentrate on new observational opportunities. At present, observational X-ray astronomy is in a rather quiet period. The *Einstein* X-ray satellite, which revolutionized the study of X-ray astronomy, is no longer operational. As this review was being written, the European Space Agency X-ray satellite EXOSAT was nearing the end of its operational life. EXOSAT is a somewhat less powerful imaging instrument than *Einstein*. What advances in the technology of X-ray astronomy are needed to answer the major questions we have about X-ray clusters, and what plans are there for the realization of these advances?

The basic data in the X-ray study of clusters consist of the surface brightness of X-rays I_ν as a function of the photon frequency ν and the position on the sky. Given these data and a suitable assumption of symmetry for the cluster, the X-ray emissivity $\varepsilon_\nu(r)$ as a function of position in the cluster can be derived by deconvolution of the surface brightness (Section 5.5.4). The emissivity varies as the square of the density ρ_g, and its frequency dependence is determined by the gas temperature T_g and by the abundances of heavy elements (equation (5.19)). At the temperatures usually found in the intracluster gas, the heavy element abundances mainly affect the emission in several narrow line features, and the temperature produces an exponential falloff in the intensity for frequencies $h\nu > kT_g$. Thus, given suitable observations of the X-ray surface brightness I_ν, one can derive the gas density, temperature, and several heavy element abundances, all as a function of position in the cluster.

In relatively nearby clusters, the required instrument for these observations must measure the X-ray surface brightness with at least moderate spatial resolution (< 1 arcmin), and modest spectral resolution (better than about 15%), and must be sensitive to X-rays with photon energies $h\nu$ of at least 7 keV. Obviously, it must also have a sufficient sensitivity to detect the clusters. Unfortunately, no past or currently existing satellite has had all these capabilities. Proportional counter systems, such as the *Uhuru* satellite, have modest spectral resolution out to high X-ray energies, but have very poor spatial resolution. The *Einstein* satellite had excellent spatial resolution, fairly poor spectral resolution in the imaging detectors, and no sensitivity for photon energies $h\nu \gtrsim 4$ keV.

The Advanced X-ray Astronomy Facility (AXAF) would provide the new observational capabilities needed for the further study of X-ray clusters (Giacconi *et al.*, 1980). AXAF is a 1.2 meter diameter X-ray telescope, which would be carried into orbit by the Space Shuttle. As currently planned, AXAF would have roughly 100 times the sensitivity of the *Einstein* telescope for point sources and a considerably increased sensitivity for extended sources as well. Its mirrors would produce images with a spatial resolution of better than one second of arc, and would be sensitive to X-rays with photon energies up to at least 8 keV. At least some of the imaging detectors being considered would have moderate spectral resolution (10–20% or better), and the satellite would have a number of higher resolution spectrometers. With its high spatial resolution, moderate spectral resolution, and sensitivity to harder X-rays, it would provide exactly the data on cluster X-ray surface brightnesses needed to derive their densities, temperatures, and abundances.

Given the run of density and temperature of the gas in a cluster or in an individual galaxy, the hydrostatic equation (5.56) allows one to determine the total mass in the galaxy or cluster as a function of position (Sections 5.5.5 and 5.8.1). These mass determinations are less uncertain than those based on the radial velocities of galaxies in clusters or stars in galaxies because the gas atoms are known to be moving isotropically. These mass distributions would provide very important information on the distribution and nature of the missing mass component in clusters and galaxies.

If measurements of the microwave diminutions of clusters can be made reliably, they can be combined with the determinations of the variation of the gas temperature and density to give distances to clusters that are independent of the Hubble constant (Section 3.5). This will provide a direct determination of the Hubble constant, independent of the usual extragalactic distance scale. If this method could be applied to high redshift clusters, it might allow the determination of the overall structure of the universe.

From the variation of the gas temperature with density and with position in a cluster, the influence of the various heating, cooling, and energy transport processes (Sections 5.3, 5.4) can be deduced. As discussed in Section 5.5.1, the surface brightness distributions in nearby clusters from the *Einstein* satellite are consistent with isothermal gas distributions, although temperatures could not be determined directly. Unfortunately, the temperatures required by the surface brightness fits are generally *not* consistent with temperatures derived from the integrated spectra of the clusters. Given this discrepancy, we cannot claim to have any real understanding of the thermal processes in intracluster gas. Direct measurements of the temperature profiles of clusters are needed to resolve this problem.

The moderate spectral resolution of AXAF's imaging detectors and sensitivity to 7 keV X-rays will allow this instrument to map out the abundance of iron and possibly other heavy elements in clusters. The distribution of heavy elements in clusters must be known if accurate abundances are to be derived for them. As noted in Sections 5.4.5 and 5.5.6, if the iron in clusters is concentrated in the core, the iron abundances may have been significantly overestimated. These abundances are used to determine the amount of gas that must have been ejected from stars in galaxies, and affect models for the origin and early evolution of galaxies (Sections 5.10.2). Moreover, the distributions of heavy elements provide information on the relative proportions of ejected galactic gas and primordial intergalactic gas in clusters.

The higher resolution spectrometers on AXAF can be used to determine the abundances of additional elements and give more precise information on the temperature structure. It is particularly useful that the 7 keV iron lines will be observable, as these are the strongest lines in the intracluster gas and have many fine structure components whose intensities are sensitive to temperature. High resolution line studies will be particularly useful in studying the physical conditions in cooling flows (Section 5.7). They may also permit the determination of redshifts for X-ray clusters for which optical data are not available, and will certainly resolve any ambiguities when several clusters at different redshifts are seen along the same line-of-sight. At the highest spectral resolution, it may be possible to determine directly the flow velocities in clusters, particularly those with cooling flows.

The high spatial resolution of AXAF will be very important to the study of cooling flows and of other gas associated with individual galaxies (Sections 5.7 and 5.8). The increased sensitivity and enhanced spectral response of AXAF should make it possible to get spectra of the gas in these individual galaxy sources, in order to test the hypothesis that the emission is from hot gas. Gas in individual galaxies in clusters has so far been studied only in relatively nearby clusters. Of particular interest is the interaction of this gas with the intracluster medium.

AXAF should detect X-ray clusters out to very high redshifts, $z \approx 1$–4. From the study of these clusters, we shall learn directly about the origin of the intracluster gas and its evolution in clusters. We may actually see the gas being ejected from galaxies. If galaxy morphologies are altered by the galaxy's environment, and the main mechanism is gas stripping by intracluster gas, the buildup of the intracluster gas should be related to the evolution of galaxy morphologies. With the Hubble Space Telescope (Hall, 1982), it should be possible to classify galaxies out to at least moderate redshifts. The variation in the heavy element abundances in clusters as a function of redshift should constrain models for the chemical evolution of galaxies. The variations in the temperature of the gas will allow us to assess the effects of heating and cooling.

It will also be interesting to see if there is any relationship between the evolution of X-ray clusters and that of quasars and other active galactic nuclei.

One problem with these studies of high redshift clusters is that few are currently known. Because AXAF is not primarily a survey instrument, it might not detect a very large number of previously unknown clusters. It is possible that deeper ground based optical surveys or studies with the Hubble Space Telescope will provide longer lists of cosmological clusters. It is also possible they may be found by studying high redshift radio galaxies and quasars with the morphological distortions normally associated with sources in clusters (Section 3.3). Another exciting possibility involves the Roentgen Satellite (ROSAT). This instrument will perform an all-sky soft X-ray survey, with a spatial resolution of about 1 minute of arc and a sensitivity limit similar to that of the *Einstein* satellite. A luminous X-ray cluster at a redshift of 1 might possibly be detected in this survey. Because an X-ray cluster at a redshift of 1 would have an angular size of about 1 minute of arc, it might appear as a resolved source. The most common extragalactic X-ray sources found in deep surveys are quasars, which are point sources. Thus most of the resolved high galactic latitude sources in the ROSAT survey should be clusters, and some of those should be at high redshifts. This survey may provide a valuable list of X-ray clusters for further study.

REFERENCES

Aarseth, S. J., and J. Binney, 1978, *Mon. Not. R. Astron. Soc.* **185,** 227.
Abell, G. O., 1958, *Astrophys. J. Suppl.* **3,** 211.
Abell, G. O., 1961, *Astron. J.* **66,** 607.
Abell, G. O., 1962, in *Problems of Extra-Galactic Research*, edited by G. C. McVittie, p. 213. Chicago: University of Chicago.
Abell, G. O., 1965, *Ann. Rev. Astron. Astrophys.* **3,** 1.
Abell, G. O., 1975, in *Stars and Stellar Systems IX: Galaxies and the Universe*, edited by A. Sandage, M. Sandage, and J. Kristian, p. 601. Chicago: University of Chicago.
Abell, G. O., 1977, *Astrophys. J.* **213,** 327.
Abell, G. O., 1982, private communication.
Abell, G. O., J. Neyman, and E. L. Scott, 1964, *Astron. J.* **69,** 529.
Abramopoulos, F., G. Chanan, and W. Ku, 1981, *Astrophys. J.* **248,** 429.
Abramopoulos, F., and W. Ku, 1983, *Astrophys. J.* **271,** 446.
Adams, M. T., K. M. Strom, and S. E. Strom, 1980, *Astrophys. J.* **238,** 445.
Adams, T. F., 1977, *Publ. Astron. Soc. Pac.* **89,** 488.
Albert, C. E., R. A. White, and W. W. Morgan, 1977, *Astrophys. J.* **211,** 309.
Allen, C. W., 1973, *Astrophysical Quantities*, p. 197. London: Athlone.
Andernach, H., J. R. Baker, A. von Kap-herr, and R. Wielebinski, 1979, *Astron. Astrophys.* **74,** 93.
Andernach, H., D. Schallwich, C. Haslam, and R. Wielebinski, 1981, *Astron. Astrophys. Suppl.* **43,** 155.
Andernach, H., H. Waldthausen, and R. Wielebinski, 1980, *Astron. Astrophys. Suppl.* **41,** 339.
Arp, H., and J. Lorre, 1976, *Astrophys. J.* **210,** 58.
Auriemma, C., G. Perola, R. Ekers, R. Fanti, C. Lari, W. Jaffe, and M. Ulrich, 1977, *Astron. Astrophys. Suppl.* **57,** 41.
Austin, T. B., J. G. Godwin, and J. V. Peach, 1975, *Mon. Not. R. Astron. Soc.* **171,** 135.
Austin, T. B., and J. V. Peach, 1974a, *Mon. Not. R. Astron. Soc.* **167,** 437.
Austin, T. B., and J. V. Peach, 1974b, *Mon. Not. R. Astron. Soc.* **168,** 591.
Avni, Y., 1976, *Astrophys. J.* **210,** 642.
Avni, Y., and N. Bahcall, 1976, *Astrophys. J.* **209,** 16.
Baan, W. A., A. D. Haschick, and B. F. Burke, 1978, *Astrophys. J.* **225,** 339.
Bahcall, J. N., and N. A. Bahcall, 1975, *Astrophys. J. Lett.* **199,** L89.
Bahcall, J. N., and C. L. Sarazin, 1977, *Astrophys. J. Lett.* **213,** L99.
Bahcall, J. N., and C. L. Sarazin, 1978, *Astrophys. J.* **219**, 781.
Bahcall, N. A., 1971, *Astron. J.* **76,** 995.

Bahcall, N. A., 1972, *Astron. J.* **77,** 550.
Bahcall, N. A., 1973a, *Astrophys. J.* **180,** 699.
Bahcall, N. A., 1973b, *Astrophys. J.* **183,** 783.
Bahcall, N. A., 1974a, *Astrophys. J.* **187,** 439.
Bahcall, N. A., 1974b, *Astrophys. J.* **193,** 529.
Bahcall, N. A., 1974c, *Nature* **252,** 661.
Bahcall, N. A., 1975, *Astrophys. J.* **198,** 249.
Bahcall, N. A., 1977a, *Ann. Rev. Astron. Astrophys.* **15,** 505.
Bahcall, N. A., 1977b, *Astrophys. J. Lett.* **217,** L77.
Bahcall, N. A., 1977c, *Astrophys. J. Lett.* **218,** L93.
Bahcall, N. A., 1979a, *Astrophys. J.* **232,** 689.
Bahcall, N. A., 1979b, *Astrophys. J. Lett.* **232,** L83.
Bahcall, N. A., 1980, *Astrophys. J. Lett.* **238,** L117.
Bahcall, N. A., 1981, *Astrophys. J.* **247,** 787.
Bahcall, N. A., D. E. Harris, and R. G. Strom, 1976, *Astrophys. J. Lett.* **209,** L17.
Bahcall, N. A., and W. L. Sargent, 1977, *Astrophys. J. Lett.* **217,** L19.
Bahcall, N. A., and R. M. Soneira, 1982, *Astrophys. J.* **262,** 419.
Baldwin, J. E., and P. F. Scott, 1973, *Mon. Not. R. Astron. Soc.* **165,** 259.
Barnes, J., 1983, *Mon. Not. R. Astron. Soc.* **203,** 223.
Basko, M. M., B. V. Komberg, and E. I. Moskalenko, 1981, *Sov. Astron.* **25,** 402.
Bautz, L. P., and G. O. Abell, 1973, *Astrophys. J.* **184,** 709.
Bautz, L. P., and W. W. Morgan, 1970, *Astrophys. J. Lett.* **162,** L149.
Bazzano, A., R. Fusco-Femiano, C. La Padula, V. Polcaro, P. Ubertini, and R. Manchanda, 1984, *Astrophys. J.* **279,** 515.
Bechtold, J., W. Forman, R. Giacconi, C. Jones, J. Schwarz, W. Tucker, and L. Van Speybroeck, 1983, *Astrophys. J.* **265,** 26.
Beers, T. C., M. J. Geller, and J. P. Huchra, 1982, *Astrophys. J.* **257,** 23.
Beers, T. C., M. J. Geller, J. P. Huchra, D. W. Latham, and R. J. Davis, 1984, *Astrophys. J.* **283,** 33.
Beers, T. C., J. P. Huchra, and M. J. Geller, 1983, *Astrophys. J.* **264,** 356.
Beers, T. C., and J. L. Tonry, 1986, *Astrophys. J.* **300,** 557.
Begelman, M. C., M. J. Rees, and R. D. Blandford, 1979, *Nature* **279,** 770.
Berthelsdorf, R. F., and J. L. Culhane, 1979, *Mon. Not. R. Astron. Soc.* **187,** 17p.
Bieging, J. H., and P. Biermann, 1977, *Astron. Astrophys.* **60,** 361.
Biermann, P., 1978, *Astron. Astrophys.* **62,** 255.
Biermann, P., and P. Kronberg, 1983, *Astrophys. J.* **268,** L69.
Biermann, P., P. P. Kronberg, and B. F. Madore, 1982, *Astrophys. J. Lett.* **256,** L37.
Biermann, P., and S. L. Shapiro, 1979, *Astrophys. J. Lett.* **230,** L33.
Bijleveld, W., and E. A. Valentijn, 1982, *Astron. Astrophys.* **111,** 50.
Bijleveld, W., and E. A. Valentijn, 1983, *Astron. Astrophys.* **125,** 217.
Binggeli, B., 1982, *Astron. Astrophys.* **107,** 338.
Binney, J., 1977, *Mon. Not. R. Astron. Soc.* **181,** 735.
Binney, J., 1980, in *X-ray Astronomy*, edited by R. Giacconi and G. Setti, p. 245. Dordrecht: Reidel.
Binney, J., and L. L. Cowie, 1981, *Astrophys. J.* **247,** 464.
Binney, J., and J. Silk, 1978, *Comm. Astrophys.* **7,** 139.
Binney, J., and O. Strimpel, 1978, *Mon. Not. R. Astron. Soc.* **185,** 473.
Birkinshaw, M., 1978, *Mon. Not. R. Astron. Soc.* **184,** 387.
Birkinshaw, M., 1979, *Mon. Not. R. Astron. Soc.* **187,** 847.
Birkinshaw, M., 1980, *Mon. Not. R. Astron. Soc.* **190,** 793.
Birkinshaw, M., and S. F. Gull, 1984, *Mon. Not. R. Astron. Soc.* **206,** 359.
Birkinshaw, M., S. F. Gull, and H. Hardebeck, 1984, *Nature* **309,** 34.

Birkinshaw, M., S. F. Gull, and A. T. Moffet, 1981a, *Astrophys. J. Lett.* **251,** L69.
Birkinshaw, M., S. F. Gull, and K. J. Northover, 1978, *Mon. Not. R. Astron. Soc.* **185,** 245.
Birkinshaw, M., S. F. Gull, and K. J. Northover, 1981b, *Mon. Not. R. Astron. Soc.* **197,** 571.
Blumenthal, G. R., S. M. Faber, J. R. Primack, and M. J. Rees, 1984, *Nature* **311,** 517.
Bohlin, R. C., R. C. Henry, and J. R. Swandic, 1973, *Astrophys. J.* **182,** 1.
Boldt, E., 1976, *Astrophys. J. Lett.* **208,** L15.
Bookbinder, J., L. L. Cowie, J. H. Krolik, J. P. Ostriker, and M. Rees, 1980, *Astrophys. J.* **237,** 647.
Bothun, G. D., M. J. Geller, T. C. Beers, and J. P. Huchra, 1983, *Astrophys. J.* **268,** 47.
Boynton, P. E., S. J. Radford, R. A. Schommer, and S. S. Murray, 1982, *Astrophys. J.* **257,** 473.
Bradt, H., W. Mayer, S. Naranan, S. Rappaport, and G. Spuda, 1967, *Astrophys. J. Lett.* **161,** L1.
Braid, M. K., and H. T. MacGillivray, 1978, *Mon. Not. R. Astron. Soc.* **182,** 241.
Branduardi-Raymont, G., D. Fabricant, E. Feigelson, P. Gorenstein, J. Grindlay, A. Soltan, and G. Zamorani, 1981, *Astrophys. J.* **248,** 55.
Brecher, K., and G. R. Burbidge, 1972, *Astrophys. J.* **174,** 253.
Bridle, A. H., and P. A. Feldman, 1972, *Nature Phys. Sci.* **235,** 168.
Bridle, A. H., and E. B. Fomalont, 1976, *Astron. Astrophys.* **52,** 107.
Bridle, A. H., E. B. Fomalont, G. K. Miley, and E. A. Valentijn, 1979, *Astron. Astrophys.* **80,** 201.
Bridle, A. H., and J. P. Vallee, 1981, *Astron. J.* **86,** 1165.
Brown, R. L., and R. J. Gould, 1970, *Phys. Rev. D* **1,** 2252.
Bruzual A, G., and H. Spinrad, 1978a, *Astrophys. J.* **220,** 1.
Bruzual A, G., and H. Spinrad, 1978b, *Astrophys. J.* **222,** 1119.
Bucknell, M. J., J. G. Godwin, and J. V. Peach, 1979, *Mon. Not. R. Astron. Soc.* **188,** 579.
Burns, J. O., 1981, *Mon. Not. R. Astron. Soc.* **195,** 523.
Burns, J. O., J. A. Eilek, and F. N. Owen, 1982, in *IAU Symposium 97: Extragalactic Radio Sources*, edited by D. Heeschen and C. Wade, p. 45. Dordrecht: Reidel.
Burns, J. O., S. A. Gregory, and G. D. Holman, 1981c, *Astrophys. J.* **250,** 450.
Burns, J. O., C. P. O'Dea, S. A. Gregory, and T. J. Balonek, 1986, *Astrophys. J.* **307,** 73.
Burns, J. O., and F. N. Owen, 1977, *Astrophys. J.* **217,** 34.
Burns, J. O., and F. N. Owen, 1979, *Astron. J.* **84,** 1478.
Burns, J. O., and F. N. Owen, 1980, *Astron. J.* **85,** 204.
Burns, J. O., F. N. Owen, and L. Rudnick, 1978, *Astron. J.* **83,** 312.
Burns, J. O., and M. P. Ulmer, 1980, *Astron. J.* **85,** 773.
Burns, J. O., R. A. White, and R. J. Hanisch, 1980, *Astron. J.* **85,** 191.
Burns, J. O., R. A. White, and M. P. Haynes, 1981a, *Astron. J.* **86,** 1120.
Burns, J. O., R. A. White, and D. H. Hough, 1981b, *Astron. J.* **86,** 1.
Burstein, D., 1979a, *Astrophys. J.* **234,** 435.
Burstein, D., 1979b, *Astrophys. J.* **234,** 829.
Butcher, H., and A. Oemler, Jr., 1978a, *Astrophys. J.* **219,** 18.
Butcher, H., and A. Oemler, Jr., 1978b, *Astrophys. J.* **226,** 559.
Butcher, H., and A. Oemler, Jr., 1984a, *Astrophys. J.* **285,** 426.
Butcher, H., and A. Oemler, Jr., 1984b, *Nature* **310,** 31.
Byram, E. T., T. A. Chubb, and H. Friedman, 1966, *Science* **152,** 66.
Cane, H. V., W. C. Erickson, R. J. Hanisch, and P. J. Turner, 1981, *Mon. Not. R. Astron. Soc.* **196,** 409.
Canizares, C. R., 1981, in *Proceedings of the HEAD Meeting on X-ray Astronomy*, edited by R. Giacconi, p. 215. Dordrecht: Reidel.
Canizares, C. R., G. W. Clark, J. G. Jernigan, and T. H. Markert, 1982, *Astrophys. J.* **262,** 33.

Canizares, C. R., G. W. Clark, T. H. Markert, C. Berg, M. Smedira, D. Bardas, H. Schnopper, and K. Kalata, 1979, *Astrophys. J.* **234,** L33.
Canizares, C. R., G. Fabbiano, and G. Trinchieri, 1987, *Astrophys. J.* **312,** 503.
Canizares, C. R., G. C. Stewart, and A. C. Fabian, 1983, *Astrophys. J.* **272,** 449.
Capelato, H. V., D. Gerbal, G. Mathez, A. Mazure, E. Salvador-Sole, and H. Sol, 1980, *Astrophys. J.* **241,** 521.
Carnevali, P., A. Cavaliere, and P. Santangelo, 1981, *Astrophys. J.* **249,** 449.
Carr, B. J., J. R. Bond, and W. D. Arnett, 1984, *Astrophys. J.* **277,** 445.
Carter, D., 1977, *Mon. Not. R. Astron. Soc.* **178,** 137.
Carter, D., 1980, *Mon. Not. R. Astron. Soc.* **190,** 307.
Carter, D., and J. G. Godwin, 1979, *Mon. Not. R. Astron. Soc.* **187,** 711.
Carter, D., and N. Metcalfe, 1980, *Mon. Not. R. Astron. Soc.* **191,** 325.
Cash, W., R. F. Malina, and R. S. Wolff, 1976, *Astrophys. J. Lett.* **209,** L111.
Catura, R. C., P. G. Fisher, M. M. Johnson, and A. J. Meyerott, 1972, *Astrophys. J. Lett.* **177,** L1.
Cavaliere, A., 1980, in *X-ray Astronomy*, edited by R. Giacconi and G. Setti, p. 217. Dordrecht: Reidel.
Cavaliere, A., L. Danese, and G. deZotti, 1977, *Astrophys. J.* **217,** 6.
Cavaliere, A., L. Danese, and G. deZotti, 1979, *Astron. Astrophys.* **75,** 322.
Cavaliere, A., G. DeBiase, P. Santangelo, and N. Vittorio, 1983, in *Clustering in the Universe*, edited by D. Gerbal and A. Mazure, p. 15. Paris: Editions Frontieres.
Cavaliere, A., and R. Fusco-Femiano, 1976, *Astron. Astrophys.* **49,** 137.
Cavaliere, A., and R. Fusco-Femiano, 1978, *Astron. Astrophys.* **70,** 677.
Cavaliere, A., and R. Fusco-Femiano, 1981, *Astron. Astrophys.* **100,** 194.
Cavaliere, A., H. Gursky, and W. H. Tucker, 1971, *Nature* **231,** 437.
Cavallo, G., and N. Mandolesi, 1982, *Astrophys. Lett.* **22,** 119.
Chamaraux, P., C. Balkowski, and E. Gerard, 1980, *Astron. Astrophys.* **83,** 38.
Chanan, G. A., and A. F. Abramopoulos, 1984, *Astrophys. J.* **287,** 89.
Chandrasekhar, S., 1939, *An Introduction to the Study of Stellar Structure*, p. 155. Chicago: University of Chicago.
Chandrasekhar, S., 1942, *Principles of Stellar Dynamics*, p. 231. Chicago: University of Chicago.
Chandrasekhar, S., 1968, *Ellipsoidal Figures of Equilibrium*, New Haven: Yale University.
Chincarini, G., 1984, *Adv. Space. Res.* **3,** 393.
Chincarini, G. L., R. Giovanelli, M. Haynes, and P. Fontanelli, 1983, *Astrophys. J.* **267,** 511.
Chincarini, G., and H. J. Rood, 1976, *Astrophys. J.* **206,** 30.
Chincarini, G., and H. J. Rood, 1977, *Astrophys. J.* **214,** 351.
Chincarini, G., M. Tarenghi, and C. Bettis, 1978, *Astrophys. J.* **221,** 34.
Chincarini, G., M. Tarenghi, and C. Bettis, 1981, *Astron. Astrophys.* **96,** 106.
Christiansen, W. A., A. G. Pacholczyk, and J. S. Scott, 1981, *Astrophys. J.* **251,** 518.
Ciardullo, R., H. Ford, F. Bartko, and R. Harms, 1983, *Astrophys. J.* **273,** 24.
Coleman, G., P. Hintzen, J. Scott, and M. Tarenghi, 1976, *Nature* **262,** 476.
Cooke, B. A., and D. Maccagni, 1976, *Mon. Not. R. Astron. Soc.* **175,** 65p.
Cooke, J. A., D. Emerson, B. D. Kelly, H. T. MacGillivray, and R. J. Dodd, 1981, *Mon. Not. R. Astron. Soc.* **196,** 397.
Corwin, H. G., 1974, *Astron. J.* **79,** 1356.
Costain, C. H., A. H. Bridle, and P. A. Feldman, 1972, *Astrophys. J. Lett.* **175,** L15.
Cowie, L. L., 1981, in *Proceedings of the HEAD Meeting on X-ray Astronomy*, edited by R. Giacconi, p. 227. Dordrecht: Reidel.
Cowie, L. L., and J. Binney, 1977, *Astrophys. J.* **215,** 723.
Cowie, L. L., A. C. Fabian, and P. E. Nulsen, 1980, *Mon. Not. R. Astron. Soc.* **191,** 399.
Cowie, L. L., M. J. Henriksen, and R. Mushotzky, 1987, *Astrophys. J.* **317,** 593.

Cowie, L., E. Hu, E. Jenkins, and D. York, 1983, *Astrophys. J.* **272,** 29.
Cowie, L. L., and C. F. McKee, 1975, *Astron. Astrophys.* **43,** 337.
Cowie, L. L., and C. F. McKee, 1977, *Astrophys. J.* **211,** 135.
Cowie, L. L., and S. C. Perrenod, 1978, *Astrophys. J.* **219,** 254.
Cowie, L. L., and A. Songaila, 1977, *Nature* **266,** 501.
Crane, P., and J. A. Tyson, 1975, *Astrophys. J. Lett.* **201,** L1.
Da Costa, L. N., and E. Knobloch, 1979, *Astrophys. J.* **230,** 639.
Dagkesamandky, R. D., A. G. Gubanov, A. D. Kuzmin, and O. B. Slee, 1982, *Mon. Not. R. Astron. Soc.* **200,** 971.
Danese, L., G. deZotti, and G. di Tullio, 1980, *Astron. Astrophys.* **82,** 322.
Davidsen, A., S. Bowyer, M. Lampton, and R. Cruddace, 1975, *Astrophys. J.* **198,** 1.
Davidsen, A., and W. Welch, 1974, *Astrophys. J. Lett.* **191,** L11.
Davies, R. D., and B. M. Lewis, 1973, *Mon. Not. R. Astron. Soc.* **165,** 231.
Davis, M., J. Huchra, D. Latham, and J. Tonry, 1982, *Astrophys. J.* **253,** 423.
Davison, P. J., 1978, *Mon. Not. R. Astron. Soc.* **183,** 39p.
Dawe, J., R. Dickens, and B. Peterson, 1977, *Mon. Not. R. Astron. Soc.* **178,** 675.
Dennison, B., 1980a, *Astrophys. J.* **236,** 761.
Dennison, B., 1980b, *Astrophys. J. Lett.* **239,** L93.
Demoulin-Ulrich, M.-H., H. R. Butcher, and A. Boksenberg, 1984, *Astrophys. J.* **285,** 527.
des Forets, G., R. Dominguez-Tenreiro, D. Gerbal, G. Mathez, Alain Mazure, and E. Salvador-Sole, 1984, *Astrophys. J.* **280,** 15.
de Vaucouleurs, G., 1948a, *Ann. d'Astrophys.* **11,** 247.
de Vaucouleurs, G., 1948b, *Contrib. Inst. Astrophys. Paris* **A,** no. 27.
de Vaucouleurs, G., 1953, *Astron. J.* **58,** 30.
de Vaucouleurs, G., 1956, *Mem. Mt. Stromlo Obs.* **15,** no. 13.
de Vaucouleurs, G., 1961, *Astrophys. J. Suppl.* **6,** 213.
de Vaucouleurs, G., 1975, in *Stars and Stellar Systems IX: Galaxies and the Universe*, edited by A. Sandage, M. Sandage, and J. Kristian, p. 557. Chicago: University of Chicago.
de Vaucouleurs, G., 1976, *Astrophys. J.* **203,** 33.
de Vaucouleurs, G., and A. de Vaucouleurs, 1970, *Astrophys. Lett.* **5,** 219.
de Vaucouleurs, G., and J.-L. Nieto, 1978, *Astrophys. J.* **220,** 449.
De Young, D. S., 1972, *Astrophys. J.* **173,** L7.
De Young, D. S., 1978, *Astrophys. J.* **223,** 47.
De Young, D. S., J. J. Condon, and H. Butcher, 1980, *Astrophys. J.* **242,** 511.
Dickens, R. J., and C. Moss, 1976, *Mon. Not. R. Astron. Soc.* **174,** 47.
Dickey, J. M., and E. E. Salpeter, 1984, *Astrophys. J.* **284,** 461.
Disney, M. J., 1974, *Astrophys. J. Lett.* **193,** L103.
Dones, L., and S. D. M. White, 1985, *Astrophys. J.* **290,** 94.
Doroshkevich, A. G., S. F. Shandarin, and E. Saar, 1978, *Mon. Not. R. Astron. Soc.* **184,** 643.
Dressler, A., 1978a, *Astrophys. J.* **222,** 23.
Dressler, A., 1978b, *Astrophys. J.* **223,** 765.
Dressler, A., 1978c, *Astrophys. J.* **226,** 55.
Dressler, A., 1979, *Astrophys. J.* **231,** 659.
Dressler, A., 1980a, *Astrophys. J. Suppl.* **42,** 565.
Dressler, A., 1980b, *Astrophys. J.* **236,** 351.
Dressler, A., 1981, *Astrophys. J.* **243,** 26.
Dressler, A., 1984, *Ann. Rev. Astron. Astrophys.* **22,** 185.
Dressler, A., and J. E. Gunn, 1982, *Astrophys. J.* **263,** 533.
Dressler, A., J. E. Gunn, and D. P. Schneider, 1985, *Astrophys. J.* **294,** 70.
Duus, A., and B. Newell, 1977, *Astrophys. J. Suppl.* **35,** 209.
Eggen, O. J., D. Lynden-Bell, and A. Sandage, 1962, *Astrophys. J.* **136,** 748.

Einasto, J., M. Joeveer, and E. Saar, 1980, *Mon. Not. R. Astron. Soc.* **193,** 353.
Einasto, J., A. Kaasik, and E. Saar, 1974, *Nature*, **250,** 309.
Ellis, R. S., W. J. Couch, I. MacLaren, and D. C. Koo, 1985, *Mon. Not. R. Astron. Soc.* **217,** 239.
Elvis, M., 1976, *Mon. Not. R. Astron. Soc.* **177,** 7p.
Elvis, M., B. A. Cooke, K. A. Pounds, and M. J. Turner, 1975, *Nature* **257,** 33.
Elvis, M., E. Schreier, J. Tonry, M. Davis, and J. Huchra, 1981, *Astrophys. J.* **246,** 20.
Erickson, W. C., T. A. Mathews, and M. R. Viner, 1978, *Astrophys. J.* **222,** 761.
Fabbri, R., F. Melchiorri, and V. Natale, 1978, *Astrophys. Space Sci.* **59,** 223.
Faber, S. M., and A. Dressler, 1976, *Astrophys. J. Lett.* **210,** L65.
Faber, S. M., and A. Dressler, 1977, *Astron. J.* **82,** 187.
Faber, S. M., and J. S. Gallagher, 1976, *Astrophys. J.* **204,** 365.
Faber, S. M., and J. S. Gallagher, 1979, *Ann. Rev. Astron. Astrophys.* **17,** 135.
Fabian, A. C., K. A. Arnaud, P. E. Nulsen, and R. F. Mushotzky, 1986a, *Astrophys. J.* **305,** 9.
Fabian, A. C., K. A. Arnaud, P. E. Nulsen, M. G. Watson, G. C. Stewart, I. McHardy, A. Smith, B. Cook, M. Elvis, and R. F. Mushotzky, 1985, *Mon. Not. R. Astron. Soc.* **216,** 923.
Fabian, A. C., K. A. Arnaud, and P. A. Thomas, 1986b, in *Proceedings of IAU Symposium 117: Dark Matter in the Universe*, edited by G. Knapp and J. Kormendy, p. 201. Dordrecht: Reidel.
Fabian, A. C., P. D. Atherton, K. Taylor, and P. E. Nulsen, 1982a, *Mon. Not. R. Astron. Soc.* **201,** L17.
Fabian, A. C., E. M. Hu, L. L. Cowie, and J. Grindlay, 1981a, *Astrophys. J.* **248,** 47.
Fabian, A. C., W. H. Ku, D. F. Malin, R. F. Mushotzky, P. E. Nulsen, and G. C. Stewart, 1981b, *Mon. Not. R. Astron. Soc.* **196,** 35p.
Fabian, A. C., and P. E. Nulsen, 1977, *Mon. Not. R. Astron. Soc.* **180,** 479.
Fabian, A. C., and P. E. Nulsen, 1979, *Mon. Not. R. Astron. Soc.* **186,** 783.
Fabian, A. C., P. E. Nulsen, and K. A. Arnaud, 1984a, *Mon. Not. R. Astron. Soc.* **208,** 179.
Fabian, A. C., P. E. Nulsen, and C. R. Canizares, 1982b, *Mon. Not. R. Astron. Soc.* **201,** 933.
Fabian, A. C., P. E. Nulsen, and C. R. Canizares, 1984b, *Nature* **310,** 733.
Fabian, A. C., and J. E. Pringle, 1977, *Mon. Not. R. Astron. Soc.* **181,** 5p.
Fabian, A. C., J. E. Pringle, and M. J. Rees, 1976, *Nature* **263,** 301.
Fabian, A. C., J. Schwarz, and W. Forman, 1980, *Mon. Not. R. Astron. Soc.* **192,** 135.
Fabricant, D., and P. Gorenstein, 1983, *Astrophys. J.* **267,** 535.
Fabricant, D., M. Lecar, and P. Gorenstein, 1980, *Astrophys. J.* **241,** 552.
Fabricant, D., G. Rybicki, and P. Gorenstein, 1984, *Astrophys. J.* **286,** 186.
Fabricant, D., K. Topka, F. R. Harnden, and P. Gorenstein, 1978, *Astrophys. J. Lett.* **226,** L107.
Fall, S. M., 1981, in *The Structure and Evolution of Normal Galaxies*, edited by M. Fall and D. Lynden-Bell, p. 1. Cambridge University Press.
Fall, S. M., and M. J. Rees, 1985, *Astrophys. J.* **298,** 18.
Falle, S. A., and P. Meszaros, 1980, *Mon. Not. R. Astron. Soc.* **190,** 195.
Fanti, C., R. Fanti, L. Feretti, I. Gioia, G. Giovannini, L. Gregorini, B. Marano, L. Padrielli, P. Parma, P. Tomasi, and V. Zitelli, 1983, *Astron. Astrophys. Suppl.* **52,** 411.
Farouki, R., G. L. Hoffman, and E. E. Salpeter, 1983, *Astrophys. J.* **271,** 11.
Farouki, R., and E. E. Salpeter, 1982, *Astrophys. J.* **253,** 1.
Farouki, R., and S. L. Shapiro, 1980, *Astrophys. J.* **241,** 928.
Farouki, R., and S. L. Shapiro, 1981, *Astrophys. J.* **243,** 32.
Feigelson, E. D., T. Maccacaro, and G. Zamorani, 1982, *Astrophys. J.* **255,** 392.
Felten, J. E., R. J. Gould, W. A. Stein, and N. J. Woolf, 1966, *Astrophys. J.* **146,** 955.
Felten, J. E., and P. Morrison, 1966, *Astrophys. J.* **146,** 686.
Field, G. E., 1980, *Astron. Gesell. Mitt.* **47,** 7.
Flannery, B. P., and M. Krook, 1978, *Astrophys. J.* **223,** 447.

Fomalont, E., and A. H. Bridle, 1978, *Astrophys. J. Lett.* **223,** L9.
Fomalont, E., and D. Rogstad, 1966, *Astrophys. J.* **146,** 52.
Ford, H. C., and H. Butcher, 1979, *Astrophys. J. Suppl.* **41,** 147.
Forman, W., J. Bechtold, W. Blair, R. Giacconi, L. Van Speybroeck, and C. Jones, 1981, *Astrophys. J. Lett.* **243,** L133.
Forman, W., and C. Jones, 1982, *Ann. Rev. Astron. Astrophys.* **20,** 547.
Forman, W., C. Jones, L. Cominsky, P. Julien, S. Murray, G. Peters, H. Tananbaum, and R. Giacconi, 1978a, *Astrophys. J. Suppl.* **38,** 357.
Forman, W., C. Jones, S. Murray, and R. Giacconi, 1978b, *Astrophys. J. Lett.* **225,** L1.
Forman, W., C. Jones, and W. Tucker, 1985, *Astrophys. J.* **293,** 102.
Forman, W., E. Kellogg, H. Gursky, H. Tananbaum, and R. Giacconi, 1972, *Astrophys. J.* **178,** 309.
Forman, W., J. Schwarz, C. Jones, W. Liller, and A. C. Fabian, 1979, *Astrophys. J. Lett.* **234,** L27.
Forster, J. R., 1980, *Astrophys. J.* **238,** 54.
Fritz, G., A. Davidsen, J. F. Meekins, and H. Friedman, 1971, *Astrophys. J. Lett.* **164,** L81.
Gaetz, T. J., and E. E. Salpeter, 1983, *Astrophys. J. Suppl.* **55,** 155.
Gaetz, T. J., E. E. Salpeter, and G. Shaviv, 1987, *Astrophys. J.* **316,** 530.
Gallagher, J. S., 1978, *Astrophys. J.* **223,** 386.
Gallagher, J. S., and J. P. Ostriker, 1972, *Astrophys. J.* **77,** 288.
Gavazzi, G., 1978, *Astron. Astrophys.* **69,** 355.
Gavazzi, G., G. C. Perola, and W. Jaffe, 1981, *Astron. Astrophys.* **103,** 35.
Geller, M. J., and T. C. Beers, 1982, *Publ. Astron. Soc. Pac.* **94,** 421.
Geller, M. J., and P. J. Peebles, 1976, *Astrophys. J.* **206,** 939.
Giacconi, R., *et al.*, 1979, *Astrophys. J.* **230,** 540.
Giacconi, R. *et al.*, 1980, *Advanced X-ray Astrophysics Facility – Science Working Group Report*, NASA Report No. TM-78285.
Giacconi, R., S. Murray, H. Gursky, E. Kellogg, E. Schreier, T. Matilsky, D. Koch, and H. Tananbaum, 1974, *Astrophys. J. Suppl.* **27,** 37.
Giacconi, R., S. Murray, H. Gursky, E. Kellogg, E. Schreier, and H. Tananbaum, 1972, *Astrophys. J.* **178,** 281.
Gingold, R. A., and S. C. Perrenod, 1979, *Mon. Not. R. Astron. Soc.* **187,** 371.
Gioia, I. M., M. J. Geller, J. P. Huchra, T. Maccacaro, J. E. Steiner, and J. Stocke, 1982, *Astrophys. J. Lett.* **255,** L17.
Giovanardi, C., G. Helou, E. E. Salpeter, and N. Krumm, 1983, *Astrophys. J.* **267,** 35.
Giovanelli, R., G. Chincarini, and M. P. Haynes, 1981, *Astrophys. J.* **247,** 383.
Giovanelli, R., M. P. Haynes, and G. L. Chincarini, 1982, *Astrophys. J.* **262,** 422.
Gisler, G. R., 1976, *Astron. Astrophys.* **51,** 137.
Gisler, G. R., 1978, *Mon. Not. R. Astron. Soc.* **183,** 633.
Gisler, G. R., 1979, *Astrophys. J.* **228,** 385.
Gisler, G. R., and G. K. Miley, 1979, *Astron. Astrophys.* **76,** 109.
Godwin, J. C., and J. V. Peach, 1977, *Mon. Not. R. Astron. Soc.* **181,** 323.
Goldstein, S. J., 1966, *Science* **151,** 3706.
Gorenstein, P., P. Bjorkholm, B. Harris, and F. Harnden, 1973, *Astrophys. J. Lett.* **183,** L57.
Gorenstein, P., D. Fabricant, K. Topka, and F. Harnden, 1979, *Astrophys. J.* **230,** 26.
Gorenstein, P., D. Fabricant, K. Topka, F. R. Harnden, and W. H. Tucker, 1978, *Astrophys. J.* **224,** 718.
Gorenstein, P., D. Fabricant, K. Topka, W. Tucker, and F. Harnden, 1977, *Astrophys. J. Lett.* **216,** L95.
Gott, J. R., 1977, *Ann. Rev. Astron. Astrophys.* **15,** 235.
Gott, J. R., and T. X. Thuan, 1976, *Astrophys. J.* **204,** 649.
Gould, R. J., and Y. Rephaeli, 1978, *Astrophys. J.* **219,** 12.
Gregory, S. A., 1975, *Astrophys. J.* **199,** 1.

Gregory, S. A., and L. A. Thompson, 1978, *Astrophys. J.* **222,** 784.
Gregory, S. A., L. A. Thompson, and W. G. Tifft, 1981, *Astrophys. J.* **243,** 411.
Gregory, S. A., and W. G. Tifft, 1976, *Astrophys. J.* **205,** 716.
Grindlay, J. A., D. R. Parsignault, H. Gursky, A. C. Brinkman, J. Heise, and D. E. Harris, 1977, *Astrophys. J. Lett.* **214,** L57.
Gudehus, D. H., 1973, *Astron. J.* **78,** 583.
Gudehus, D. H., 1976, *Astrophys. J.* **208,** 267.
Guindon, B., 1979, *Mon. Not. R. Astron. Soc.* **186,** 117.
Guindon, B., and A. H. Bridle, 1978, *Mon. Not. R. Astron. Soc.* **184,** 221.
Gull, S. F., and K. J. Northover, 1975, *Mon. Not. R. Astron. Soc.* **173,** 585.
Gull, S. F., and K. J. Northover, 1976, *Nature* **263,** 572.
Gunn, J. E., 1977, *Astrophys. J.* **218,** 592.
Gunn, J. E., 1978, in *Observational Cosmology*, edited by A. Maeder, L. Martinet, and G. Tammann. Geneva: Geneva Observatory.
Gunn, J. E., and J. R. Gott, 1972, *Astrophys. J.* **176,** 1.
Gunn, J. E., and J. B. Oke, 1975, *Astrophys. J.* **195,** 255.
Gunn, J. E., and B. M. Tinsley, 1976, *Astrophys. J.* **210,** 1.
Gursky, H., E. Kellogg, C. Leong, H. Tananbaum, and R. Giacconi, 1971a, *Astrophys. J. Lett.* **165,** L43.
Gursky, H., E. M. Kellogg, S. Murray, C. Leong, H. Tananbaum, and R. Giacconi, 1971b, *Astrophys. J. Lett.* **167,** L81.
Gursky, H., and D. Schwartz, 1977, *Ann. Rev. Astron. Astrophys.* **15,** 541.
Gursky, H., A. Solinger, E. Kellogg, S. Murray, H. Tananbaum, R. Giacconi, and A. Cavaliere, 1972, *Astrophys. J. Lett.* **173,** L99.
Guthie, B. N., 1974, *Mon. Not. R. Astron. Soc.* **168,** 15.
Hall, A., and D. Sciama, 1979, *Astrophys. J. Lett.* **228,** L15.
Hall, D. N., 1982, *The Space Telescope Observatory*, NASA Report No. CP-2244.
Hamilton, A. J., C. L. Sarazin, and R. A. Chevalier, 1983, *Astrophys. J. Suppl.* **51,** 115.
Hanisch, R. J., 1980, *Astron. J.* **85,** 1565.
Hanisch, R. J., 1982a, *Astron. Astrophys.* **111,** 97.
Hanisch, R. J., 1982b, *Astron. Astrophys.* **116,** 137.
Hanisch, R. J., and W. C. Erickson, 1980, *Astron. J.* **85,** 183.
Hanisch, R. J., T. A. Matthews, and M. M. Davis, 1979, *Astron. J.* **84,** 946.
Hanisch, R. J., and R. A. White, 1981, *Astron. J.* **86,** 806.
Harris, D. E., P. Dewdney, C. Costain, H. Butcher, and A. Willis, 1983a, *Astrophys. J.* **270,** 39.
Harris, D. E., V. K. Kapahi, and R. D. Ekers, 1980, *Astron. Astrophys. Suppl.* **39,** 215.
Harris, D. E., and G. K. Miley, 1978, *Astron. Astrophys. Suppl.* **34,** 117.
Harris, D. E., J. G. Robertson, P. E. Dewdney, and C. H. Costain, 1982, *Astron. Astrophys.* **111,** 299.
Harris, D. E., and W. Romanishin, 1974, *Astrophys. J.* **188,** 209.
Harris, E. W., M. G. Smith, and E. S. Myra, 1983b, *Astrophys. J.* **272,** 456.
Hartwick, F. D., 1973, *Astrophys. J.* **219,** 345.
Hartwick, F. D., 1976, *Astrophys. J. Lett.* **208,** L13.
Haslam, C., P. Kronberg, H. Waldthausen, R. Wielebinski, and D. Schallwich, 1978, *Astron. Astrophys. Suppl.* **31,** 99.
Hausman, M. A., and J. P. Ostriker, 1978, *Astrophys. J.* **224,** 320.
Havlen, R. J., and H. Quintana, 1978, *Astrophys. J.* **220,** 14.
Haynes, M. P., R. L. Brown, and M. S. Roberts, 1978, *Astrophys. J.* **221,** 414.
Heckman, T. M., 1981, *Astrophys. J. Lett.* **250,** L59.
Helfand, D., W. Ku, and F. Abramopoulos, 1980, *Highlights Astron.* **5,** 747.
Helmken, H., J. P. Delvaille, A. Epstein, M. J. Geller, H. W. Schnopper, and J. G. Jernigan, 1978, *Astrophys. J. Lett.* **221,** L43.

Helou, G., and E. E. Salpeter, 1982, *Astrophys. J.* **252,** 75.
Helou, G., E. E. Salpeter, and N. Krumm, 1979, *Astrophys. J. Lett.* **228,** L1.
Henriksen, M. J., 1985, Ph.D. Thesis, University of Maryland.
Henriksen, M. J., and R. F. Mushotzky, 1985, *Astrophys. J.* **292,** 441.
Henriksen, M. J., and R. F. Mushotzky, 1986, *Astrophys. J.* **302,** 287.
Henry, J. P., G. Branduardi, U. Briel, D. Fabricant, E. Feigelson, S. Murray, A. Soltan, and H. Tananbaum, 1979, *Astrophys. J. Lett.* **239,** L15.
Henry, J. P., J. T. Clarke, S. Bowyer, and R. J. Lavery, 1983, *Astrophys. J.* **272,** 434.
Henry, J. P., M. J. Henriksen, P. A. Charles, and J. R. Thorstensen, 1981, *Astrophys. J. Lett.* **243,** L137.
Henry, J. P., and R. J. Lavery, 1984, *Astrophys. J.* **280,** 1.
Henry, J. P., A. Soltan, U. Briel, and J. E. Gunn, 1982, *Astrophys. J.* **262,** 1.
Henry, J. P., and W. Tucker, 1979, *Astrophys. J.* **229,** 78.
Hesser, J. E., H. C. Harris, S. van den Bergh, and G. L. Harris, 1984, *Astrophys. J.* **276,** 491.
Hickson, P., 1977, *Astrophys. J.* **217,** 16.
Hickson, P., 1982, *Astrophys. J.* **255,** 382.
Hickson, P., and P. J. Adams, 1979a, *Astrophys. J. Lett.* **234,** L87.
Hickson, P., and P. J. Adams, 1979b, *Astrophys. J. Lett.* **234,** L91.
Hill, J. M., J. R. Angel, J. S. Scott, D. Lindley, and P. Hintzen, 1980, *Astrophys. J. Lett.* **242,** L69.
Hill, J. M., and M. S. Longair, 1971, *Mon. Not. R. Astron. Soc.* **154,** 125.
Himmes, A., and P. Biermann, 1980, *Astron. Astrophys.* **86,** 11.
Hintzen, P., G. O. Boeshaar, and J. S. Scott, 1981, *Astrophys. J. Lett.* **246,** L1.
Hintzen, P., J. M. Hill, D. Lindley, J. S. Scott, and J. R. Angel, 1982, *Astron. J.* **87,** 1656.
Hintzen, P., and J. S. Scott, 1978, *Astrophys. J. Lett.* **224,** L47.
Hintzen, P., and J. S. Scott, 1979, *Astrophys. J. Lett.* **232,** L145.
Hintzen, P., and J. S. Scott, 1980, *Astrophys. J.* **239,** 765.
Hintzen, P., J. S. Scott, and J. D. McKee, 1980, *Astrophys. J.* **242,** 857.
Hintzen, P., J. S. Scott, and M. Tarenghi, 1977, *Astrophys. J.* **212,** 8.
Hintzen, P., J. Ulvestad, and F. Owen, 1983, *Astron. J.* **88,** 709.
Hirayama, Y., 1978, *Prog. Theor. Phys.* **60,** 724.
Hirayama, Y., Y. Tanaka, and T. Kogure, 1978, *Prog. Theor. Phys.* **59,** 751.
Hoessel, J. G., 1980, *Astrophys. J.* **241,** 493.
Hoessel, J. G., K. D. Borne, and D. P. Schneider, 1985, *Astrophys. J.* **293,** 94.
Hoessel, J. G., J. E. Gunn, and T. X. Thuan, 1980, *Astrophys. J.* **241,** 486.
Hoffman, A. A., and P. Crane, 1977, *Astrophys. J.* **215,** 379.
Hoffman, G. L., and E. E. Salpeter, 1982, *Astrophys. J.* **263,** 485.
Holman, G. D., J. A. Ionson, and J. S. Scott, 1979, *Astrophys. J.* **228,** 576.
Holman, G. D., and J. D. McKee, 1981, *Astrophys. J.* **249,** 35.
Holmberg, E. B., A. Lauberts, H. E. Schuster, and R. M. West, 1974, *Astron. Astrophys. Suppl.* **18,** 463.
Holt, S., and R. McCray, 1982, *Ann. Rev. Astron. Astrophys.* **20,** 323.
Hooley, T., 1974, *Mon. Not. R. Astron. Soc.* **166,** 259.
Hu, E. M., L. L. Cowie, P. Kaaret, E. B. Jenkins, D. G. York, and F. L. Roesler, 1983, *Astrophys. J. Lett.* **275,** L27.
Hu, E. M., L. L. Cowie, and Z. Wang, 1985, *Astrophys. J. Suppl.* **59,** 447.
Hubble, E. P., 1930, *Astrophys. J.* **71,** 231.
Huchra, J. P., and M. F. Geller, 1982, *Astrophys. J.* **257,** 423.
Huchtmeier, W. K., G. A. Tammann, and H. J. Wendker, 1976, *Astron. Astrophys.* **46,** 381.
Humason, M. L., N. U. Mayall, and A. R. Sandage, 1956, *Astron. J.* **61,** 97.
Humason, M. L., and A. Sandage, 1957, *Carnegie Inst. Washington Yearb.* **56,** 62.
Hunt, R., 1971, *Mon. Not. R. Astron. Soc.* **154,** 141.

Ikeuchi, S., 1977, *Prog. Theor. Phys.* **58,** 1742.
Ikeuchi, S., and Y. Hirayama, 1979, *Prog. Theor. Phys.* **61,** 881.
Ikeuchi, S., and Y. Hirayama, 1980, *Prog. Theor. Phys.* **64,** 81.
Ives, J. C., and P. W. Sanford, 1976, *Mon. Not. R. Astron. Soc.* **176,** 13p.
Jaffe, W. J., 1977, *Astrophys. J.* **212,** 1.
Jaffe, W. J., 1980, *Astrophys. J.* **241,** 924.
Jaffe, W. J., 1982, *Astrophys. J.* **262,** 15.
Jaffe, W. J., and G. C. Perola, 1973, *Astron. Astrophys.* **26,** 423.
Jaffe, W. J., and G. Perola, 1975, *Astron. Astrophys. Suppl.* **21,** 137.
Jaffe, W. J., and G. Perola, 1976, *Astron. Astrophys.* **46,** 273.
Jaffe, W. J., G. C. Perola, and E. A. Valentijn, 1976, *Astron. Astrophys.* **49,** 179.
Jaffe, W. J., and L. Rudnick, 1979, *Astrophys. J.* **233,** 453.
Jenner, D. C., 1974, *Astrophys. J.* **191,** 55.
Joeveer, M., J. Einasto, and E. Tago, 1978, *Mon. Not. R. Astron. Soc.* **185,** 357.
Johnson, H. M., 1981, *Astrophys. J. Suppl.* **47,** 235.
Johnson, H. M., H. W. Schnopper, and J. P. Delvaille, 1980, *Astrophys. J.* **236,** 738.
Johnson, M. W., R. G. Cruddace, G. Fritz, S. Shulman, and H. Friedman, 1979, *Astrophys. J. Lett.* **231,** L45.
Johnson, M. W., R. G. Cruddace, M. P. Ulmer, M. P. Kowalski, and K. S. Wood, 1983, *Astrophys. J.* **266,** 425.
Johnston, M. D., H. V. Bradt, R. E. Doxsey, B. Margon, F. E. Marshall, and D. A. Schwartz, 1981, *Astrophys. J.* **245,** 799.
Jones, C., and W. Forman, 1978, *Astrophys. J.* **224,** 1.
Jones, C., and W. Forman, 1984, *Astrophys. J.* **276,** 38.
Jones, C., E. Mandel, J. Schwarz, W. Forman, S. S. Murray, and F. R. Harnden, 1979, *Astrophys. J. Lett.* **234,** L21.
Jones, T. W., and F. N. Owen, 1979, *Astrophys. J.* **234,** 818.
Jura, M., 1977, *Astrophys. J.* **212,** 634.
Just, K., 1959, *Astrophys. J.* **129,** 268.
Karachentsev, I. D., and A. I. Kopylov, 1980, *Mon. Not. R. Astron. Soc.* **192,** 109.
Karzas, W., and R. Latter, 1981, *Astrophys. J. Suppl.* **6,** 167.
Kato, T., 1976, *Astrophys. J. Suppl.* **30,** 397.
Katz, J. I., 1976, *Astrophys. J.* **207,** 25.
Kellogg, E. M., 1973, in *X- and Gamma-Ray Astronomy*, edited by H. Bradt and R. Giacconi, p. 171. Dordrecht: Reidel.
Kellogg, E. M., 1974, in *X-Ray Astronomy*, edited by R. Giacconi and H. Gursky, p. 321. Dordrecht: Reidel.
Kellogg, E. M., 1975, *Astrophys. J.* **197,** 689.
Kellogg, E. M., 1977, *Astrophys. J.* **218,** 582.
Kellogg, E. M., 1978, *Astrophys. J. Lett.* **220,** L63.
Kellogg, E. M., J. R. Baldwin, and D. Koch, 1975, *Astrophys. J.* **199,** 299.
Kellogg, E. M., H. Gursky, C. Leong, E. Schreier, H. Tananbaum, and R. Giacconi, 1971, *Astrophys. J. Lett.* **165,** L49.
Kellogg, E. M., H. Gursky, H. Tananbaum, R. Giacconi, and K. Pounds, 1972, *Astrophys. J. Lett.* **174,** L65.
Kellogg, E. M., and S. Murray, 1974, *Astrophys. J. Lett.* **193,** L57.
Kellogg, E. M., S. Murray, R. Giacconi, H. Tananbaum, and H. Gursky, 1973, *Astrophys. J. Lett.* **185,** L13.
Kent, S. M., and J. E. Gunn, 1982, *Astron. J.* **87,** 945.
Kent, S. M., and W. L. Sargent, 1979, *Astrophys. J.* **230,** 667.
Kent, S. M., and W. L. Sargent, 1983, *Astron. J.* **88,** 697.
Kiang, T., 1961, *Mon. Not. R. Astron. Soc.* **122,** 263.

King, I. R., 1962, *Astron. J.* **67,** 471.
King, I. R., 1966, *Astron. J.* **71,** 64.
Kirshner, R. P., A. Oemler, and P. L. Schechter, 1978, *Astron. J.* **83,** 1549.
Kirshner, R. P., A. Oemler, P. L. Schechter, and S. A. Schechtman, 1981, *Astrophys. J. Lett.* **248,** L57.
Klemola, A. R., 1969, *Astron. J.* **74,** 804.
Knobloch, E., 1978a, *Astrophys. J.* **222,** 779.
Knobloch, E., 1978b, *Astrophys. J. Suppl.* **38,** 253.
Koo, D. C., 1981, *Astrophys. J. Lett.* **251,** L75.
Kotanyi, C., J. H. van Gorkom, and R. D. Ekers, 1983, *Astrophys. J. Lett.* **273,** L7.
Kowalski, M., M. P. Ulmer, and R. Cruddace, 1983, *Astrophys. J.* **268,** 540.
Kowalski, M., M. P. Ulmer, R. Cruddace, and K. S. Wood, 1984, *Astrophys. J. Suppl.* **56,** 403.
Kraan-Korteweg, R. C., 1981, *Astron. Astrophys.* **104,** 280.
Kriss, G. A., C. R. Canizares, J. E. McClintock, and E. D. Feigelson, 1980, *Astrophys. J. Lett.* **235,** L61.
Kriss, G. A., C. R. Canizares, J. E. McClintock, and E. D. Feigelson, 1981, *Astrophys. J. Lett.* **245,** L51 erratum for above.
Kriss, G. A., D. F. Cioffi, and C. R. Canizares, 1983, *Astrophys. J.* **272,** 439.
Kristian, J., A. Sandage, and J. A. Westphal, 1978, *Astrophys. J.* **221,** 383.
Kritsuk, A. G., 1983, *Astrophys.* **19,** 263.
Kron, R. G., H. Spinrad, and I. R. King, 1977, *Astrophys. J.* **217,** 951.
Krumm, N., and E. E. Salpeter, 1976, *Astrophys. J. Lett.* **208,** L7.
Krumm, N., and E. E. Salpeter, 1979, *Astrophys. J.* **227,** 776.
Krupp, E. C., 1974, *Publ. Astron. Soc. Pac.* **86,** 385.
Ku, W., F. Abramopoulos, P. Nulsen, A. Fabian, G. Stewart, G. Chincarini, and M. Tarenghi, 1983, *Mon. Not. R. Astron. Soc.* **203,** 253.
Lake, G., and R. B. Partridge, 1977, *Nature* **270,** 502.
Lake, G., and R. B. Partridge, 1980, *Astrophys. J.* **237,** 378.
Larson, R. B., and H. L. Dinerstein, 1975, *Publ. Astron. Soc. Pac.* **87,** 911.
Larson, R. B., B. M. Tinsley, and C. N. Caldwell, 1980, *Astrophys. J.* **237,** 692.
Lasenby, A. N., and R. D. Davies, 1983, *Mon. Not. R. Astron. Soc.* **203,** 1137.
Lawler, J. M., and B. Dennison, 1982, *Astrophys. J.* **252,** 81.
Lawrence, A., 1978, *Mon. Not. R. Astron. Soc.* **185,** 423.
Lea, S. M., 1975, *Astrophys. Lett.* **16,** 141.
Lea, S. M., 1976, *Astrophys. J.* **203,** 569.
Lea, S. M., and D. S. De Young, 1976, *Astrophys. J.* **210,** 647.
Lea, S. M., and G. D. Holman, 1978, *Astrophys. J.* **222,** 29.
Lea, S. M., K. Mason, G. Reichert, P. Charles, and G. Reigler, 1979, *Astrophys. J. Lett.* **227,** L67.
Lea, S. M., R. Mushotzky, and S. Holt, 1982, *Astrophys. J.* **262,** 24.
Lea, S. M., G. Reichert, R. Mushotzky, W. A. Baity, D. E. Gruber, R. Rothschild, and F. A. Primini, 1981, *Astrophys. J.* **246,** 369.
Lea, S. M., J. Silk, E. Kellogg, and S. Murray, 1973, *Astrophys. J. Lett.* **184,** L105.
Lecar, M., 1975, in *Dynamics of Stellar Systems*, edited by A. Hayli, p. 161. Dordrecht: Reidel.
Leir, A. A., and S. van den Bergh, 1977, *Astrophys. J. Suppl.* **34,** 381.
Limber, D. N., and W. G. Mathews, 1960, *Astrophys. J.* **132,** 286.
Livio, M., O. Regev, and G. Shaviv, 1978, *Astron. Astrophys.* **70,** L7.
Livio, M., O. Regev, and G. Shaviv, 1980, *Astrophys. J. Lett.* **240,** L83.
Lugger, P. M., 1978, *Astrophys. J.* **221,** 745.
Lynden-Bell, D., 1967, *Mon. Not. R. Astron. Soc.* **136,** 101.

Lynds, R., 1970, *Astrophys. J. Lett.* **159,** L151.
Maccacaro, T., B. A. Cooke, M. J. Ward, M. V. Penston, and R. F. Haynes, 1977, *Mon. Not. R. Astron. Soc.* **180,** 465.
Maccagni, D., and M. Tarenghi, 1981, *Astrophys. J.* **243,** 42.
MacDonald, R., S. Kenderdine, and A. Neville, 1968, *Mon. Not. R. Astron. Soc.* **138,** 259.
MacGillivray, H. T., and R. J. Dodd, 1979, *Mon. Not. R. Astron. Soc.* **186,** 743.
MacGillivray, H. T., R. Martin, N. M. Pratt, V. C. Reddish, H. Seddon, L. W. Alexander, G. S. Walker, and P. R. Williams, 1976, *Mon. Not. R. Astron. Soc.* **176,** 649.
Malina, R., M. Lampton, and S. Bowyer, 1976, *Astrophys. J.* **209,** 678.
Malina, R. F., S. M. Lea, M. Lampton, and C. S. Bowyer, 1978, *Astrophys. J.* **219,** 795.
Malumuth, E. M., and R. P. Kirshner, 1981, *Astrophys. J.* **251,** 508.
Malumuth, E. M., and R. P. Kirshner, 1985, *Astrophys. J.* **291,** 8.
Malumuth, E. M., and G. A. Kriss, 1986, *Astrophys. J.* **308,** 10.
Malumuth, E. M., and D. O. Richstone, 1984, *Astrophys. J.* **276,** 413.
Marchant, A. B., and S. L. Shapiro, 1977, *Astrophys. J.* **215,** 1.
Margon, B., M. Lampton, S. Bowyer, and R. Cruddace, 1975, *Astrophys. J.* **197,** 25.
Markert, T. H., C. R. Canizares, G. W. Clark, F. K. Li, P. L. Northridge, G. F. Sprott, and G. F. Wargo, 1976, *Astrophys. J.* **206,** 265.
Markert, T., P. Winkler, F. Laird, G. Clark, D. Hearn, G. Sprott, F. Li, V. Bradt, W. Lewin, and H. Schnopper, 1979, *Astrophys. J. Suppl.* **39,** 573.
Marshall, F. E., E. A. Boldt, S. S. Holt, R. F. Mushotzky, S. H. Pravdo, R. E. Rothschild, and P. J. Serlemitsos, 1979, *Astrophys. J. Suppl.* **40,** 657.
Mason, K. O., H. Spinrad, S. Bowyer, G. Reichert, and J. Stauffer, 1981, *Astron. J.* **86,** 803.
Materne, J., G. Chincarini, M. Tarenghi, and U. Hopp, 1982, *Astron. Astrophys.* **109,** 238.
Mathews, T. A., W. W. Morgan, and M. Schmidt, 1964, *Astrophys. J.* **140,** 35.
Mathews, W. G., 1978a, *Astrophys. J.* **219,** 408.
Mathews, W. G., 1978b, *Astrophys. J.* **219,** 413.
Mathews, W. G., and J. C. Baker, 1971, *Astrophys. J.* **170,** 241.
Mathews, W. G., and J. N. Bregman, 1978, *Astrophys. J.* **224,** 308.
Mathieu, R. D., and H. Spinrad, 1981, *Astrophys. J.* **251,** 485.
Matilsky, T., C. Jones, and W. Forman, 1985, *Astrophys. J.* **291,** 621.
McGlynn, T. A., and A. C. Fabian, 1984, *Mon. Not. R. Astron. Soc.* **208,** 709.
McGlynn, T. A., and J. P. Ostriker, 1980, *Astrophys. J.* **241,** 915.
McHardy, I. M., 1974, *Mon. Not. R. Astron. Soc.* **169,** 527.
McHardy, I. M., 1978a, *Mon. Not. R. Astron. Soc.* **184,** 783.
McHardy, I. M., 1978b, *Mon. Not. R. Astron. Soc.* **185,** 927.
McHardy, I. M., 1979, *Mon. Not. R. Astron. Soc.* **188,** 495.
McHardy, I. M., A. Lawrence, J. P. Pye, and K. A. Pounds, 1981, *Mon. Not. R. Astron. Soc.* **197,** 893.
McKee, C. F., and L. L. Cowie, 1977, *Astrophys. J.* **215,** 213.
McKee, J. D., R. F. Mushotzky, E. A. Boldt, S. S. Holt, F. E. Marshall, S. H. Pravdo, and P. J. Serlemitsos, 1980, *Astrophys. J.* **242,** 843.
Meekins, J. F., F. Gilbert, T. A. Chubb, H. Friedman, and R. C. Henry, 1971, *Nature* **231,** 107.
Melnick, J., and H. Quintana, 1975, *Astrophys. J. Lett.* **198,** L97.
Melnick, J., and H. Quintana, 1981a, *Astron. Astrophys. Suppl.* **44,** 87.
Melnick, J., and H. Quintana, 1981b, *Astron. J.* **86,** 1567.
Melnick, J., and W. L. Sargent, 1977, *Astrophys. J.* **215,** 401.
Melnick, J., S. D. M. White, and J. Hoessel, 1977, *Mon. Not. R. Astron. Soc.* **180,** 207.
Merritt, D., 1983, *Astrophys. J.* **264,** 24.
Merritt, D., 1984a, *Astrophys. J.* **276,** 26.
Merritt, D., 1984b, *Astrophys. J. Lett.* **280,** L5.

Merritt, D., 1985, *Astrophys. J.* **289**, 18.
Mewe, R., and E. H. Gronenschild, 1981, *Astron. Astrophys. Suppl.* **45**, 11.
Meyer, S. S., A. D. Jeffries, and R. Weiss, 1983, *Astrophys. J.* **271**, L1.
Miley, G., 1973, *Astron. Astrophys.* **26**, 413.
Miley, G., 1980, *Ann. Rev. Astron. Astrophys.* **18**, 165.
Miley, G., and D. E. Harris, 1977, *Astron. Astrophys.* **61**, L23.
Miley, G., G. Perola, P. van der Kruit, and H. van der Laan, 1972, *Nature* **237**, 269.
Miller, G. E., 1983, *Astrophys. J.* **274**, 840.
Mills, B. Y., 1980, *Aust. J. Phys.* **13**, 550.
Minkowski, R., 1960, *Astrophys. J.* **132**, 908.
Minkowski, R., 1961, *Astron. J.* **66**, 558.
Minkowski, R., and G. O. Abell, 1963, in *Basic Astronomical Data*, edited by K. A. Strand, p. 481. Chicago: University of Chicago.
Mitchell, R. J., P. A. Charles, J. L. Culhane, P. J. Davison, and A. C. Fabian, 1975, *Astrophys. J. Lett.* **200**, L5.
Mitchell, R. J., and J. L. Culhane, 1977, *Mon. Not. R. Astron. Soc.* **178**, 75p.
Mitchell, R. J., J. L. Culhane, P. J. Davison, and J. C. Ives, 1976, *Mon. Not. R. Astron. Soc.* **176**, 29p.
Mitchell, R. J., R. J. Dickens, S. J. Bell Burnell, and J. L. Culhane, 1979, *Mon. Not. R. Astron. Soc.* **189**, 329.
Mitchell, R. J., J. C. Ives, and J. L. Culhane, 1977, *Mon. Not. R. Astron. Soc.* **131**, 25p.
Mitchell, R. J., and R. Mushotzky, 1980, *Astrophys. J.* **236**, 730.
Morgan, W. W., 1981, *Proc. Natl. Acad. Sci. U.S.A.* **47**, 905.
Morgan, W. W., S. Kayser, and R. A. White, 1975, *Astrophys. J.* **199**, 545.
Morgan, W. W., and J. Lesh, 1965, *Astrophys. J.* **142**, 1364.
Moss, C., and R. J. Dickens, 1977, *Mon. Not. R. Astron. Soc.* **178**, 701.
Mottmann, J., and G. O. Abell, 1977, *Astrophys. J.* **218**, 53.
Murray, S. S., W. Forman, C. Jones, and R. Giacconi, 1978, *Astrophys. J. Lett.* **219**, L89.
Mushotzky, R. F., 1980, in *X-ray Astronomy*, edited by R. Giacconi and G. Setti, p. 171. Dordrecht: Reidel.
Mushotzky, R. F., 1984, *Phys. Scripta* **T7**, 157.
Mushotzky, R. F., 1985, in *Proceedings of the Conference on Non-thermal and Very High Temperature Phenomena in X-ray Astronomy*, edited by G. Perola and M. Salvati.
Mushotzky, R. F., W. A. Baity, and L. E. Peterson, 1977, *Astrophys. J.* **212**, 22.
Mushotzky, R. F., S. S. Holt, B. W. Smith, E. A. Boldt, and P. J. Serlemitsos, 1981, *Astrophys. J. Lett.* **244**, L47.
Mushotzky, R. F., P. J. Serlemitsos, B. W. Smith, E. A. Boldt, and S. S. Holt, 1978, *Astrophys. J.* **225**, 21.
Nepveu, M., 1981a, *Astron. Astrophys.* **98**, 65.
Nepveu, M., 1981b, *Astron. Astrophys.* **101**, 362.
Noonan, T. W., 1973, *Astron. J.* **78**, 26.
Noonan, T. W., 1974, *Astron. J.* **79**, 775.
Noonan, T. W., 1975a, *Astrophys. J.* **202**, 551.
Noonan, T. W., 1975b, *Astrophys. J.* **196**, 683.
Noonan, T. W., 1980, *Astrophys. J.* **238**, 793.
Noonan, T. W., 1981, *Astrophys. J. Suppl.* **45**, 613.
Noordam, J. E., and A. G. de Bruyn, 1982, *Nature* **299**, 597.
Norman, C., and J. Silk, 1979, *Astrophys. J. Lett.* **233**, L1.
Nugent, J., K. Jensen, J. Nousek, G. Garmire, K. Mason, F. Walter, C. Bowyer, R. Stern, and G. Riegler, 1983, *Astrophys. J. Suppl.* **51**, 1.
Nulsen, P. E., 1982, *Mon. Not. R. Astron. Soc.* **198**, 1007.
Nulsen, P. E., and A. C. Fabian, 1980, *Mon. Not. R. Astron. Soc.* **191**, 887.

Nulsen, P. E., A. C. Fabian, R. F. Mushotzky, E. A. Boldt, S. S. Holt, F. J. Marshall, and P. J. Serlemitsos, 1979, *Mon. Not. R. Astron. Soc.* **189,** 183.
Nulsen, P. E., G. C. Stewart, and A. C. Fabian, 1984, *Mon. Not. R. Astron. Soc.* **208,** 185.
Nulsen, P. E., G. C. Stewart, A. C. Fabian, R. F. Mushotzky, S. S. Holt, W. Ku, and D. Malin, 1982, *Mon. Not. R. Astron. Soc.* **109,** 1089.
O'Connell, R. W., B. McNamara, and C. L. Sarazin, 1987, preprint.
O'Dea, C. P., and F. N. Owen, 1986, *Astrophys. J.* **301,** 841.
Oemler, A. Jr., 1973, *Astrophys. J.* **180,** 11.
Oemler, A. Jr., 1974, *Astrophys. J.* **194,** 1.
Oemler, A. Jr., 1976, *Astrophys. J.* **209,** 693.
Omer, G. C., T. L. Page, and A. G. Wilson, 1965, *Astron. J.* **70,** 440.
Oort, J. H., 1983, *Ann. Rev. Astron. Astrophys.* **21,** 373.
Osmer, P. S., 1978, *Phys. Scripta* **17,** 357.
Osterbrock, D. E., 1974, *Astrophysics of Gaseous Nebulae.* San Francisco: Freeman.
Ostriker, J. P., 1980, *Comm. Astrophys.* **8,** 177.
Ostriker, J. P., and M. A. Hausman, 1977, *Astrophys. J. Lett.* **217,** L125.
Ostriker, J. P., P. J. Peebles, and A. Yahil, 1974, *Astrophys. J. Lett.* **193,** L1.
Ostriker, J. P., and S. D. Tremaine, 1975, *Astrophys. J. Lett.* **202,** L113.
Owen, F. N., 1974, *Astrophys. J. Lett.* **189,** 155.
Owen, F. N., 1975, *Astrophys. J.* **195,** 593.
Owen, F. N., J. O. Burns, and L. Rudnick, 1978, *Astrophys. J. Lett.* **226,** L119.
Owen, F. N., J. O. Burns, L. Rudnick, and E. W. Greison, 1979, *Astrophys. J. Lett.* **229,** L59.
Owen, F. N., C. P. O'Dea, M. Inoue, and J. A. Eilek, 1985, *Astrophys. J. Lett.* **294,** L85.
Owen, F. N., and L. Rudnick, 1976a, *Astrophys. J.* **203,** 307.
Owen, F. N., and L. Rudnick, 1976b, *Astrophys. J. Lett.* **205,** L1.
Owen, F. N., R. A. White, K. C. Hilldrup, and R. J. Hanisch, 1982, *Astron. J.* **87,** 1083.
Paal, G., 1964, in *Mitt. d. Sternwarted. Ungarischen Akad. d. Wissenschaften,* no. 54.
Pacholczyk, A. G., and J. S. Scott, 1976, *Astrophys. J.* **203,** 313.
Pariiskii, Y. N., 1973, *Sov. Astron. AJ* **16,** 1048.
Peach, J. V., 1969, *Nature* **223,** 1140.
Peebles, P. J., 1974, *Astron. Astrophys.* **32,** 197.
Perola, G. C., and M. Reinhardt, 1972, *Astron. Astrophys.* **17,** 432.
Perrenod, S. C., 1978a, *Astrophys. J.* **224,** 285.
Perrenod, S. C., 1978b, *Astrophys. J.* **226,** 566.
Perrenod, S. C., 1980, *Astrophys. J.* **236,** 373.
Perrenod, S. C., and J. P. Henry, 1981, *Astrophys. J. Lett.* **247,** L1.
Perrenod, S. C., and C. J. Lada, 1979, *Astrophys. J. Lett.* **234,** L173.
Perrenod, S. C., and M. P. Lesser, 1980, *Publ. Astron. Soc. Pac.* **92,** 764.
Peterson, B. M., 1978, *Astrophys. J.* **223,** 740.
Peterson, B. M., S. E. Strom, and K. M. Strom, 1979, *Astron. J.* **84,** 735.
Piccinotti, G., R. F. Mushotzky, E. A. Boldt, S. S. Holt, F. E. Marshall, P. J. Serlemitsos, and R. A. Shafer, 1982, *Astrophys. J.* **253,** 485.
Pravdo, S., E. Boldt, F. Marshall, J. McKee, R. Mushotzky, and B. Smith, 1979, *Astrophys. J.* **234,** 1.
Press, W. H., 1976, *Astrophys. J.* **203,** 14.
Press, W. H., and P. Schechter, 1974, *Astrophys. J.* **187,** 425.
Primini, F., E. Basinska, S. Howe, F. Lang, A. Levine, W. Lewin, R. Rothschild, W. Baity, D. Gruber, F. Knight, J. Matteson, S. Lea, and G. Reichert, 1981, *Astrophys. J. Lett.* **243,** L13.
Pye, J. P., and B. A. Cooke, 1976, *Mon. Not. R. Astron. Soc.* **177,** 21p.
Quintana, H., 1979, *Astron. J.* **84,** 15.

Quintana, H., and D. G. Lawrie, 1982, *Astron. J.* **87,** 1.

Quintana, H., and J. Melnick, 1982, *Astron. J.* **87,** 972.

Raymond, J. C., D. P. Cox, and B. W. Smith, 1976, *Astrophys. J.* **204,** 290.

Raymond, J. C., and B. W. Smith, 1977, *Astrophys. J. Suppl.* **35,** 419.

Rees, M. J., 1978, *Phys. Scripta* **17,** 193.

Reichert, G., K. Mason, S. Lea, P. Charles, S. Bowyer, and S. Pravdo, 1981, *Astrophys. J.* **247,** 803.

Rephaeli, Y., 1977a, *Astrophys. J.* **212,** 608.

Rephaeli, Y., 1977b, *Astrophys. J.* **218,** 323.

Rephaeli, Y., 1978, *Astrophys. J.* **225,** 335.

Rephaeli, Y., 1979, *Astrophys. J.* **227,** 364.

Rephaeli, Y., 1980, *Astrophys. J.* **241,** 858.

Rephaeli, Y., 1981, *Astrophys. J.* **245,** 351.

Rephaeli, Y., and E. E. Salpeter, 1980, *Astrophys. J.* **240,** 20.

Richstone, D. O., 1975, *Astrophys. J.* **200,** 535.

Richstone, D. O., 1976, *Astrophys. J.* **204,** 642.

Richstone, D. O., and E. M. Malumuth, 1983, *Astrophys. J.* **268,** 30.

Richter, O. E., J. Materne, and W. K. Huchtmeier, 1982, *Astron. Astrophys.* **111,** 193.

Ricketts, M. J., 1978, *Mon. Not. R. Astron. Soc.* **183,** 51p.

Riley, J. M., 1975, *Mon. Not. R. Astron. Soc.* **170,** 53.

Robertson, J. G., 1983, *Proc. Astron. Soc. Aust.* **5,** 144.

Roland, J., H. Sol, I. Pauliny-Toth, and A. Witzel, 1981, *Astron. Astrophys.* **100,** 7.

Roland, J., P. Veron, I. Pauliny-Toth, E. Preuss, and A. Witzel, 1976, *Astron. Astrophys.* **50,** 165.

Rood, H. J., 1965, Ph.D. Thesis, University of Michigan.

Rood, H. J., 1969, *Astrophys. J.* **158,** 657.

Rood, H. J., 1975, *Astrophys. J.* **201,** 551.

Rood, H. J., 1976, *Astrophys. J.* **207,** 16.

Rood, H. J., 1981, *Rep. Prog. Phys.* **44,** 1077.

Rood, H. J., and G. O. Abell, 1973, *Astrophys. Lett.* **13,** 69.

Rood, H. J., and A. A. Leir, 1979, *Astrophys. J. Lett.* **231,** L3.

Rood, H. J., T. L. Page, E. C. Kintner, and I. R. King, 1972, *Astrophys. J.* **175,** 627.

Rood, H. J., V. C. Rothman, and B. E. Turnrose, 1970, *Astrophys. J.* **162,** 411.

Rood, H. J., and G. N. Sastry, 1971, *Publ. Astron. Soc. Pac.* **83,** 313.

Rood, H. J., and G. N. Sastry, 1972, *Astron. J.* **77,** 451.

Roos, N., and C. A. Norman, 1979, *Astron. Astrophys.* **95,** 349.

Rose, J. A., 1976, *Astron. Astrophys. Suppl.* **23,** 109.

Rothenflug, R., L. Vigroux, R. Mushotzky, and S. Holt, 1984, *Astrophys. J.* **279,** 53.

Rothschild, R. E., W. A. Baity, A. P. Marscher, and W. A. Wheaton, 1981, *Astrophys. J. Lett.* **243,** L9.

Rowan-Robinson, M., and A. C. Fabian, 1975, *Mon. Not. R. Astron. Soc.* **170,** 199.

Rubin, V. C., W. K. Ford, C. J. Peterson, and C. R. Lynds, 1978, *Astrophys. J. Suppl.* **37,** 235.

Rubin, V. C., W. K. Ford, C. J. Peterson, and J. H. Oort, 1977, *Astrophys. J.* **211,** 693.

Ruderman, M. A., and E. A. Spiegel, 1971, *Astrophys. J.* **165,** 1.

Rudnick, L., 1978, *Astrophys. J.* **223,** 37.

Rudnick, L., and F. N. Owen, 1976a, *Astron. J.* **82,** 1.

Rudnick, L., and F. N. Owen, 1976b, *Astrophys. J. Lett.* **203,** L107.

Ryle, M., and M. D. Windram, 1968, *Mon. Not. R. Astron. Soc.* **138,** 1.

Salpeter, E. E., 1955, *Astrophys. J.* **121,** 161.

Salpeter, E. E., and J. M. Dickey, 1985, *Astrophys. J.* **292,** 426.

Sandage, A., 1972, *Astrophys. J.* **178,** 1.

Sandage, A., 1976, *Astrophys. J.* **205,** 6.
Sandage, A., K. C. Freeman, and N. R. Stokes, 1970, *Astrophys. J.* **160,** 831.
Sandage, A., J. Kristian, and J. A. Westphal, 1976, *Astrophys. J.* **205,** 688.
Sandage, A., and G. A. Tammann, 1975, *Astrophys. J.* **197,** 265.
Sandage, A., and N. Visvanathan, 1978, *Astrophys. J.* **225,** 742.
Sarazin, C. L., 1979, *Astrophys. Lett.* **20,** 93.
Sarazin, C. L., 1980, *Astrophys. J.* **236,** 75.
Sarazin, C. L., 1986a, *Rev. Mod. Phys.* **58,** 1.
Sarazin, C. L., 1986b, in *Proceedings of the Greenbank Workshop on Gaseous Haloes of Galaxies*, edited by J. Bregman and F. Lockman, p. 223.
Sarazin, C. L., 1986c, in *Proceedings of IAU Symposium 117: Dark Matter in the Universe*, edited by G. Knapp and J. Kormendy, p. 183. Dordrecht: Reidel.
Sarazin, C. L., and J. N. Bahcall, 1977, *Astrophys. J. Suppl.* **34,** 451.
Sarazin, C. L., and R. W. O'Connell, 1983, *Astrophys. J.* **268,** 552.
Sarazin, C. L., and H. Quintana, 1987, preprint.
Sarazin, C. L., H. J. Rood, and M. F. Struble, 1982, *Astron. Astrophys.* **108,** L7.
Sargent, W. L., 1973, *Publ. Astron. Soc. Pac.* **85,** 281.
Sastry, G. N., 1968, *Publ. Astron. Soc. Pac.* **80,** 252.
Sastry, G. N., and H. J. Rood, 1971, *Astrophys. J. Suppl.* **23,** 371.
Schallwich, D., and R. Wielebinski, 1978, *Astron. Astrophys.* **71,** L15.
Schechter, P. L., 1976, *Astrophys. J.* **203,** 297.
Schechter, P. L., and P. J. E. Peebles, 1976, *Astrophys. J.* **209,** 670.
Schechter, P. L., and W. H. Press, 1976, *Astrophys. J.* **203,** 557.
Scheepmaker, A., G. Ricker, K. Brecher, S. Ryckman, J. Ballantine, J. Doty, P. Downey, and W. Lewin, 1976, *Astrophys. J. Lett.* **205,** L65.
Schild, R., and M. Davis, 1979, *Astron. J.* **84,** 311.
Schipper, L., 1974, *Mon. Not. R. Astron. Soc.* **168,** 21.
Schipper, L., and I. R. King, 1978, *Astrophys. J.* **220,** 798.
Schmidt, M., 1978, *Phys. Scripta* **17,** 329.
Schneider, D. P., and J. E. Gunn, 1982, *Astrophys. J.* **263,** 14.
Schneider, D. P., J. E. Gunn, and J. G. Hoessel, 1983, *Astrophys. J.* **264,** 337.
Schnopper, H. W., J. P. Delvaille, A. Epstein, H. Helmken, D. E. Harris, R. G. Strom, G. W. Clark, and J. G. Jernigan, 1977, *Astrophys. J. Lett.* **217,** L15.
Schreier, E., P. Gorenstein, and E. Feigelson, 1982, *Astrophys. J.* **261,** 42.
Schwartz, D. A., 1976, *Astrophys. J. Lett.* **206,** L95.
Schwartz, D. A., 1978, *Astrophys. J.* **220,** 8.
Schwartz, D. A., M. Davis, R. E. Doxsey, R. E. Griffiths, J. Huchra, M. D. Johnston, R. F. Mushotzky, J. Swank, and J. Tonry, 1980a, *Astrophys. J. Lett.* **238,** L53.
Schwartz, D. A., J. Schwartz, and W. Tucker, 1980b, *Astrophys. J. Lett.* **238,** L59.
Schwarz, J., U. Briel, R. Doxsey, G. Fabbiano, R. Griffiths, M. Johnston, D. Schwartz, J. McKee, and R. Mushotzky, 1979, *Astrophys. J. Lett.* **231,** L105.
Schwarzschild, M., 1954, *Astron. J.* **59,** 273.
Schwarzschild, M., 1979, *Astrophys. J.* **232,** 236.
Scott, E. L., 1962, in *Problems in Extra-Galactic Research*, edited by G. C. McVittie, p. 269. Chicago: University of Chicago.
Scott, J. S., G. D. Holman, J. A. Ionson, and K. Papadopoulos, 1980, *Astrophys. J.* **239,** 769.
Serlemitsos, P. J., B. W. Smith, E. A. Boldt, S. S. Holt, and J. H. Swank, 1977, *Astrophys. J. Lett.* **211,** L63.
Sersic, J. L., 1974, *Astrophys. Space Sci.* **28,** 365.
Shapiro, P. R., and J. N. Bahcall, 1980, *Astrophys. J.* **241,** 1.
Shaviv, G., and E. E. Salpeter, 1982, *Astron. Astrophys.* **110,** 300.

Shectman, S. A., 1982, *Astrophys. J.* **262,** 9.

Shibazaki, N., R. Hoshi, F. Takahara, and S. Ikeuchi, 1976, *Prog. Theor. Phys.* **56,** 1475.

Shostak, G., D. Gilra, J. Noordam, H. Nieuwenhuijzen, T. deGraauw, and J. Vermne, 1980, *Astron. Astrophys.* **81,** 223.

Shu, F. H., 1978, *Astrophys. J.* **225,** 83.

Shull, J. M., 1981, *Astrophys. J. Suppl.* **46,** 27.

Shull, J. M., and M. Van Steenberg, 1982, *Astrophys. J. Suppl.* **48,** 95.

Silk, J., 1976, *Astrophys. J.* **208,** 646.

Silk, J., 1978, *Astrophys. J.* **220,** 390.

Silk, J., and S. D. M. White, 1978, *Astrophys. J. Lett.* **226,** L103.

Simon, A. J., 1978, *Mon. Not. R. Astron. Soc.* **184,** 537.

Simon, A. J., 1979, *Mon. Not. R. Astron. Soc.* **188,** 637.

Slingo, A., 1974a, *Mon. Not. R. Astron. Soc.* **166,** 101.

Slingo, A., 1974b, *Mon. Not. R. Astron. Soc.* **168,** 307.

Smith, B., R. Mushotzky, and P. Serlemitsos, 1979a, *Astrophys. J.* **227,** 37.

Smith, H. Jr., 1980, *Astrophys. J.* **241,** 63.

Smith, H. Jr., P. Hintzen, G. Holman, W. Oegerle, J. Scott, and S. Sofia, 1979b, *Astrophys. J. Lett.* **234,** L97.

Smith, S., 1936, *Astrophys. J.* **83,** 23.

Smyth, R. J., and R. S. Stobie, 1980, *Mon. Not. R. Astron. Soc.* **190,** 631.

Snow, T. P., 1970, *Astron. J.* **75,** 237.

Sofia, S., 1973, *Astrophys. J. Lett.* **179,** L35.

Solinger, A. B., and W. H. Tucker, 1972, *Astrophys. J. Lett.* **175,** L107.

Soltan, A., and J. P. Henry, 1983, *Astrophys. J.* **271,** 442.

Sparke, L. S., 1983, *Astrophys. Lett.* **23,** 113.

Spinrad, H., 1985, private communication.

Spinrad, H., J. Stauffer, and H. Butcher, 1985, *Astrophys. J.* **296,** 784.

Spitzer, L. Jr., 1956, *Physics of Fully Ionized Gases.* New York: Interscience.

Spitzer, L. Jr., 1978, *Physical Processes in the Interstellar Medium.* New York: Wiley.

Spitzer, L. Jr., and W. Baade, 1951, *Astrophys. J.* **113,** 413.

Spitzer, L. Jr., and J. Greenstein, 1951, *Astrophys. J.* **114,** 407.

Spitzer, L. Jr., and R. Harm, 1958, *Astrophys. J.* **127,** 544.

Stauffer, J., 1983, *Astrophys. J.* **264,** 14.

Stauffer, J., and H. Spinrad, 1978, *Publ. Astron. Soc. Pac.* **90,** 20.

Stauffer, J., and H. Spinrad, 1979, *Astrophys. J. Lett.* **231,** L51.

Stauffer, J., and H. Spinrad, 1980, *Astrophys. J.* **235,** 347.

Stauffer, J., H. Spinrad, and W. Sargent, 1979, *Astrophys. J.* **228,** 379.

Steiner, J. E., J. E. Grindlay, and T. Maccacaro, 1982, *Astrophys. J.* **259,** 482.

Stewart, G. C., C. R. Canizares, A. C. Fabian, and P. E. Nulsen, 1984a, *Astrophys. J.* **278,** 536.

Stewart, G. C., A. C. Fabian, C. Jones, and W. Forman, 1984b, *Astrophys. J.* **285,** 1.

Stimpel, O., and J. Binney, 1979, *Mon. Not. R. Astron. Soc.* **188,** 883.

Strom, K. M., and S. E. Strom, 1978a, *Astron. J.* **83,** 73.

Strom, K. M., and S. E. Strom, 1978b, *Astron. J.* **83,** 1293.

Strom, S. E., and K. M. Strom, 1978c, *Astron. J.* **83,** 732.

Strom, S. E., and K. M. Strom, 1978d, *Astrophys. J. Lett.* **225,** L93.

Strom, S. E., and K. M. Strom, 1979, *Astron. J.* **84,** 1091.

Struble, M. F., and H. J. Rood, 1981, *Astrophys. J.* **251,** 471.

Struble, M. F., and H. J. Rood, 1982, *Astron. J.* **87,** 7.

Struble, M. F., and H. J. Rood, 1985, *Astron. J.* **89,** 1487.

Sulentic, J. W., 1980, *Astrophys. J.* **241,** 67.

Sullivan, W. T., G. D. Bothun, B. Bates, and R. A. Schommer, 1981, *Astron. J.* **86,** 919.

Sullivan, W. T., and P. E. Johnson, 1978, *Astrophys. J.* **225,** 751.
Sunyaev, R. A., 1981, *Sov. Astron. Lett.* **6,** 213.
Sunyaev, R. A., and Y. B. Zel'dovich, 1972, *Comm. Astrophys. Space Phys.* **4,** 173.
Sunyaev, R. A., and Y. B. Zel'dovich, 1980a, *Ann. Rev. Astron. Astrophys.* **18,** 537.
Sunyaev, R. A., and Y. B. Zel'dovich, 1980b, *Mon. Not. R. Astron. Soc.* **190,** 413.
Sunyaev, R. A., and Y. B. Zel'dovich, 1981, *Astrophys. Space Sci. Rev.* **1,** 1.
Takahara, F., and S. Ikeuchi, 1977, *Prog. Theor. Phys.* **58,** 1728.
Takahara, F., S. Ikeuchi, N. Shibazaki, and R. Hoshi, 1976, *Prog. Theor. Phys.* **56,** 1093.
Takahara, M., and F. Takahara, 1979, *Prog. Theor. Phys.* **62,** 1253.
Takahara, M., and F. Takahara, 1981, *Prog. Theor. Phys.* **65,** 1.
Takeda, H., P. Nulsen, and A. Fabian, 1984, *Mon. Not. R. Astron. Soc.* **208,** 461.
Tarenghi, M., G. Chincarini, H. Rood, and L. Thompson, 1980, *Astrophys. J.* **235,** 724.
Tarenghi, M., and J. S. Scott, 1976, *Astrophys. J. Lett.* **207,** L9.
Tarter, J. C., 1975, Ph.D. Thesis, University of California, Berkeley.
Tarter, J. C., 1978, *Astrophys. J.* **220,** 749.
Thomas, J. C., and D. Batchelor, 1978, *Astron. J.* **83,** 1160.
Thompson, L. A., 1976, *Astrophys. J.* **209,** 22.
Thompson, L. A., 1986, *Astrophys. J.* **300,** 639.
Thompson, L. A., and S. A. Gregory, 1978, *Astrophys. J.* **220,** 809.
Thompson, L. A., and S. A. Gregory, 1980, *Astrophys. J.* **242,** 1.
Thuan, T. X., 1980, in *Physical Cosmology: Proceedings of the Les Houches Summer School XXXII*, edited by R. Balian, J. Audouze, and D. N. Schramm, p. 278. Amsterdam: North-Holland.
Thuan, T. X., and J. Kormendy, 1977, *Publ. Astron. Soc. Pac.* **89,** 466.
Thuan, T. X., and W. Romanishin, 1981, *Astrophys. J.* **248,** 439.
Tifft, W. G., 1978, *Astrophys. J.* **222,** 54.
Tifft, W. G., and S. A. Gregory, 1976, *Astrophys. J.* **205,** 696.
Tifft, W. G., and M. Tarenghi, 1975, *Astrophys. J. Lett.* **198,** L7.
Tifft, W. G., and M. Tarenghi, 1977, *Astrophys. J.* **217,** 944.
Tinsley, B. M., and A. G. Cameron, 1974, *Astrophys. Space Sci.* **31,** 31.
Tonry, J. L., 1984, *Astrophys. J.* **279,** 13.
Tonry, J. L., 1985a, *Astrophys. J.* **291,** 45.
Tonry, J. L., 1985b, *Astron. J.* **90,** 2431.
Toomre, A., and J. Toomre, 1972, *Astrophys. J.* **178,** 623.
Toyama, K., and S. Ikeuchi, 1980, *Prog. Theor. Phys.* **64,** 831.
Tremaine, S., 1981, in *The Structure and Evolution of Normal Galaxies*, edited by M. Fall and D. Lynden-Bell, p. 67. Cambridge University Press.
Tremaine, S. D., and D. O. Richstone, 1977, *Astrophys. J.* **212,** 311.
Trinchieri, G., and G. Fabbiano, 1985, *Astrophys. J.* **296,** 447.
Trinchieri, G., G. Fabbiano, and C. R. Canizares, 1986, *Astrophys. J.* **310,** 637.
Tucker, W. H., and R. Rosner, 1983, *Astrophys. J.* **267,** 547.
Turner, E. L., and J. R. Gott, 1976a, *Astrophys. J.* **209,** 6.
Turner, E. L., and J. R. Gott, 1976b, *Astrophys. J. Suppl.* **32,** 409.
Turner, E. L., and W. L. Sargent, 1974, *Astrophys. J.* **194,** 587.
Tytler, D., and N. V. Vidal, 1979, *Mon. Not. R. Astron. Soc.* **182,** 33p.
Ulmer, M. P., and R. G. Cruddace, 1981, *Astrophys. J. Lett.* **246,** L99.
Ulmer, M. P., and R. G. Cruddace, 1982, *Astrophys. J.* **258,** 434.
Ulmer, M. P., R. G. Cruddace, and M. P. Kowalski, 1985, *Astrophys. J.* **290,** 551.
Ulmer, M. P., R. G. Cruddace, K. Wood, J. Meekins, D. Yentis, W. D. Evans, H. Smathers, E. Byram, T. Chubb, and H. Friedman, 1980a, *Astrophys. J.* **236,** 58.
Ulmer, M. P., and J. G. Jernigan, 1978, *Astrophys. J. Lett.* **222,** L85.
Ulmer, M. P., R. Kinzer, R. G. Cruddace, K. Wood, W. Evans, E. Byram, T. Chubb, and H. Friedman, 1979, *Astrophys. J. Lett.* **227,** L73.

Ulmer, M. P., M. Kowalski, R. Cruddace, M. Johnson, J. Meekins, H. Smathers, D. Yentis, K. Wood, D. McNutt, T. Chubb, E. Byram, and H. Friedman, 1981, *Astrophys. J.* **243,** 681.
Ulmer, M. P., S. Shulman, W. D. Evans, W. N. Johnson, D. McNutt, J. Meekins, G. H. Share, D. Yentis, K. Wood, E. T. Byram, T. A. Chubb, and H. Friedman, 1980b, *Astrophys. J.* **235,** 351.
Ulrich, M., 1978, *Astrophys. J.* **221,** 422.
Uson, J. M., and D. T. Wilkinson, 1985, private communication.
Valentijn, E. A., 1978, *Astron. Astrophys.* **68,** 449.
Valentijn, E. A., 1979a, *Astron. Astrophys. Suppl.* **38,** 319.
Valentijn, E. A., 1979b, *Astron. Astrophys.* **78,** 362.
Valentijn, E. A., 1979c, *Astron. Astrophys.* **78,** 367.
Valentijn, E. A., 1983, *Astron. Astrophys.* **118,** 123.
Valentijn, E. A., and W. Bijleveld, 1983, *Astron. Astrophys.* **125,** 223.
Valentijn, E. A., and R. Giovanelli, 1982, *Astron. Astrophys.* **114,** 208.
Valentijn, E. A., and G. C. Perola, 1978, *Astron. Astrophys.* **63,** 29.
Vallee, J. P., 1981, *Astrophys. Lett.* **22,** 193.
Vallee, J. P., A. H. Bridle, and A. S. Wilson, 1981, *Astrophys. J.* **250,** 66.
Vallee, J. P., and A. S. Wilson, 1976, *Nature* **259,** 451.
Vallee, J. P., A. S. Wilson, and H. van der Laan, 1979, *Astron. Astrophys.* **77,** 183.
Valtonen, M., and G. Byrd, 1979, *Astrophys. J.* **230,** 655.
van Albada, T. S., 1982, *Mon. Not. R. Astron. Soc.* **201,** 939.
van Breugel, W. J., 1980, *Astron. Astrophys.* **88,** 248.
van Breugel, W. J., T. Heckman, and G. Miley, 1984, *Astrophys. J.* **276,** 79.
van den Bergh, S., 1961a, *Zeit Astrophys.* **53,** 219.
van den Bergh, S., 1961b, *Astrophys. J.* **134,** 970.
van den Bergh, S., 1976, *Astrophys. J.* **206,** 883.
van den Bergh, S., 1977a, *Publ. Astron. Soc. Pac.* **89,** 746.
van den Bergh, S., 1977b, *Vistas Astron.* **21,** 71.
van den Bergh, S., 1983a, *Publ. Astron. Soc. Pac.* **95,** 275.
van den Bergh, S., 1983b, *Astrophys. J.* **265,** 606.
van den Bergh, S., and J. de Roux, 1978, *Astrophys. J.* **219,** 352.
van Gorkom, J. H., and R. D. Ekers, 1983, *Astrophys. J.* **267,** 528.
Vestrand, W. T., 1982, *Astron. J.* **87,** 1266.
Vidal, N. V., 1975a, *Astron. Astrophys.* **42,** 145.
Vidal, N. V., 1975b, *Publ. Astron. Soc. Pac.* **87,** 625.
Vidal, N. V., 1980, in *IAU Symp. No. 92: Objects of High Redshift*, edited by G. O. Abell and P. J. E. Peebles, p. 69. Dordrecht: Reidel.
Vidal, N. V., and B. A. Peterson, 1975, *Astrophys. J. Lett.* **196,** L95.
Vidal, N. V., and D. T. Wickramasinghe, 1977, *Mon. Not. R. Astron. Soc.* **180,** 305.
Vigroux, L., 1977, *Astron. Astrophys.* **56,** 473.
Villumsen, J. V., 1982, *Mon. Not. R. Astron. Soc.* **199,** 493.
Waldthausen, H., C. Haslam, R. Wielebinski, and P. Kronberg, 1979, *Astron. Astrophys. Suppl.* **36,** 237.
Weinberg, S., 1972, *Gravitation and Cosmology: Principles and Applications of the General Theory of Relativity*. New York: Wiley.
Wellington, K., G. Miley, and H. van der Laan, 1973, *Nature* **244,** 502.
West, R. M., 1974, *Euro. South. Obs. Bull.* **10,** 25.
West, R. M., and S. Frandsen, 1981, *Astron. Astrophys. Suppl.* **44,** 329.
Westphal, J. A., J. Kristian, and A. R. Sandage, 1975, *Astrophys. J. Lett.* **197,** L95.
White, R. A., 1978a, *Astrophys. J.* **226,** 591.
White, R. A., and J. O. Burns, 1980, *Astron. J.* **85,** 117.
White, R. A., and H. Quintana, 1985, private communication.

White, R. A., C. L. Sarazin, and H. Quintana, 1987, preprint.
White, R. A., C. L. Sarazin, H. Quintana, and W. J. Jaffe, 1981a, *Astrophys. J. Lett.* **245,** L1.
White, R. E., and C. L. Sarazin, 1987a, *Astrophys. J.* **318,** 612.
White, R. E., and C. L. Sarazin, 1987b, *Astrophys. J.* **318,** 629.
White, S. D. M., 1976a, *Mon. Not. R. Astron. Soc.* **174,** 19.
White, S. D. M., 1976b, *Mon. Not. R. Astron. Soc.* **174,** 467.
White, S. D. M., 1976c, *Mon. Not. R. Astron. Soc.* **177,** 717.
White, S. D. M., 1977a, *Comm. Astrophys.* **7,** 95.
White, S. D. M., 1977b, *Mon. Not. R. Astron. Soc.* **179,** 33.
White, S. D. M., 1978b, *Mon. Not. R. Astron. Soc.* **184,** 185.
White, S. D. M., 1979, *Mon. Not. R. Astron. Soc.* **189,** 831.
White, S. D. M., 1982, in *Morphology and Dynamics of Galaxies*, edited by L. Martinet and M. Mayor, p. 289. Geneva: Geneva Observatory.
White, S. D. M., 1985, private communication.
White, S. D. M., and M. J. Rees, 1978, *Mon. Not. R. Astron. Soc.* **183,** 341.
White, S. D. M., and J. Silk, 1980, *Astrophys. J.* **241,** 864.
White, S. D. M., J. Silk, and J. P. Henry, 1981b, *Astrophys. J. Lett.* **251,** L65.
Wilkinson, A., and J. B. Oke, 1978, *Astrophys. J.* **220,** 376.
Willson, M., 1970, *Mon. Not. R. Astron. Soc.* **151,** 1.
Wilson, A. S., and J. P. Vallee, 1977, *Astron. Astrophys.* **58,** 79.
Wirth, A., S. J. Kenyon, and D. A. Hunter, 1983, *Astrophys. J.* **269,** 102.
Wolf, M., 1906, *Astron. Nachr.* **170,** 211.
Wolf, R. A., and J. N. Bahcall, 1972, *Astrophys. J.* **176,** 559.
Wolff, R. S., H. Helava, T. Kifune, and M. C. Weisskopf, 1974, *Astrophys. J. Lett.* **193,** L53.
Wolff, R. S., H. Helava, and M. C. Weisskopf, 1975, *Astrophys. J. Lett.* **197,** L99.
Wolff, R. S., R. J. Mitchell, P. A. Charles, and J. L. Culhane, 1976, *Astrophys. J.* **208,** 1.
Wood, K., J. Meekins, D. Yentis, H. Smathers, D. McNutt, R. Bleach, E. Byron, T. Chubb, H. Friedman, and M. Meidav, 1984, *Astrophys. J. Suppl.* **56,** 507.
Yahil, A., and J. P. Ostriker, 1973, *Astrophys. J.* **185,** 787.
Yahil, A., and N. V. Vidal, 1977, *Astrophys. J.* **214,** 347.
Young, P. J., 1976, *Astron. J.* **81,** 807.
Zel'dovich, Y. B., and R. A. Sunyaev, 1969, *Astrophys. Space Sci.* **4,** 301.
Zel'dovich, Y. B., and R. A. Sunyaev, 1981, *Sov. Astron. Lett.* **6,** 285.
Zwicky, F., 1933, *Helv. Phys. Acta* **6,** 110.
Zwicky, F., 1957, *Morphological Astronomy*. Berlin: Springer.
Zwicky, F., E. Herzog, P. Wild, M. Karpowicz, and C. T. Kowal, 1961–1968, *Catalogues of Galaxies and Clusters of Galaxies*, Vol. 1–6. Pasadena: Caltech.